Immunochemistry of Solid-Phase Immunoassay

Editor

John E. Butler, Ph.D.

Professor
Department of Microbiology
College of Medicine
University of Iowa
Iowa City, Iowa

CRC Press
Boca Raton Ann Arbor Boston London

Library of Congress Cataloging-in-Publication Data

Immunochemistry of solid-phase immunoassay / editor, John E. Butler
p. cm.
Includes bibliographical references.
ISBN 0-8493-5394-7
1. Immunochemistry. 2. Immunoassay. I. Butler, John E.
[DNLM: 1. Immunoassay—methods. 2. Immunochemistry. QW 570
I335]
QR183.6.I454 1991
616.0756—dc20
DNLM/DLC 90-2592
for Library of Congress CIP

Direct all inquiries to CRC Press, Inc., 2000 Corporate Blvd., N.W., Boca Raton, Florida 33441.

International Standard Book Number 0-8493-5394-7

Library of Congress Card Number 90-2592
Printed in the United States

PREFACE

The *Immunochemistry of Solid-Phase Immunoassay* is designed to fill a nichè in the field of immunology. Although solid-phase immunoassay, e.g., ELISA, RIA, etc., are used worldwide, there is no comprehensive source of information which focuses on the immunochemical principles underlying this technology. This book attempts to fill that role.

Solid-phase ligand-receptor (antigen-antibody) reactions offer numerous challenges to our understanding when compared with conventional, solution-phase antigen-antibody interactions because the interaction of ligand and receptor occur on an interface, the exact reaction volume which is difficult to determine. This means that functional reactant concentrations are different than the concentrations of these reactants in the total test volume. Immobilization of one reactant also means that the time needed to reach equilibrium is longer than in solution but perhaps even more important, immobilization itself can alter the function of antibodies or the epitopes of antigens.

Solid-phase ligand-receptor interactions should not be considered *in vitro* peculiarities, as many important *in vivo* biological interactions occur at interfaces, especially those involving antigens and antibodies. Hence, the understanding of interfacial reactions *in vitro* could serve as a useful model for better understanding interfacial interactions *in vivo.*

Solid-phase systems not only present certain problems that must be overcome in assay design, but they also provide the opportunity for developing new concepts in immunoassay, e.g., multianalyte systems. New as well as traditional methods of solid-phase immunoassay require a reliable means of data analyses and quantitation. Of special concern is the quantitation of specific antibodies. Antibody responses are considered to be of great importance in medical diagnostics of, e.g., HIV, allergy, and bacterial infection. Hence, the proper design, use, and interpretation of data from solid-phase immunoassays used for such diagnoses are critical to medical science.

The *Immunochemistry of Solid-Phase Immunoassay* is composed of 20 chapters; five concern general principles, four focus on practical application, and the remainder are short contributions dedicated to specific, highly relevant issues in the field of solid-phase immunoassay. A glossary of terms and an index are also provideed.

The book is designed to assemble the ideas, data, and concepts of investigators at the cutting edge of an important area of biotechnology as well as serving as an advanced textbook in the field of molecular immunology. The *Immunochemistry of Solid-Phase Immunoassay* integrates the knowledge and opinions of an international panel of over 25 investigators who develop, study, and extensively use this popular category of immunoassays. Scientific concepts which are influenced by data generated from solid-phase immunoassays will be strengthened when users of the technology gain a better understanding of its underlying chemistry and immunochemistry. *Immunochemistry of Solid-Phase Immunoassay* is not a laboratory methods manual, but a textbook of principles; practical aspects are nevertheless considered by nearly every contributor.

THE EDITOR

John E. Butler, Ph.D., is Professor of Microbiology in the College of Medicine at the University of Iowa, where he teaches molecular immunology. Dr. Butler is a native of Rice Lake, Wisconsin and received his B.S. degree from the University of Wisconsin, River Falls in 1961 where he majored in chemistry and biology. He received his PhD. from the University of Kansas in Zoology/Biochemistry in 1966. Dr. Butler has had a diverse background including professional experience as a Ranger-naturalist of the National Park Service, Research Biologist at the USDAs Allergens Investigations in Washington, DC, Visiting Scientist, Southeast Pacific Oceanographic Expedition, and as a visiting scientist on several occasions to West Germany. He has conducted research in the areas of mucosal immunity, immunochemistry and veterinary immunology. Dr. Butler was appointed Assistant Professor of Microbiology at the University of Iowa in 1971: Associate Professor in 1974 and Professor in 1980.

Dr. Butler has been the recipient of a National Science Foundation Fellowship, an USPHS Traineeship in Immunology, a Fellowship for the Max-Planck Society of Germany, a Fogarty International Fellowship as well as receiving a Citation Classic Award from the Institute of Scientific Information. He is a member of the American Association of Immunologists, the New York Academy of Sciences, and the Society of Mucosal Immunity. Dr. Butler has been the recipient of grants from the NIH and USDA, has published more than 90 peer-reviewed articles, >30 reviews and invited chapters and has previously edited a book on the ruminant immune system. He has organized two international symposia, chaired sessions at national and international meetings, and has trained >20 graduate and post doctoral students from nine different countries.

The current interests of the editor are solid-phase immunoassay and the regulation of immunoglobulin andFe receptor genes of swine and cattle. He is chairman of an I.U.I.S. committee on the nomenclature and standardization of veterinary animal immunoglobulins.

CONTRIBUTORS

Ole Jannik Bjerrum, M.D., Ph.D.
Professor
Department of Biochemistry and Nutrition
The Technical University of Denmark
and
Director of Research
Assay Technology
Novo Nordisk
Bagsvaerd, Denmark

Ole Børmer, M.D.
Consultant
Central Laboratory
The Norwegian Radium Hospital
Oslo, Norway

William R. Brown, B.S.
Pre-Doctoral Fellow
Department of Microbiology
University of Iowa Medical School
Iowa City, Iowa

John E. Butler, Ph.D.
Professor
Department of Microbiology
University of Iowa
Iowa City, Iowa

Robert Chalifour
Diagnostics Research Group
IAF BioChem International, Inc.
Laval, Quebec, Canada

Steven E. Dierks, Ph.D.
Postdoctoral Research Associate
Department of Microbiology
Virginia Commonwealth University
Richmond, Virginia

Gervais Dionne
Diagnostics Research Group
IAF BioChem International, Inc.
Laval, Quebec, Canada

Lisa Djavadi-Ohaniance, D.Sc.
Director of Research
Centre National de la Recherche Scientifique
Biochimie et Génétique Moléculaire
Institut Pasteur
Paris, France

Robert J. Dwyer
Diagnostics Research Group
IAF BioChem International, Inc.
Laval, Quebec, Canada

Roger Ekins, Ph.D., D.Sc.
Professor
Department of Molecular Endocrinology
Thorn Institute of Clinical Science
University College and Middlesex School of Medicine
London, England

Martha Fesler
Durham, North Carolina

Burkart Franz, D.V.M.
Institute for Microbiology and Infectious Diseases
Veterinary School
Hannover, Germany

Bertrand Friguet, D.Univ.
Assistant Professor
Université Paris 7
Departement de Biochimie et Génétique Moléculaire
Institut Pasteur
Paris, France

J. M. Gershoni
National Cancer Institute
National Institutes of Health
Bethesda, Maryland

K. S. Joshi
Department of Microbiology
University of Iowa
Iowa City, Iowa

Robert G. Hamilton, Ph.D.
Associate Professor of Medicine
Department of Medicine
and
Director
DACI Reference Laboratory
Johns Hopkins University School of Medicine
Baltimore, Maryland

William E. Keating
Research Assistant
Diagnostic Chemistry
Becton Dickinson Research Center
Research Triangle Park, North Carolina

Omar S. Khalil, Ph.D.
Associate Research Fellow
Diagnostics Division
Abbott Laboratories
Abbott Park, Illinois

Martial Lacroix, Ph.D.
Director
Diagnostics Research Group
IAF BioChem International, Inc.
Laval, Quebec, Canada

Patrick D. Mize, Ph.D.
Senior Scientist
Molecular Biology
Becton Dickinson Research Center
Research Triangle Park, North Carolina

Ronald G. Myatich
UNC Medical School
University of North Carolina at Chapel Hill
Chapel Hill, North Carolina

Ali Naqui, Ph.D.
Scientist
Becton Dickinson Advanced Diagnostics
Baltimore, Maryland

That T. Ngo, Ph.D.
Chairman and CEO
BioProbe International, Inc.
Tustin, CA

Kvell Nustad, M.D., Ph.D.
Senior Consultant
Central Laboratory
The Norwegian Radium Hospital
Oslo, Norway

Håkan Nygren, Ph.D.
Associate Professor
Department of Histology
University of Göteborg
Göteborg, Sweden

Catherine M. O'Connell, Ph.D.
Manager of Undergraduate Studies
Department of Chemistry
Duke University
Durham, North Carolina

Elisabeth Paus, Ph.D.
Biochemist
Central Laboratory
The Norwegian Radium Hospital
Oslo, Norway

Jeffrey H. Peterman, Ph.D.
Research Immunochemist
Igen, Inc.
Rockville, Maryland
and
Research Associate
Department of Microbiology
University of Alabama at Birmingham
Birmingham, Alabama

Fritz Poulsen, Ph.D.
Research Fellow
Assay Technology
Novo Nordisk
Bagsvaerd, Denmark

Michael Stegemann, D.V.M.
Institute for Microbiology and Infectious Diseases
Veterinary School
Hannover, Germany

Manne Stenberg, Ph.D.
Department of Histology
University of Göteberg
Göteborg, Sweden

Fred J. Stevens, Ph.D.
Biophysicist
Biological and Medical Research Division
Argonne National Laboratory
Argonne, Illinois

Mark Suter, D.V.M.
Head
Molecular Immunology
Swiss Institute for Allergy and Asthma Research
Davos, Switzerland

Peter J. Tarcha, Ph.D.
Section Head
and
Senior Polymer Chemist
Diagnostics Division
Abbott Laboratories
North Chicago, Illinois

Jess N. Waller
Department of Polymer Chemistry
University of Akron
Akron, Ohio

Maan Zrein
Diagnostics Research Group
IAF BioChem International, Inc.
Laval, Quebec, Canada

TABLE OF CONTENTS

General Principles

Chapter 1

PERSPECTIVES, CONFIGURATIONS AND PRINCIPLES

John E. Butler

TABLE OF CONTENTS

I. PERSPECTIVE

A. SEROLOGY, IMMUNOCHEMISTRY, AND RADIOIMMUNOASSAY

The first century of immunology* saw the emergence of at least three disciplines concerned with the reaction of antigens and antibodies which have gradually amalgamated in the last 15 years. *Serology* assesses the antibody activity to infectious agents, erythrocyte and leukocyte antigens, autoantibodies, and allergens. Serological *activity* is usually expressed as titer, e.g., agglutination, C′ fixation, or neutralization titer. The emergence of *immunochemistry* in the manner of Landsteiner, Heidelberger, Kabat, and others, focused on experiments designed to understand the kinetics and stoichiometry of antigen-antibody reactions and described a means of expressing antibody in micrograms per milliliter, not merely as titer. Despite advancements made by these pioneers, serologists did not immediately convert their "titers to micrograms". It must be remembered that the classic work of Heidelberger and colleagues was done with what today would be regarded as specialized and/or contrived systems, precipitating IgG antibodies from hyperimmune rabbits to soluble proteins and bacterial polysaccharides. Subsequently, we have learned that serum antibodies can belong to five different classes, that precipitating antibodies are practically confined to the IgG class, and that subclasses differ in precipitability.[1-3] Fortunately for Heidelberger, rabbits are unusual in having only one IgG subclass.[4] When attempts were made to understand the immunochemistry of other types of antibody-antigen reactions, e.g., agglutination or C′ fixation, the substantial differences in the activity of different isotypes made the conversion of titer into micrograms awkward if not impossible. One mole of IgM can be as effective as 20 to 100 mol of IgG in C′ fixation, neutralization, and agglutination assays,[5-7] and IgE can initiate *in vivo* activities that are out of proportion to its mere nanogram level in serum. Serological measurements of titer are therefore alive and well, and remain the most frequently encountered, and perhaps honest method, of expressing antibody activity (Chapter 9).

The immunochemical studies of Heidelberger and colleagues made two longer-lasting contributions. First, they demonstrated that antigen-antibody reactions were not a form of "20th-century alchemy", but followed nearly predictable rules of biochemical interaction. Second, they showed that an antigen-antibody reaction could be used to accurately determine the quantity of antigen in a sample. The practicability of this principle received wide attention when a simple gel version of the precipitin assay was developed to allow technicians with little immunochemical experience to quantify multivalent antigens in clinical and basic science laboratories. Hence, the antigen-antibody reaction became a quantitative tool in biochemistry, immunology, and medicine.

The pioneer efforts of Yalow and Berson and Ekins 30 years ago ushered in the era of *radioimmunoassay* (RIA); this constitutes the third discipline in which antigen-antibody reactions were used and studied. Methods like agglutination-inhibition and immunoprecipitation lacked the sensitivity and applicability to the quantitation of low molecular weight substances afforded by RIA. RIA was originally designed as a direct competitive immunoassay in which the unlabeled antigen being measured competed with known amounts of iodinated antigen for binding sites on a limited quantity of preferably high-affinity antibody. The field of RIA did foster numerous immunochemical studies which increased our knowledge of the kinetics of solution-phase antigen-antibody interactions, the role of affinity, the half-life of the immune complexes, and the nature and liability of antigen epitopes. Despite these advances, "the slowest ship in the RIA convoy" remained the separation of bound and free antigen, a process necessary in all heterogenous immunoassays.** It was the need to make

* 1890 is arbitrarily regarded as the beginning of the science of immunology although certain historic observations were made before this date.

** Heterologous immunoassays are those usually involving multiple reaction steps and eventually requiring that bound reactants be physically separated from one of the free reactants to obtain the result of the test.

this process more efficient and cost-effective which spawned the era of solid-phase immunoassay.

B. SOLID-PHASE BIOCHEMISTRY: A REVOLUTION IN THE SEPARATION OF BOUND AND FREE REACTANTS

The measurement of the analyte (either antibody or antigen) in heterologous immunoassays requires that the immune complex be somehow separated from the uncomplexed reactants, irrelevant serum, or test fluid macromolecules. Richard Farr[8] was able to separate free and antibody-complexed antigen provided that the free antigen remained soluble in concentrations of neutral salt sufficient to precipitate antibodies and immune complexes, e.g., 50% ammonium sulfate. Unfortunately, these "Farr Assays" were only applicable to proteins like serum albumins and haptens that are soluble in 50% ammonium sulfate. Charcoal, dextran particles, and secondary antibody co-precipitation were among the separation agents or methods used initially in RIAs; these were effective but cumbersome. Hence, the observations, pioneered by Catt and Tregear,[9] that proteins could be readily adsorbed on plastic surfaces and membranes like nitrocellulose revolutionized the field of immunoassay. Plastic test tubes or microtiter plates could now be simultaneously used as reaction vessels and as a solid phase permitting rapid separation of the bound and free analyte by washing the tube. Numerous modifications of this basic principle, involving various plasticware, membranes, and beads, have emerged in the last 15 years.

Solid-phase biochemistry did not begin with immunoassay. Campbell and co-workers[10,11] made the first antibody-antigen solid-phase affinity matrix and in 1968 Goldstein and Katchalski[12] introduced solid-phase enzymes. By 1970, affinity chromatography had become a routine laboratory tool, with hydrophilic supports used to covalently immobilize reactants.[13] Modern solid-phase immunoassays utilize both adsorption- and covalent-based technology in the immobilization process.

C. PERMUTATIONS, PURPOSE, AND PROBLEMS

Twenty years of solid-phase immunoassay has been witness to various permutations or extensions of the basic principle. Each has its purpose, each provided advantages over predecessor methodology and each brought awareness of new problems for the assayist. In the 1970's, enzymes were introduced to replace radionuclides as signal generators. With them came greater safety and convenience which fostered the appearance of enzyme immunoassay (EIA) in laboratories that would have formerly never considered using RIA. Enzymes, however, introduced steric problems; substituting an enzyme label for an atom of iodine can have major consequences for molecular structure.

At the same time at which enzymes and EIA* became popular, alternative solid phases began to appear. The test tubes used by Engvall and Perlmann[14] gave way to microtiter plates, beads, and membranes. Microtiter plates offered the advantage of automation, and, except for assays in clinical labs designed to measure a particular analyte, remain the most widely used.

The use of membranes as a solid phase opened the field of immunoblotting (Chapter 8). Immunoblots replaced the many attempts to immunochemically identify proteins and peptides directly in electrophoretic gels by instead transferring them to nitrocellulose (NC) prior to analysis. The membranes were much easier to work with than the gels, replicate blots could be made for multiple analysis, and some argued that the transferred molecules were now more immunoreactive.

In a quest for greater sensitivity, investigators configured a variety of technical advancements which could amplify detection. Some involved the use of multiple antibodies, other amplified the enzyme product by fluorescence while others reverted to combined RIA/EIA

* Throughout this book, the terms EIA and ELISA are used synonymously and interchangeably.

methodology to achieve greater sensitivity. Multiple antibodies create steric problems, fluorescence involved additional equipment, and adding radioactivity seems to counter the advantage gained in the switch from RIA to EIA.

A practical permutation has been the production of "yes/no" assays which everyone from clinicians to farmers could use in field applications. "Yes/no" assays must have forgiving tolerances to overcome the many sources of error which assays run under such conditions inevitably encounter. Perhaps uppermost is the inherent danger of "single-point determinations" (Chapter 9) which, for practical reasons, is necessarily a built-in liability.

While the list of permutations could be lengthened, one final example is the appearance of systems utilizing covalent attachment of the solid-phase reactant. This approach is believed to overcome the capacity limitation and molecular alteration events associated with the popular surface adsorption method. In spite of the theoretical promise of this approach, most solid-phase immunoassays are still conducted with adsorbed reactant either for convenience or in lieu of a truly successful and reproducible means of covalent attachment.

Most of the permutations described for solid-phase immunoassay have usually achieved their purpose in a specific application, although few, save the replacement of radionucleotides with enzymes, have been a universal panacea. Many permutations are more technical than scientific. For example, using a fluorescent product in an EIA no doubt amplifies the signal, but it will also amplify the background. The better solution would be to improve the immunochemistry of the system. Similarly, assays utilizing covalently bound antigen may be thought to function better because of increased antigen density when in fact the real reason may be that fewer epitopes are lost than during adsorption.

D. IS SOLID-PHASE IMMUNOASSAY A MODEL FOR ANTIGEN-ANTIBODY REACTIONS *IN VIVO?*

Biologists are continuously confronted with the challenge of designing *in vitro* models that represent the *in vivo* situation. With the exception of neutralizing and "blocking" antibodies in serum and elsewhere, few antibodies exert their *in vivo* effect in free solution. Rather, they act in concert with cells, i.e., they are involved in interfacial reactions. There are more than a few similarities between the potential immunochemical behavior of surface-bound reactants *in vivo* and those attached to a solid phase *in vitro*. Beyond the fact that immobilized components are no longer freely diffusible, the reaction volume is also confined to the interface. Molecules which are concentrated in a small surface volume, such as is in lymphocyte capping, should, according to the Mass Law, promote ligand binding. There are data showing that the kinetics of cell surface antibody interactions[15] are more similar to those of antibodies binding solid-phase antigen than for solution-phase antigen-antibody interactions[16] (see Section III.B). However, the simple comparison of *in vivo* receptor-ligand interactions with solid-phase receptor-ligand interactions overlooks many significant differences. Living membranes are dynamic, not chemically fixed as is presumed to be the case with *in vitro* solid phases. While *in vitro* solid phase are often hydrophobic, the solvent exposed surfaces of cell membranes are the opposite. The dynamic fluidity of living membranes also means that such interfacial reactions can be proteolytically altered, endocytosed, or processed in some other manner.

Although enough differences exist between *in vitro* solid-phase and *in vivo* membrane interactions to make direct interpolation difficult, certain principles learned from studies of the *in vitro* solid phases could be instructive for the living system.

E. PROGNOSIS AND PREDICTION

There is little doubt that the popularity enjoyed by solid-phase immunoassay will not only continue but intensify in the future. The microtiter format will probably be retained because equipment manufacturers and laboratories have such hardware in place and will adapt and automate it rather than replace it; the advantages of microparticular systems (Chapter 14) or membrane dot-blot systems (Chapter 8) may result in their adaptation to the microtiter format.

The growing interest among some users and many manufacturers of the need to understand the immunochemistry of the system, not just to produce more specialized kits, is gradually having an impact on innovation in this field of biotechnology. The increasing awareness of the molecular alterations taking place during surface adsorption (Chapters 10, 11, 13, and 17) seems to be the driving force behind new methods of immobilization. The diffusion dependence of especially micrototiter systems (Chapters 18 and 19) will necessitate techniques and equipment so that the time required for reaction equilibrium can be reduced by an order of magnitude.

The need to reduce technician work loads will promote the development of assays with a wider dynamic range and which require fewer reaction steps. The labor-intensive aspects of solid-phase immunoassay will be reduced through the use of additional automation and robotics. Field kits will retain their popularity, but better immunochemistry will be employed to increase reliability in testing, and flow-through systems will replace static systems for analysis of dilute solutions such as the search for pesticides and herbicides in water.

Recombinant DNA technology is already used to produce chimeric antibodies for reference standards (Chapter 9); however, the future will also see it used to engineer capture antibodies (CAb), antigens, and peptides for attachment to solid phases. A greater awareness of how to generate and use monoclonal antibodies in solid-phase immunoassay will emerge; international standardization will become a reality and both recombinant and monoclonal technology will play an important role in this development.

One of the purposes of this volume is to bring an awareness of solid-phase immunochemistry not just to those who study and develop such assays, but to the many biologists and immunobiologists who use this form of bio-technology, and who might use it more effectively when better informed of the principles governing these immunoassays.

II. THE ACTORS: ANTIBODIES, ANTIGENS, SOLID-PHASES AND SIGNAL GENERATORS

A. THE ANTIBODY MOLECULE

1. Structure, Heterogeneity, and Specificity

Antibodies are a family of glycoproteins synthesized by the β-lymphocyte lineage. Antibodies belong to the immunoglobulin supergene family which is composed of many immunologically important molecules that are structurally characterized by their repetitive "immunoglobulin fold" (Chapter 9). The term immunoglobulin describes a protein structure while an antibody refers to an immunoglobulin with specific affinity for a particular antigen; antibodies are defined functionally by the antigens they bind. Immunoglobulins are biochemically heterogeneous; an individual can probably synthesize antibodies with $>10^8$ different antigen specificities. This vast specificity potential contributes substantially to the biochemical heterogeneity and sets them apart from other members of the immunoglobulin supergene family not involved with antigen recognition.

Antibodies recognize determinant groups on antigens known as epitopes (see Section II.B.2) which can occupy the space equivalent to at least a molecule the size of the *xanthene* ring system of fluorescein. Landsteiner, though not defining epitope size, demonstrated that antibodies recognize simple chemical grouping; antibodies recognize large molecules only because they display such simple groups. Antibodies can be more or less specific; a single antibody can recognize a series of related epitopes or only one. Some antibodies can even distinguish between optical isomers of single compounds.[17] Antibody specificity is biased toward conformational epitopes and they bind their antigens non-covalently, utilizing both hydrophilic and, for some, hydrophobic bond forces.

2. Antibodies as Diagnostic Indicators of Immune Function

Specific antibodies are an arm of the vertebrate organism's specific defense mechanism so that the appearance of an antibody specific for a particular pathogen is usually evidence that the organism has encountered the pathogen. Similarly, the presence of antibodies to allergens suggest an encounter with the allergen while IgG antibodies to a non-self blood group antigen usually suggests an earlier nonmatched blood transfusion. Hence, antibody titers have diagnostic significance both in human and veterinary medicine, and the ability to reliably measure antibodies lies at the heart of clinical immunology. Reliability in the measurement of specific antibodies is compromised by a plethora of variables; the lack of standardization among laboratories, the form in which antibody responses are expressed, the cross-reactivity of antibodies for related and sometimes unrelated organisms, the identification of the class of antibodies involved, the specificity of the detection antibody and variations in the performance of the assays employed. These are discussed in detail in Chapter 9.

3. Antibodies as Tools in Biotechnology

The potentially exquisite specificity of antibodies, their potentially high affinity for antigens, and the ease by which they can be obtained either as polyclonal reagents raised in animals or as monoclonals produced by hybridoma technology has made them popular tools in biology and medicine. It has been repeatedly shown that differences in proteins detected by antibodies eventually can be traced to subtle or major differences in structure; the degree of cross-reactivity of an antiserum generally parallels the degree of structural similarity. Nuttall, >80 years ago and using relatively crude methods, used antisera prepared in rabbits to demonstrate that serological cross-reactivity among >500 species of animals reflected their taxonomic and/or phylogenetic relationship.[18] Numerous studies conducted since that time and using purified proteins and tissue extracts from various types of organisms have confirmed his basic observations.

Modern immunological techniques have refined this principle, probably reaching a pinnacle with the introduction of hybridoma technology. Monoclonal antibodies (mAbs) produced by fusion of a specific splenic B-cell to an immortal B-cell tumor are able to make very fine structural distinctions in contrast to animal sera containing a spectrum of antibodies. mAbs can "find the needle *in* the haystack" whereas polyclonals "find the needle *and* the haystack". Nevertheless, monoclonal technology has by no means made polyclonal technology obsolete, a point which will no doubt be emphasized by various contributors to this book.

Biological and medical science has taken advantage of antibody specificity as a means of identifying pathogens, histological structures, and through the use of RIA and EIA, to quantify hormones, pharmaceuticals, environmental contaminants, and allergens. Antibodies made in one species can also be used to recognize those from another so that immunoglobulin levels in dilute solutions can be quantitatively measured (see Section II.B.3). Anti-antibodies are also used in assays configured to measure serum antibody responses according to the class of reactive antibody. mAbs further extend this technology by recognizing even the idiotype (see Section II.B3 and Chapters 9 and 13) of the reactive antibody! Specific antibodies are widely used in molecular biology for such solid-phase techniques as immunoblotting (Chapter 8) and, when immobilized on column supports, i.e., affinity columns, can be used to purify antigens from complex mixtures.

Antibodies can also be used *in vivo*. When tumor-specific antibodies are coupled to toxins, they can "home" to and bind tumor cells; the toxin may then kill the target. Anti-idiotypic antibodies have been used as non-pathogenic alternatives to vaccines and anti-lymphocyte antibodies have been used to delay graft rejection by blocking or killing effector cells.

B. THE ANATOMY OF AN ANTIGEN

1. Immunogenicity, Antigenicity, and Epitopes

The immune system recognizes and responds to substances (immunogens) for reasons intrinsic to the immune system and because of properties of the immunogens. These properties include such things as size, the degree of foreignness, and unique chemical structures. Antibodies which appear in response to immunogens can recognize a portion of the substance which is called the *antigen.* Low molecular weight substances, e.g., aromatic dyes, are nonimmunogenic (i.e., haptenic); however, when coupled to an immunogenic protein "carrier", some of the antibodies formed in response to these immunogenic conjugates are specific for the conjugated dye molecule. As mentioned earlier, the size of the antibody site is such that it recognizes only small chemical groupings, in the example given here, the haptenic, organic dye. These groupings are currently called epitopes or, by some, "antigenic determinants". The low molecular weight nonconjugated and nonimmunogenic dye, is called a *hapten.*

2. Factors Influencing Antigenicity

The topic of antigenicity has intrigued immunologists for some time and has been often reviewed.[19-22] Most agree that antibodies are most likely to recognize chemical grouping exposed to the solvent on the surface of an antigen. The determinant groupings are conformational and may be either continuous (formed by a series of adjacent amino acid sides) or discontinuous (formed as noncontinuous amino acids that are brought into juxtaposition by protein folding). Antibodies have also been prepared against linear peptides, either free or conjugated to a carrier (Chapter 16); whether such antibodies truely recognize linear sequences or rather sequences which form conformational structures in the test system is controversial. The size of an antigenic determinant was first estimated by Kabat and Berg.[23] These studies, which used carbohydrates, as well as later experiments with silk fibroin peptides[24] and high-affinity antibodies to fluorescein,[25] suggested that the antibody binding site was large enough to accommodate a pentasaccharide, a hexapeptide, or a fluorescein molecule. The current view that the antibody binding site can exhibit flexibility or show conformational adaptability suggests that an antigenic epitope need not be rigidly restricted in size.

Flexibility may also be a factor in antigenicity. Rigid configurations may present stable configurations whereas flexibility may allow the epitope to more readily assume the thermodynamically best and most avid configuration in the antibody binding site. Recognition is likely to depend both on the conformation of the epitope and its ability to form electrostatic, hydrophilic or even hydrophobic interaction with residues within the binding site. Chemical modifications of either antigens or antibodies are likely to alter their ability to recognize or be recognized. Such substitution procedures as biotinylation, iodination, or fluoresceination may directly change exposed epitopes or indirectly alter them through conformation changes elsewhere in the molecule. Modification of tryptophan is known to seriously alter antigenicity.[26] Conjugation with enzymes or covalent attachment to solid phases can block epitopes entirely. Surface adsorption of protein offers numerous possibilities for the inadvertent alteration of structure as evidenced by the loss of epitopes exposed in solution and the appearance of new ones[27-30] (Chapters 10 and 13).

3. The Antibody as an Antigen

Administered to an appropriately foreign recipient, the immunoglobulin molecule is of suitable size and complexity to serve as a potent immunogen and to be recognized as an antigen. The antigenic determinants of immunoglobulins have been classified according to their location on the immunoglobulin and the antibody is identified according to the epitope recognized. Immunoglobulins possess both constant and variable domains; the former determine the class, subclass, or light chain type while the latter control antibody specificity (see Chapter 9). Anti-antibodies (called antiglobulins by some) can be prepared which are specific for a single class of antibodies, e.g., IgG. These are regarded as isotype (class or subclass) specific, and the determinants recognized are called *isotopes.* Anti-antibodies can also be prepared which recognize genetic variant between individuals of a population, i.e., *allotypes,*

TABLE 1
Types of Solid Phases for Immunoassays

Type	Capacity	Bonding force[a]
	Membraneous forms	
Nitrocellulose	High	Hydrophobic (hydrophilic?)[b]
PVDF[c]	High	Hydrophobic
Nylon	High	Hydrophobic
Modified membranes	High	Covalent, hydrophilic, hydrophobic
	Plastic plate and tubes	
Polystyrene (PS)	Low	Hydrophobic[b]
Polyvinyl	Low	Hydrophobic
Derivatized microtiter plates	Low	Covalent, hydrophilic, hydrophobic
	Beads and microparticles	
PS beads	Moderate	Hydrophobic
Derivatized PS beads	High	Covalent, hydrophobic, hydrophilic
Microparticles[d]	High	Covalent and hydrophobic

[a] Information regarding bonding forces are indirect and general and the topic represents an area for fertile investigation.
[b] Solid-phase immunoassays based on immobilization of the solid-phase reactant are successful because the adsorptive process is hydrophobic and the subsequent immunochemical reactions with the immobilized reactant are largely hydrophilic. Hence, solid phases, which exhibit hydrophilic character, require the more extensive use of blocking agents.
[c] PVDF = polyvinylidene difluoride.
[d] Microparticles are <1 μm in diameter and, in contrast to beads, remain as a colloidal suspension through the assay.

by recognizing determinants called *allotopes*. Finally, anti-antibodies can also recognize determinants in the variable portion of antibodies which can also be markers for the specificity of the antibody. These determinants are called *idiotopes;* idiot means unusual, "he who occurs infrequently" (see Chapter 13). Because of the extensive heterogeneity encoded in the V-regions of antibodies vs. the constant regions, an IgG isotope could be present on >80% of all antibodies of an individual while an idiotope may only be present on 1 in 10,000 antibodies.

The antigenic properties of the antibody molecule have considerably simplified the task of immunoassay development permitting the anti-antibody or "antiglobulin" to become a powerful reagent in immunoglobulin quantitation and in identifying the isotype or allotype of the antibodies involved in the immune response to the pathogen or antigen in question.

C. THE SOLID-PHASE

1. Categories and Characteristics

Solid phases come in a variety of forms and chemistries as will be discussed in Chapters 2, 7, 8, 14, and 15. Table 1 summarizes the solid phases in common use. The three categories in use are amphipathic with highly hydrophilic surface excluded to reduce nonspecific binding. Microtiter plates are extensively used for various types of ELISA systems, especially in research labs. Membranes are used in immunoblotting procedures, and beads and microparticles are used in many automated clinical assays, in RIA, and for separation and analysis of cells. Major differences in the application of these solid phases have to do with convenience, "laboratory tradition", and specialized conditions. For example, immunoblots could hardly be made using a microtiter plate, and derivatization of microtiter plates to permit covalent immobilization of reactants can be costly in comparison to derivatization of beads and membranes. Membranes have a greater functional surface area (Chapter 8) and are amenable to flow-through sample application which accelerates the antigen-antibody reaction (Chapter 15) and have potential for multianalyte application. Selection of the solid phase is also influenced by the instrument industry which may have recently focused their efforts on the use of microtiter plates and associated robatic systems. Microparticles offer many advantages for automated, fluorescence-based analyses and are gaining widespread acceptance in clinical immunology and biochemistry. Magnetic beads are gaining acceptance in leukocyte-separation technology and offer potential advantages for immunoassays including those utilizing microtiter plates.

2. The Immobilization Process

Catt and Tregear were among the first to demonstrate that, because plastic surfaces firmly adsorbed proteins, plastic test tubes could be both a solid-phase and a reaction vessel. At present, solid-phase reactants are immobilized in one of three ways: (1) adsorption to predominantly hydrophobic surfaces, (2) covalent attachment to activated surface groups, or (3) noncovalent, electrostatic, and hydrophilic bonds between a molecule immobilized by either method (1) or (2), which in turn is coupled to a ligand by, for example, an antibody-antigen or biotin-avidin bond. The extent to which these methods are chemically discrete may not be absolute. Studies on covalent immobilization of protein to microparticles show that, while the initial bonding is >80% covalent, the effective bonds after incubation for 16 h can be > 80% noncovalent (Nustad, K., personal communication). Most material listed in Table 1 are mildly to strongly hydrophobic, thus providing the potential for secondary hydrophobic interaction even when the primary interaction is covalent. Direct covalent coupling to hydrophobic surfaces might therefore be a means of facilitating secondary hydrophobic interactions. The immobilization of solid-phase reactants is chemically heterogeneous and varies with the specific surface, ligand chemistry and conditions of immobilization.

3. Activities of Immobilized Reactants

Were there no evidence to suggest that immobilized reactants behave any differently than soluble reactants, or that the reaction of molecules in solution did not differ from those in which one reactant is immobilized, there would be little reason to write this book! Bull, in 1955,[31] presented data suggesting that adsorption of albumin on glass resulted in a conformation change of the protein; the thermo-dynamic findings of Nyilas et al.[32] and Lyklema and Norde[33] support his findings as do data obtained by other methods.[27-30,34-39] Unfortunately, most companies anxious to sell their reagents, assays, and instruments, as well as the many "how to do" laboratory manuals on ELISAs and immunoblotting, do little to warn users of this and other potential problems associated with solid-phase immunoassay; the field represent a good example of getting the "technological cart ahead of the scientific horse". Figure 1 depicts the various conformational changes, changes in epitope or antibody binding site availability and concentration which can occur as a result of immobilization on the solid phase.

The following scenario may describe protein adsorption on plastic. Molecules incubated with plastic, in an effort to achieve thermodynamic stability, orient their hydrophobic region toward the adsorptive surface. Achieving the most energetically favorable conformation on the surface may thus alter or destroy epitopes that are normally expressed by these molecules in solution (Figure 1D). Molecules which undergo little conformational change may nevertheless express some of their epitopes cryptically after adsorption (Figure 1B). The loss of epitopes during immobilization need not be restricted to adsorptive methods but some might also become sterically hindered or destroyed by covalent attachment (Figure 1A). The adsorptive process appears complex; it appears to be biphasic with differences in binding constants

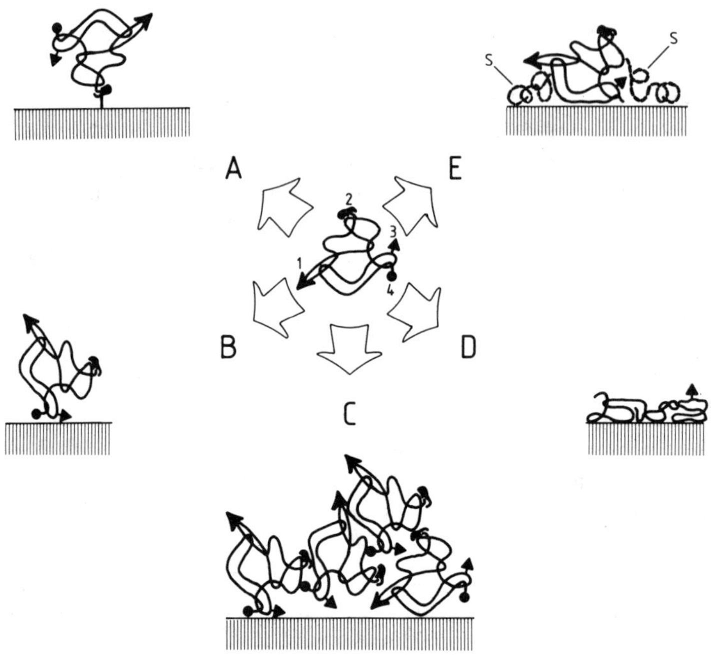

FIGURE 1. The effects on a soluble protein of immobilization on a solid phase. The hypothetical soluble globular protein at the center of the diagram expresses four functional groups, antigenic epitopes, enzyme active sites, etc., indicated by different symbols 1 to 4. Five different examples of conformational or steric effects resulting from immobilization on a solid phase (shaded, vertical bar) are illustrated on the peripheral diagrams A to E. (A). Covalent attachment resulting in loss of functional group #2, and if the hypothetical protein were an antigen, possible steric hindrance of epitope #3. (B) Immobilization by adsorption without conformational change but with functional groups #3 and #4 becoming cryptic and nonfunctional. (C) Adsorptive immobilization which proceeds through the process of aggregation resulting in the steric hindrance of certain functional groups but an increase in the local concentration of others. (D) Adsorptive immobilization resulting in conformational change and loss of all but one functional group. (E) Immobilization by adsorption in the presence of a stabilizing protein(s) which prevents the formation of secondary surface interaction and thus reduces conformational deformation.

associated with each phase.[40,41] The relationship between the amount of antigen bound vs. the amount added is initially linear, but thereafter becomes logarithmic[42] (Figure 2A). The concentration dependence of the binding plot may reflect the concentration-dependent thermodynamic changes reported by Nyilas et al.;[32] protein adsorbed at low concentration (corresponding to the log-log linear region, Figure 2A) undergoes the most conformational change. Kochwa et al.[36] reached a similar conclusion reporting that IgG adsorbed on polystyrene latex at low concentration behaved immunogenically like heat-denatured material. A progressive loss of antigenicity was also observed when progressively lower amounts of rabbit IgG were adsorbed on microtiter wells; this could be overcome if albumin was coadsorbed[33] (Figure 1E).

Within the linear binding region, the ratio of bound:free protein is constant but the percentage adsorbed is protein specific, i.e., each protein has its own unique affinity for the hydrophobic polymer surface (Figure 2B). A similar difference in affinity was observed by others[43,44] and fibrinogen (high affinity), can even displace previously adsorbed albumin.[45] High molecular weight bio-molecules may adsorb best probably due to multisite attachment.

The changing ratio of bound:free protein in the logarithmic phase of the binding plot (Figure 2A) suggests that the equilibrium constant changes with concentration as shown by

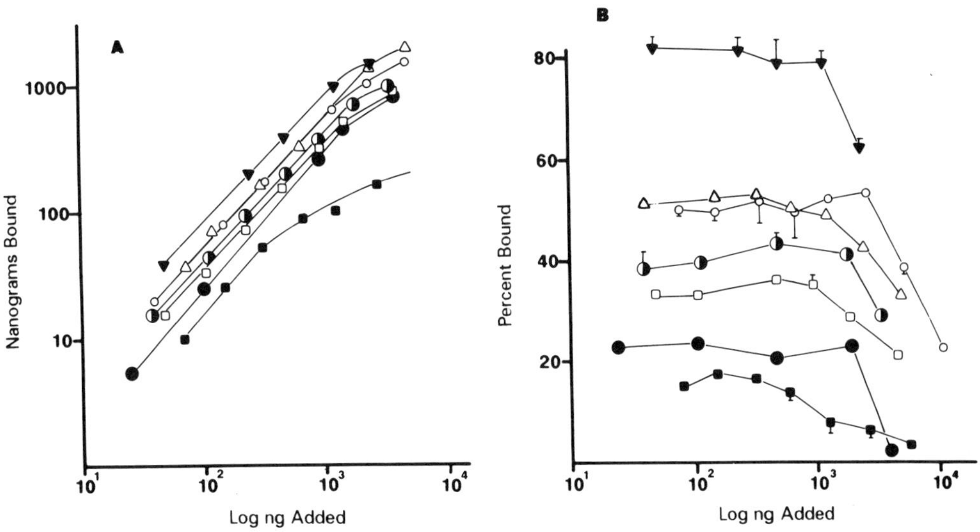

FIGURE 2. The adsorption characteristics of six different proteins on polystyrene.[42] (A) Data are plotted in the form of a log-log saturation plot whereas in B, data are presented in terms of the percent bound as a function of the amount added (see Figure 3 for details concerning the interpretation of Figure 2B). Legend for proteins: ▼ bovine IgM, Δ bovine SIgA, ○ ovalbumin, ◑ bovine IgG1, □ bovine IgG2, ● bovine serum albumin, ■ α-lactalbumin.

the biphasic Langmiur plots of Oreskes and Singer.[39] One explanation is that the adsorbing protein can exist in different conformation, i.e., it is allosteric, that these conformations are concentration dependent and that each has a different affinity for plastic. If one equates adsorption to solubility chemistry, increasing the concentration of protein usually increases the tendency for aggregation which in turn increases the tendency toward insolubility. Insolubility or aggregation might be a prelude to adsorption such that adsorbed proteins occur as aggregate clusters (Figure 1C). Because aggregation is typically concentration dependent, molecules adsorbed at different concentrations might be expected to be more or less aggregated and hence express different conformations. This would be consistent with the fact that most proteins are least soluble at their Ip and that various investigators have reported that hydrophobic adsorption is maximal near the Ip of the protein.[31,41,46] EM data show that surface-adsorbed protein and antibody bound to surface adsorbed antigen are not uniformly distributed and not all of the surface is covered.[43,47]* The cooperative binding of monoclonals reported by Greenspan et al.[48] would require clustering of unrelated antigens as would the activation of complement by adsorbed IgG.[49]

The influence of pH on adsorption on plastic remains controversial; early studies indicated a pH dependence such that the "classical", pH 9.6 carbonate buffer continues to be widely used in ELISAs. There is evidence that antibodies exposed to low pH prior to adsorption, display higher activity than when stored at neutral pH.[50-53] The higher-antigen capture capacity of polyclonal capture antibodies prepared using caprylic acid is probably not due to acidity.[54] Blotting membranes, especially NC, interact hydrophilically as well as hydrophobically with protein, so that pH is more important for adsorption on this solid-phase than it is for polystyrene (see Chapter 8).

There is a paucity of quantitative data regarding differences in capacity and stability of molecules adsorbed on different surfaces. Desorption can occur and this can be significant when high concentrations of protein are used.[55-57] While some studies report little differences in adsorption of several proteins to a spectrum various hydrophobic surfaces[43] (Chapter 12),

* Unpublished data by the author, obtained with the Hitachi S-4000 scanning EM confirm the absence of any monolayer and suggest the formation of clusters comprising many antibody molecules.

capacity differences among microtiter plates and nitrocellulose membrane with differing chemistries and/or from different manufacturers are known[58] (Chapter 8). Mobility and/or flexibility may also vary among adsorptive surfaces; it is known that albumin adsorbed on glass continues to creep until inhibited by other molecules.[39]

Covalent attachment of the solid-phase reactant may permit a higher concentration of reactants to be immobilized than by adsorption, although covalent bond formation may also cause conformational change, block epitopes, or chemically alter them (Figure 1A). Adsorption on microtiter plates permits equal or greater amounts of antibody to be immobilized than either covalent attachment or attachment via a streptavidin linkage (see Chapters 11 and 12). Monoclonal antibodies immobilized via a streptavidin linkage show the highest functional activity when compared to covalent and hydrophobically immobilized antibodies.[59]

The literature suggest that not all solid-phase reactants are created equal, that they can be expected to behave differently from each other and from reactants which interact in the fluid phase. Hopefully, the biochemistry of the various immobilization processes will become better understood so that the conformation of solid-phase reactants can be predicted and thus their functional behavior optimized.

D. THE SIGNAL GENERATORS*

1. Fluorescence, Radioactivity, and Enzyme Systems

Gamma radiation from radioactive iodine was the signal generated by antigen in classical RIA;[60] indirect competitive RIA or immunoradiometric assays (IRMA) utilizing radio-iodinated antibodies. Iodinated reagents have a comparatively large number of disadvantages when judged by 1990 standards. First, universities and clinical laboratories utilize far more immunoassays than 2 decades before, and with the more rigorously enforced Health Physics standards,[61] the widespread use of ^{125}Iodine in crowded, multi-use laboratories, especially those involving students where personnel changes are frequent, is becoming prohibitive. Second, ^{125}I and ^{131}I have half-lives of 57 and 9 d, respectively, which makes it necessary to continuously generate or purchase fresh signal generators. Radioactive iodine still remains a fundamental tool of immunochemists for studying reaction kinetics and stoichiometry. The introduction of one mole of iodine per mole of protein provides a strong detection signal and usually does not alter the biochemical or conformational properties of proteins; iodinated peptides may suffer some alteration by this modification.[62] By contrast, fluoresceination, biotinylation, and conjugation with enzymes can alter the biochemistry of the labeled compound. The other common radionuclides used for proteins, e.g., ^{3}H, ^{14}C, ^{35}S, must be introduced metabolically. ^{99m}Tc may also be used for proteins but its very brief half-life is impractical in immunoassay. The signals generated by ^{3}H, and particularly ^{14}C, are weak compared to iodine, making them poor choices for use in modern immunoassays in which picogram-nanogram of analyte or reactant must be detected.

Fluorescent labeling of proteins, especially antibodies, is a method of wide use for antigen localization. Initially confined to use with bacteria, cell suspensions, and tissue sections, fluorescein-labeled antibody technology had a rebirth with the growth of flow cytometry. Fluorescent signals are also in widespread use in clinical immunoassays. The physical chemical aspects of fluorescence and its use in immunoassays is discussed in Chapter 4. The lack of any biohazard associated with fluorescence has, together with major advances in instrumentation, been responsible for its widespread use.

Enzymes are complex signal generators in contrast to radionuclides and fluors. In contrast to direct fluorescent** and radioactive labels which generate either a constant or diminishing

* Signal generators or "reporter" reagents have become vernacular in current biotech jargon.

** Some EIAs are designed to generate a fluorescent product; these are nevertheless defined as EIAs, not fluorescent immunoassays, because the signal-generating label is an enzyme, not a simple, direct fluorescent label.

signal with time, enzymes generate progressively more signal with time providing both amplification and the opportunity for measuring immunoreactivity as a function of the enzyme reaction rate. The use of enzymes as signal generators was pioneered in the field of immunohistochemistry where peroxidase was used to convert substrates like diaminobenzoate to an insoluble product which remained at the site of enzyme localization. A variety of methods involving single enzymes,[63] soluble enzyme-antibody amplification systems,[64] and even small peptides in electron microscopy with peroxidase activity,[65] have been described. Avrameas and Guilbert[66] and Engvall and Perlmann[14] were first to apply this form of signal generator to quantitative immunoassay. Thereafter, a parade of modifications followed, some involving different enzymes, new substrates, amplified systems and configurational changes. Collectively these became known as either ELISA or EIA. In fact, ELISA has come to refer to not only a method of detection, but rather a complete system involving solid-phase immunoassay as an integral part. Currently, the advantages of enzymes and fluors have been married to yield a number of hybrids such as the generation of fluorescent enzyme products (Chapter 4).

Enzymes as signal generators overcome all the disadvantages of radioactive labels but when used in heterogeneous immunoassay are more cumbersome than fluorescent labels because of the additional enzyme-substrate step involved. Hence, enzymes as signal generators have been largely confined to heterogeneous solid-phase immunoassay conducted in, for example, microtiter plates or to homogeneous EIA.* Their macromolecular and substrate-dependent character has virtually kept them out of the field of flow cytometry. Reasons for the popularity of enzymes as signal generators in solid-phase immunoassay are (1) their ease of use, (2) the substantial amplification in signal which they can generate, (3) their applicability to measurement of reactions by rate analyses, (4) the fact that the user views a visual reaction, and (5) the considerably less expensive instrumentation needed for detection. The last statement requires some qualification. EIA can be done with only the naked *eye* as a detection unit. These assays are only qualitative but are useful in field applications. At the opposite extreme, an automated "ELISA reader" complete with computer and other gadgetry, can be as expensive as a radioactivity counter. The importance of enzymes as signal generators in solid-phase immunoassay justified devoting Chapter 5 to this topic.

The student or first-time user of immunoassay should not assume that conversion from RIA to EIA is a simple matter of changing the signal generator. Classical RIA is a direct competitive immunoassay and substituting a protein enzyme for an atom of iodine is sure to change the biochemistry, and perhaps reactivity, of the labeled reactant. Typically, assays formerly done by RIA and later converted to ELISA, usually involve either a switch to an indirect competitive assay (see Section A) or to a homogeneous enzyme immunoassay.

2. Novel Detection Systems

The widely used detection systems which generate radioactivity, color, or fluorescence may give way to newer technology in future solid-phase immunoassays. For example, the binding of antibody to solid-phase antigen causes charge redistribution that can be used to generate electrical impulses which can in turn be converted to an optical signal. Chemiluminescence and time-resolved and internal-reflection fluorescence (Chapter 4) may become more popular as well as surface-enhanced Raman spectroscopy.[67] Reaction rate analyses of EIAs, although widely used, might still be considered as novel by some.

3. Sensitivity and the Detection System

It is interesting to read the claims of manufacturers about the sensitivity of their immunoassay or to read textbook dogma stating "ELISA systems are very popular but lack the sensitivity

* Homogenous EIA is a solution-phase system dependent on modification of enzymic activity as a result of antibody-antigen interaction. Homogenous immunoassays do not fall into the scope of solid-phase immunoassay.

of RIA and fluorescent-based systems." The mid 1970's to mid 1980's saw reports of ELISA claiming ever-improved sensitivity. While some of these reports did provide a logical basis for their improved sensitivity, e.g., signal amplification, the remainder probably did not improve sensitivity when used in an application other than in the specialized system for which they were initially described. It is common to hear an investigator say that an assay "works great"; "works great" can usually be distilled to (1) the enzyme-substrate reaction occurs more rapidly, (2) the antiglobulin reactant can be used at a much higher dilution, (3) the linear binding region is longer, or (4) lower amounts of the analyte can be detected. The latter is the desirable meaning of the jargon and could result from (1) a higher signal:noise ratio, i.e., less background, (2) a greater quantity of the signal generator bound per molecule of analyte measured, or (3) a greater activity of the signal generator per molecule of bound analyte. Only investigators directly measuring the binding of the analyte, as in the case of ^{125}I-antigen, have *direct evidence* that increased signal means an increase in bound analyte. For most ELISA systems, such evidence is only indirect which then raises the issue as to *what* is most important in increasing the sensitivity of an assay? Improvements in sensitivity are most convincing when investigators directly measure the stoichiometry at every step in the assay. This approach can explain *how* the same amount of analyte can produce a greater signal in one system than in another. Indirect evidence to explain improvement in detection can be gained by showing that changing one reagent dramatically improves or decreases detection of the analyte. As typical examples, using a different CAb in a sandwich assay or different antiglobulin in specific antibody immunoassays (SpAbI) can greatly increase detectability. Hence, it is often not the novel configuration of a solid-phase immunoassay which improves performance, but rather the higher affinity of a critical antibody used. This lesson was learned early in the field of RIA when stories of famous rabbits producing high-affinity antibodies and living in gold cages on an *ad libitum* diet of carrots and lettuce, were legend. A common reason why the acclaimed sensitivity of a published immunoassay cannot be duplicated in a remote laboratory, even though the actual detection system is identical (having been purchased from, for example, Sigma Chemical Co.), is often due to an antibody. The goal of improved sensitivity is best achieved by increasing the affinity of the relevant antigen-antibody or receptor-ligand interaction so that a greater portion of receptor (antibody sites) are occupied by ligand (Chapter 6 and Figure 3). The diffusion dependence of macromolecular interactions means that longer times will be required to reach equilibrium in nonagitated vs. agitated system. Hence, the claims of "increased sensitivity in ELISAs" by some companies are merely achieved by reducing the time to equilibrium through agitation without per sec changing the sensitivity of the immunochemical reaction. In other words, improvement is not the result of the signal generator itself, but is rather a configurational and kinetic issue. The claim that RIA is more sensitive than EIA or one ELISA is more sensitive than another, is equivalent to saying that all Chevrolets go faster or cost more than Fords without specifying age, model, or driver. *The Immunochemistry of Solid-Phase Immunoassay* is written with the goal of helping students and users of this technology to understand the *immunochemical principles* of these assays so that "the slow Ford in your future can be made to go faster or at least you can understand why not".

III. IMMUNOCHEMISTRY AND SOLID-PHASE IMMUNOASSAY

A. SOLID-PHASE IMMUNOASSAY AND THE LAW OF MASS ACTION

Solution-phase, biomolecular noncovalent reactions are known to proceed according to the Law of Mass Action with similar forward rate constant (k_1) but with very

$$\text{Ab} + \text{Ag} \underset{k_2}{\overset{k_1}{\rightleftharpoons}} \text{AgAb} \qquad (1)$$

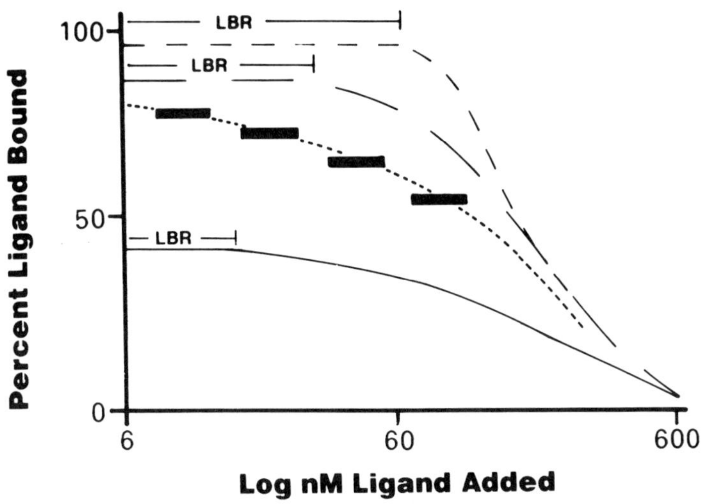

FIGURE 3. The Law of Mass Action for receptor-ligand interactions expressed as the percentage of bound ligand as a function of the amount added. LBR = linear binding region, where the receptor is in large excess and percentage of bound ligand remains constant. The extent, i.e., length, of the LBR is proportional to receptor concentration. Short dashes = high-affinity homogeneous receptor-ligand interaction; large dashes = moderate affinity, homogeneous interaction; solid line = low-affinity homogeneous interaction; dotted line with dark bars = polyclonal (heterogeneous) receptor-ligand interaction. In the latter, the dark bars represent homogeneous subpopulations of progressively lower affinity (left to right); the mixture of these describe the curve dotted line characteristic of a polyclonal receptor performance.

different dissociation rates (Table 2) so that κ_2 largely determines the equilibrium constant ($K_{eq.}$; Equation 2). Hence, the K_{eq} can be estimated by assuming an "average" κ_1, e.g. 5×10^7 m^{-1} s^{-1}, and then determining κ_2, which is slower and therefore easier to measure. The relatively similar association constants, whether involving antibodies or enzymes, suggests that encounter is diffusion controlled. It has been calculated that the diffusion-controlled encounter rate constant is 10^9 s^{-1}, m^{-1} suggesting that the 1- to 2-log lower association rates observed (Table 2) are due to two-step processes that may involve dissociation of intermediate encounter complexes[74] (see Chapter 19). Solid-phase antibody-antigen reactions appear to associate more slowly (Table 2), and their dissociation, as well as the dissociation of antibodies from cell surfaces, may also proceed slowly and not by simple first-order rates.[15,16]

The equilibrium constant of an antigen-antibody reaction can also be

$$K_{eq.} = \frac{k_1}{k_2} \tag{2}$$

TABLE 2
The Kinetics of Antibody-Antigen and Enzyme-Substrate Reactions

System	Association rate (k_{-1}) (m^{-1} s^{-1})	Dissociation rate (k_2) (s^{-1})	$K_{eq.}$	Ref.
	Solution Systems			
MOPC-315 (IgA)	1.3×10^8	5.3×10^1	2.5×10^6	68
Sheep antidigoxin (IgG?)	1.7×10^7	3.4×10^{-4}	5.0×10^{10}	69
Rabbit anti-DNP (IgG)	8.4×10^7	1.1×10^1	7.7×10^6	70
mAb 4-4-20 (IgG2b anti-FLU)	5.6×10^6	1.11×10^{-2}	5.0×10^8	71
mAb 93.6	1.4×10^6	5.0×10^{-4}	2.8×10^9	72
Rabbit anti-hemoglobin S(Val)	0.6×10^6	2.7×10^{-5}	2.2×10^{10}	73
Chymotrypsin × proflavin	1.2×10^8	8.3×10^3	1.45×10^4	74
Tyrosyl - tRNA synthetase × $tRNA^{tyr}$	2.2×10^8	1.5	1.47×10^8	74
Trypsin × pancreatic trypsin inhibitor	1.1×10^6	6.6×10^{-8}	1.67×10^{13}	74
Alcohol dehydrogenase × NADH	2.5×10^7	9	2.78×10^6	74
	Interfacial Systems			
Anti-lymphocyte antigens	0.8×10^5 to 1.1×10^6	6.0×10^{-4a}	1.3×10^8 2×10^9	15
Various mAb to immobilized haptens	0.57 to 5.1 $\times 10^3$	1.7 to 6.4 $\times 10^{-4}$	3.4×10^6 to 7.9×10^6	75
Anti-arsanilate mAb to immobilized arsanilate-BSA	8.7×10^3	2.47×10^{-4}	3.52×10^7	76

[a] Dissociation was biphasic, value given is for initial dissociation rate.

empirically determined by incubation of a known amount of antibody with various concentrations of excess antigen. At equilibrium, the concentration of bound and free antigen is determined and K_{eq} is calculated (Equation 3).

$$\text{Keq.} = \frac{[\text{AbAg}]}{[\text{Ab}_f]\,[\text{Ag}_f]} \tag{3}$$

where [AbAg] = molar concentration of bound antibody binding sites or bound antigenic epitopes; $[\text{Ab}_f]$ = molar concentration of free antibody binding sites and $[\text{Ag}_f]$ = molar concentrations of free antigenic epitopes. As these reactions are generally conducted using monovalent antigens, i.e., low molecular weight ligands, the complications of multivalent binding effects are excluded. Hence, $K_{eq.}$ becomes an estimate of the intrinsic binding site $K_{eq.}$, or the affinity constant, which could be much lower than the $K_{eq.}$ for reactions involving polyvalent antibodies and multivalent antigen with repeating epitopes.* Equation 3 can be modified to provide an estimate of valence (n) as shown below.

$$\text{Keq.} = \frac{[\text{AbAg}]}{[\text{nAb}_f]\;[\text{Ag}_f]} = \frac{[\text{Ab}_b]}{[\text{nAb}_t\text{–Ab}_b]\;[\text{Ag}_f]} \tag{4}$$

* Multivalent interactions lead to slower dissociation rates, i.e., "sticky interactions". The K_{eq} of such interactions therefore reflects these longer off rates and for such interactions *avidity* rather than *intrinsic affinity* is a more appropriate term.

where $[Ab_b]$ = molar concentration of occupied binding sites which numerically equals the molar concentration of the complex and bound epitopes. n = antibody valence and Ab_t denotes the total molar concentration of antibody.

Equation 4 can be algebraically converted into a number of widely accepted graphic versions of the Law of Mass Action including those developed by Scatchard,[77] Langmuir,[78] and Steward and Petty.[79] The Steward-Petty plot can be used to estimate the number of functional binding sites which for IgG is twice the $[Ab_t]$. The $[Ab_t]$ estimated by this method or determined independently, can then be used in Scatchard or Langmiur equations to determine K_{eq} and n.

An instructive derivation of the Mass Law is the one graphically presented in Figure 3, discussed in Chapter 3, and applied to solid-phase CAb in Chapter 12. The equation of this graph predicts that for a homogeneous system *in the region of excess solid-phase CAb* (Figure 3) the percentage of Ag bound at equilibrium remains constant and its numerical value is a function of the affinity of the interaction. The range over which the percent bound remains constant, i.e., "linear binding region" (LBR), depends on the functional concentration of the solid-phase receptor. The equation is also applicable to studies on the immobilization of protein on a solid phase. The data in Figure 2B therefore indicate that polystyrene has a higher affinity for IgM than α-lactalbumin but suggest that, on a molar basis, the concentration of chemical groups on the plastic which are able to interact with both proteins is approximately equal. Percent bound plots are especially applicable for the study of solid-phase interactions in which the vagueries about actual reaction volume prohibit accurate determination of concentrations as are needed for saturation, Scatchard, and Langmuir plots.

Another useful derivation of the Mass Law is discussed by Ekins (Chapter 6). This quantitation predicts that at equilibrium the proportion of occupied sites becomes a constant which is determined by the affinity of the interaction. While increasing the concentration of the solid-phase reactant would increase the amount of bound ligand, occupancy (fraction of filled sites) would remain constant. This "occupancy principle" predicts that increasing the sensitivity of an assay through addition of more solid-phase reactant would only occur provided the "noise" generated by the additional solid-phase reactant *did not* increase in parallel with the specific signal. Because this is unlikely, increasing sensitivity would be best achieved by increasing occupancy which means increasing the equilibrium constant of the reaction.

The various equations reviewed above have been derived from studies of solution-phase ligand-receptor interactions. As discussed in Chapter 19, these may not be directly applicable to interfacial antigen-antibody interactions. Equation 5 presents a hypothetical scheme which may more accurately reflect interfacial kinetics.

$$Ag_{SLD}{}^{*} + Ab_{SOL} \underset{k_2D}{\overset{k_1D}{\rightleftharpoons}} Ag_{SLD} \text{ --- } Ab_{SOL} \underset{k_2R}{\overset{k_1R}{\rightleftharpoons}} Ag_{SLD} - Ab_{SOL} \underset{k_2A}{\overset{k_1A}{\rightleftharpoons}} (Ag_{SLD}Ab_{SOL}) \quad (5)$$

Equation 5 includes a set of rate constants (k_1D and k_2D) which govern the mass transfer of reactants to the site of the initial interfacial ligand-receptor interaction. This diffusion-dependent phase can be overcome by various methods: (1) a vigorous agitation, (2) confinement of the reaction volume, and (3) use of microparticle volume (see Chapter 18). The reaction rate dependent phase of the interaction (k_1R, k_2R) is perhaps most similar to fluid-phase kinetics, although, as stated above, both forward and reverse rates may be 1 to 2 logs slower than those reported for fluid-phase systems (Table 2). Equation 5 also includes a set of rate constants (k_1A, k_2A) which govern the possible interfacial coagulation or aggregation of passively adsorbed reactants via translational diffusion on the synthetic interface. There is evidence from a number of sources that interfacial surface clusters are a reality in SPI.[43,47-49]

* SLD = solid phase, SOL = soluble.

B. DO SOLID-PHASE IMMUNOASSAYS FOLLOW THE RULES OF SOLUTION PHASE ASSAYS?

Our concept of antibody-antigen reactions has been almost exclusively based on the study of antigen-antibody reactions in solution; application of this concept to solid-phase systems raises various interrelated questions.

1. What is the molar concentration of the immobilized receptor (antibody or antigen) and its reaction volume?
2. Is the functional concentration of immobilized antibody or antigen the same as their actual concentration or is functionality lost or changed during immobilization?
3. Are immobilized reactant uniformly distributed or do they exist as aggregate clusters?
4. Are solid-phase reactants stable or do their actual or functional concentrations change during the course of the immunoassay?
5. When and under what conditions is equilibrium attained?
6. What is the dissociation rate of the primary antibody-antigen reaction and how is it affected by the subsequent reactions involved in heterologous solid-phase immunoassays?
7. How does solvent constituency, i.e., the presence of detergents, blocking proteins, and salts, affect receptor-ligand interaction in solid-phase immunoassays?
8. Is antigen cross-linking and/or cooperative binding more or less likely in solid-phase than in solution-phase immunoassays?
9. Does the high concentration of solid-phase reactant (due to a very small reaction volume) substantially reduce the dissociation rate of the ligand-receptor complex?
10. Is the performance of reagents used in fluid phase immunoassays the same as in solid-phase immunoassays?

The Mass Law equations (Equations 3 and 4) require that values be entered as concentrations. Without knowing the answers to the first four questions, values are only guesses. In addition, such values refer to that obtained at equilibrium (the fifth and sixth questions), and in solid-phase immunoassay this may be different than in a solution-phase system and may differ considerably among different types of solid-phase immunoassays. Furthermore, application of the Mass Law to antibodies and haptens in solution assumes independent binding site behavior which is probably not the case for most solid-phase systems in which multisite binding stabilizes the interaction (the eighth question). This may partially explain the slow dissociation rate (Table 2) and perhaps complex reversibility of solid-phase antibody-antigen interaction.[16,80]

"Solid phase" is in itself a generic term encompassing a broad spectrum of technologies. Solid phases differ in adsorptive capacity (see Table 1 and Chapter 8), adsorbed biomolecules may differ in their distribution and functional properties, and covalently attached molecules may be differently distributed than adsorbed biomolecules. Some molecules may undergo considerable conformational change during immobilization while others may not. Therefore, questions may have different answers in different systems. In other words, the broad spectrum of solid-phase systems and configurations in use suggests that the behavior of a reactant may not be extrapolated from one system to the next. This consideration is more relevant for immobilized antigen than for antibodies because the latter are closely related members of the same family of proteins. Nevertheless, the affinity expressed by an immobilized monoclonal CAb (1) may be different than when the same monoclonal binds its antigen in solution and (2) may differ significantly depending on the type of solid phase, concentration, and method of immobilization. Hence, the term functional affinity, used by Steward,[81] may be especially meaningful in solid-phase systems.

TABLE 3
The Quantitative Relationship between Antibodies Adsorbed on Plastic and Those used as Solution Phase Reagents

Type of antibody preparation	Antibody (ng) per 200 μl of solution phase[a]			Antibody (ng) Adsorbed as CAb[a,b] from			
	1/50	1/500	1/5000	Whole serum	SAS[c] Ppt	Pro A[d] Pure	Affi[d] Pure
Whole serum (WS)	6,000	600	60	10.7	41	62	500
Specific[e] IgA in WS	5,400	540	54	9.7	37	56	450
Contaminating anti-L chain in WS	300	30	3	0.5	2	3	25
Contaminating anti-γ chain in WS	300	30	3	0.5	2	3	25

[a] Calculated for a microtiter system using 200 μl per well.
[b] CAb (capture antibody) is applied at the optimal concentration of 5 μg/ml or 1 μg per well based on total protein. Values are for specific antibody of which 50% is adsorbed on Immulon 2 (Dynatech), Maxisorp, or Polysorp (NUNC) plates (see Chapter 12).
[c] The IgG content of a 35% saturated ammonium sulfate (SAS) precipitate is 65% which contains all the antibody. Many companies supply reagents in this form.
[d] Assumes 100% IgG and 100% specific antibody, respectively.
[e] In this example, 90% of the antibody is IgA specific.

C. TESTING PROTOCOLS

The protocols for solid-phase immunoassay are often "hand-me-downs" from technician to technician, few of whom ever investigate their validity or study the variables involved. Such technicians may work in laboratories supervised by investigators who have little or no immunochemical experience and often heavy clinical or other responsibilities which places control of assay quality and validity on the shoulders of the technician. How many protocols and procedures for data analyses (1) have their origin in scenarios like the one above, (2) are accepted on faith from a published procedure, or (3) are the end product of optimization through knowledgeable testing by the current user? For example, how many investigators still adsorb proteins on plastic in carbonate buffer, pH 9.6 because this was done in a 20-year-old classic publication despite the fact that pH may not affect adsorption or that adsorption would be maximal at the pI of the protein in question? Has the scientific community fallen victim to the belief that all published methods are a product of judicious application of the scientific method? Perhaps the current structure of scientific support places too much pressure on scientists to publish on biological or immunobiological discoveries, while regarding methods validity as second-rate science; investigators may feel that they dare not take time from experiments for which a grant was awarded, to validate the assay used. Associated with this "catch 22" is the absence of a federal agency which regulates the quality of marketed immunological reagents. A product labeled "anti-human IgA, heavy chain specific" can be marketed by an individual or company without meeting any specificity guidelines. Some companies may sell two products with the same designation but specify one as highly specific and "ELISA quality"; does this imply that the other reagent *is not specific?* Perhaps a more important question is "ELISA specific but in what type of ELISA"? Theoretically and empirically, it can be demonstrated that an anti-Ig, e.g., anti-IgA, will appear more specific when adsorbed as a CAb on a solid phase than when tested as a solution phase reagent against IgG and IgA on a solid phase. Assuming no loss of activity, adsorption of the globulin fraction of a polyclonal anti-IgA would result in 37 ng of specific anti-IgA being immobilized but negligible amounts of the contaminating cross-reactive antibodies (Table 3). On the other hand, 200 μl of unfractionated whole anti-IgA serum when used as the solution phase anti-IgA reagent to measure IgA bound on a solid phase would contain a comparable amount of anti-IgA only if used at a dilution of 1:5000 (Table 3). More important is that a 1:500 dilution of whole serum would contain *as much contaminating antibodies* as the adsorbed globulin contains anti-IgA (Table 3)! Hence, when used as a solution-phase reagent the concentration of the added specific anti-IgA or contaminating antibodies is controlled by the user so that a manufacturer can claim specificity for a reagent either by the contrived experiment of making it a CAb on plastic or by testing it at high dilution against adsorbed antigen so that cross-reactive antibodies are at such low concentration that their effect is not observed. A purchaser and user, not knowing at what dilution or in which manner specificity was established, might easily observe considerable cross-reactivity if the same reagent was used at low dilution.

A fundamental rule for all immunological tests, applicable to the example in Table 3, is that "the specificity of an immunochemical reagent should be demonstrated in the same system in which it will be used". Furthermore, specificity follows the Law of Mass Action so that decreasing concentrations (controlled by the users or as a consequence of adsorption) increases apparent specificity. As more assays are offered in the form of kits with fixed reagent concentrations, more reproducible results should be obtained among laboratories.

Incubation time is a protocol-related, Mass Law consideration. How many protocols suggest incubating the primary serum with the solid-phase antigen for only 1 to 3 h even though equilibrium may require 16 to 24 h (Chapter 18)? Supplier "A" may be able to claim greater sensitivity in an assay merely by suggesting a longer incubation time than supplier "B"; supplier "C" may claim greater sensitivity than supplier "A" because his system utilizes vigorous agitation to facilitate attainment of equilibrium. The use of higher reagent concentrations can also increase apparent sensitivity but this must be weighed against (1) cost and (2) the possibility that higher reagent concentrations will expose unwanted cross-reactivity or increase nonspecific binding that will manifest as reduced signal:noise ratios (see Section II.D and Table 3).

The same Mass Law considerations which apply to the primary antibody-antigen reaction also apply to the detection phase, e.g., a third antibody, or antibody-enzyme conjugate, or even a bacterial Ig receptor protein such as Protein-A. Ideally, the $K_{eq.}$ of interaction involving the detection system should be high. High-affinity monoclonal or polyclonal antiglobulins are most effective and the high affinity of the avidin-biotin reaction makes it an effective and popular part of any detection system (Chapter 17).

The advent of enzymes as signal generators also increased the probability that steric factors can play important roles in solid-phase immunoassay.[82,83] Testing protocols must therefore also consider this factor in their design.

D. SPECIFIC IMMUNITY AND IMMUNOASSAY

A major application of solid-phase immunoassay is the measurement of humoral immune responses in man and animals. Investigators undertaking such measurements need to be mindful of (1) the degree of binding site heterogeneity of antibodies even to a single, defined epitope, (2) the isotype heterogeneity of these antibodies even when binding site heterogeneity is limited, and (3) the dependence of all immunoassays designed to measure antibodies, i.e., specific antibody immunoassays (SpAbI), on the affinity spectrum of the antibody population being studied. Chapter 9 is devoted to SpAbI and also reviews the origin of antibody heterogeneity in more detail.

A popular misconception is that at equilibrium *all antibody* is bound to the immobilized antigen provided that the amount of added antibody falls in the titratable range and that the antigen is in excess. The mistruth in this concept is illustrated theoretically (Figure 3) and can also be demonstrated empirically by transferring the supernatant from a well that has reached equilibrium to a new well containing antigen.[84] Figure 3 illustrates how different percentages of the antibodies present in a specimen are bound to their antigen at equilibrium as a function of their affinity so that they cannot be easily quantified in absolute, gravimetric units, e.g. μg/ml. At most, the detected antibody represents only the *bound* antibodies, not the total antibody in the sample. Hence, SpAbI are more realistically measures of *antibody activity*. The affinity dependence of SpAbIs is a predictable experimental reality, effecting all immunoassays as well as being influenced by the manner in which they are performed.[81,85-87] Affinity differences are a biological reality, not an *in vitro* serological artifact.

The objectives of *Immunochemistry of Solid-Phase Immunoassay* are multiple. First, an attempt is made to integrate the ideas and data of investigators from diverse disciplines who share the common denominator of using solid-phase immunoassays in a manner that brings clarity to some of the major and often confusing immunochemical aspects of these assays. Second, this volume collects the data and views of over 25 scientists thus giving readers the opportunity to obtain in one volume a collection of ideas from some of the most knowledgeable scientists in the field. Finally, the book attempts to stimulate the current and potential users of solid-phase immunoassays to more carefully consider the underlying principles involved in these assays as an aid in the design of current and future assays to optimally measure the immunobiological parameter of greatest importance in their respective scientific endeavors.

REFERENCES

1. **Klinman, N. R., Rockey, J. H., Frauenberger, G., and Karush, F.,** Equine anti-hapten antibody. III. The comparative properties of γG and γA antibodies, *J. Immunol.*, 96, 587, 1966.
2. **McGuire, T., Musoke, A. J., and Kurtti, T.,** Functional properties of bovine IgG1 and IgG2: interaction with complement, macrophages, neutrophils and skin, *Immunology*, 38, 249, 1979.
3. **van der Zee, J. S., van Swieten, P., and Aalberse, R. C.,** Serologic aspects of IgG4 antibodies. II. IgG4 antibodies form small non-precipitating immune complexes due to functional monovalency, *J. Immunol.*, 137, 3566, 1986.
4. **Knight, K. L., Burnett, R. C., and McNicholas, J. M.,** Organization and polymorphism of rabbit immunoglobulin heavy chain genes, *J. Immunol.*, 143, 1245, 1985.
5. **Greenbury, C. L., Moore, D. H., and Nunn, L. A. C.,** Reaction of 7S and 19S components of immune rabbit antisera with human group A and AB red cells, *Immunology*, 6, 421, 1963.
6. **Sarvas, H. and Mäkelä, Ö.,** Haptenated bacteriophage in the assay of antibody quantity and affinity: maturation of an immune response, *Immunochemistry*, 7, 933, 1970.
7. **Hornick, C. L. and Karash, F.,** Antibody affinity. III. The role of multivalency, *Immunochemistry*, 9, 325, 1972.
8. **Farr, R. S.,** A quantitative immunochemical measure of the primary interaction between I*BSA and antibody, *J. Infect. Dis.*, 103, 239, 1958.
9. **Catt, K. and Tregear, G. W.,** Solid-phase radioimmunoassay in antibody-coated tubes, *Science*, 158, 1570, 1967.
10. **Campbell, D. H., Leuscher, E., and Lerman, L. S.,** Immunologic adsorbents. I. Isolation of antibody by means of cellulose-protein antigen, *Proc. Natl. Acad. Sci. U.S.A.*, 37, 575, 1961.
11. **Sutherland, G. B. and Campbell, D. H.,** The use of antigen-coated glass as a specific adsorbent for antibody, *J. Immunol.*, 80, 294, 1958.
12. **Goldstein, L. and Katchalski, E.,** Use of water-insoluble enzyme derivatives in biochemical analysis and separation, *Z. Anal. Chem.*, 243, 375, 1968.
13. **Cuatrecasas, P. and Anfinsen, C. B.,** Affinity chromatography, *Annu. Rev. Biochem.*, 40, 259, 1971.
14. **Engvall, E., Jonsson, K., and Perlmann, P.,** Enzyme-linked immunosorbent assay. II. Quantitative assay of protein antigen, immunoglobulin G, by means of enzyme labelled antigen and antibody coated tubes, *Biochim. Biophys. Acta*, 251, 427, 1971.
15. **Mason, D. W. and Williams, A. F.,** The kinetics of antibody binding to membrane antigens in solution and at the cell surface, *Biochem. J.*, 187, 1, 1980.
16. **Nygren, H., Czerkinsky, C., and Stenberg, M.,** Dissociation of antibodies bound to surface-immobilized antigen, *J. Immunol. Meth.*, 85, 87, 1985.

17. **Voss, E. W., Jr. and Watt, R. W.,** Steric orientation of hapten groups and the effect on antibody reactivity, *Immunochemistry,* 14, 741, 1977.
18. **Nuttal, G. H. F.,** A demonstration of certain blood-relationships amongst animals by means of the precipitin test, in *Blood Immunity and Blood Relationships,* Cambridge University Press, Oxford, 1904.
19. **Crumpton, M. H.,** Protein antigens: the molecular bases of antigenicity and immunogenicity, in *The Antigen,* Sela, M., Ed., Academic Press, New York, 1974, 1.
20. **Getzoff, E. D., Tainer, J. A., Lerner, R. A., and Geysen, H. M.,** The chemistry and mechanism of antibody binding to protein antigens, *Adv. Immunol.,* 43, 1, 1988.
21. **Bezofsky, J. A.,** Intrinsic and extrinsic factors in protein antigen structure, *Science,* 229, 932, 1985.
22. **Benjamin, D. C., Bezofsky, J. A., East, I. J., Gurd, F. R. N., Hannum, C., Leach, S. J., Margoliash, E., Michael, J. G., Miller, A., Prager, E. M., Reichlin, M., Sercarz, E. E., Smith-Gill, S. J., Todd, P. E., and Wilson, A. C.,** The antigenic structure of proteins: a reappraisal, *Annu. Rev. Immunol.,* 2, 67, 1984.
23. **Kabat, E. A. and Berg, D.,** Dextran-antigen in man, *J. Immunol.,* 70, 514, 1953.
24. **Cebra, J. J.,** Studies on the combining sites of the protein antigen silk fibroin. III. Inhibition of the silk fibroin-antifibroin system by peptides derived from the antigen, *J. Immunol.,* 86, 205, 1961.
25. **Voss, E. W., Jr., Eschenfeldt, W., and Roo, T.,** Fluorescein: a complete antigenic group, *Immunochemistry,* 13, 447, 1976.
26. **Atassi, M. Z. and Caruso, D. R.,** Immunochemistry of sperm whale myoglobin. II. Modification of the two tryptophan residues and their role in the conformation and antigen-antibody reaction, *Biochemistry,* 7, 699, 1968.
27. **Kennel, S.,** Binding of monoclonal antibody to protein in fluid phase and bound to solid supports, *J. Immunol. Meth.,* 55, 1, 1982.
28. **Soderquist, M. E. and Walton, A. G.,** Structural changes in proteins adsorbed on polymer surfaces, *J. Coll. Interface Sci.,* 75, 386, 1980.
29. **Friguet, B., Djavadi-Ohaniance, L., and Goldberg, M. E.,** Some monoclonal antibodies raised with a native protein bind preferentially to the denatured antigen, *Mol. Immunol.,* 21, 673, 1984.
30. **Hollander, Z. and Katchalski-Katzir, E.,** Use of monoclonal antibodies to detect conformational alterations in lactate dehydrogenase isoenzyme 5 on heat denaturation and on adsorption to polystyrene plates, *Mol. Immun.,* 23, 927, 1986.
31. **Bull, H. B.,** Adsorption of bovine serum albumin on glass, *Biochem. Biophys. Acta,* 19, 464, 1956.
32. **Nyilas, E., Chiu, T.-H., and Herzlinger, G. A.,** Thermodynamics of native protein/foreign surface interactions. I. Colorimetry of the human γ-globulin/glass system, *Trans Am. Soc. Artif. Intern. Organs,* 20, 480, 1974.
33. **Lyklema, J. and Norde, W.,** Biopolymer adsorption with special reference to the serum albumin-polystyrene latex system, *Croatica Chem. Acta,* 45, 67, 1973.
34. **Suter, M. and Butler, J. E.,** The immunochemistry of sandwich ELISAs. II. A novel system prevents the denaturation of capture antibodies, *Immun. Lett.,* 13, 313, 1986.
35. **Dierks, S. E., Butler, J. E., and Richerson, H. B.,** Altered recognition of surface-adsorbed compared to antigen-bound antibodies in the ELISA, *Mol. Immunol.,* 23, 403, 1986.
36. **Kochwa, S., Brownell, M., Rosenfield, R. E., and Wasserman, L. R.,** Adsorption of proteins by polystyrene particles. I. Molecular unfolding and acquired immunogenicity of IgG, *J. Immunol.,* 99, 981, 1967.
37. **Morrissey, B. W. and Fenstermaker, C. A.,** Conformation of adsorbed γ-globulin and β-lactoglobulin. Effect of surface concentration, *Trans Am. Soc. Artif. Intern. Organs,* 22, 278, 1976.
38. **Stevens, F. J., Iwo, J., Carperos, W., Kohler, H., and Schiffer, M.,** Relationships between liquid and solid-phase antibody association characteristics: implications for the use of competitive ELISA techniques to map the spatial location of idiotopes, *J. Immunol.,* 137, 1937, 1986.
39. **Michaeli, I., Absolon, D. R., and van Oss, C. J.,** Diffusion of adsorbed protein within the plane of adsorption, *J. Coll. Interfacial Sci.,* 77, 586, 1980.
40. **Pesce, A. J., Ford, D. J., Gaizutis, M., and Pollak, V. W.,** Binding of protein to polystyrene in solid-phase immunoassay, *Biochim. Biophys. Acta,* 492, 399, 1977.
41. **Oreskes, I. and Singer, J. M.,** The mechanism of particulate carrier reactions. Adsorption of human γ-globulin to polystyrene latex particles, *J. Immunol.,* 86, 338, 1961.
42. **Cantarero, L. A., Butler, J. E., and Osborne, J. W.,** The adsorptive characteristics of proteins for polystyrene and their significance in solid-phase immunoassays, *Analyt. Biochem.,* 105, 375, 1980.
43. **Brash, J. L. and Lyman, D. J.,** Adsorption of plasma proteins in solution to uncharged hydrophobic polymer surfaces, *J. Biomed. Mater. Res.,* 3, 175, 1969.
44. **Lee, R. G., Adamson, C., and Kim, S. W.,** Competitive adsorption of plasma proteins onto polymer surfaces, *Thrombosis Res.,* 4, 485, 1974.
45. **Fenstermaker, C. A., Grant, W. H., Morrissey, B. W., Smith, L. E., and Stromberg, R. R.,** Interaction of plasma proteins with surfaces. Cited in Morrissey, B. W., 1977, *Ann. N.Y. Acad. Sci.,* 283, 50, 1974.
46. **Morrissey, B. W. and Stromberg, R. R.,** The conformation of adsorbed blood proteins by infrared bound fraction measurements, *J. Colloid Interface Sci.,* 46, 152, 1974.

47. **Nygren, H.,** Experimental demonstration of lateral cohesion in a layer of adsorbed protein and in layers of gold-antibody complexes bound to surface immobilized antigen, *J. Immunol. Meth.,* 114, 107, 1988.
48. **Greenspan, N. S., Dacek, D. A., and Cooper, L. J. N.,** Cooperative binding of two antibodies to independent antigens by an Fc-dependent mechanism, *FASEB J.,* 3, 2203, 1989.
49. **Børmer, O. P.,** Interference of complement with the binding of carcino-embryonic antigen to solid-phase monoclonal antibodies, *J. Immunol. Meth.,* 121, 85, 1989.
50. **Vanderbranden, M., DeCoen, J. L., Jenner, R., Kanarek, L., and Ruyschaert, J. M.,** Interaction of γ-immunoglobulins with lipid mono- or bilayers and liposomes. Existence of two conformations of γ-immunoglobulins of different hydrophobicities, *Mol. Immunol.,* 18, 621, 1981.
51. **Vanderbranden, M., Kayser, G., Banerjee, S., and Ruysschaert, J. M.,** Immunoglobulin-lipid interactions, *Biochem. Biophys. Acta.,* 685, 177, 1982.
52. **Ishikawa, E., Hamaguchi, Y., Imagawa, M., et al.,** An improved preparation of antibody-coated polystyrene beads for sandwich enzyme immunoassays, *J. Immunoassay,* 1, 385, 1980.
53. **Conradie, J. E., Gouender, M., and Visser, L.,** ELISA solid phase: partial denaturation of coating antibody yields a more efficient solid phase, *J. Immunol. Meth.,* 59, 289, 1983.
54. **Joshi, K. S. and Butler, J. E.,** The immunochemistry of sandwich ELISAs. V. The antibody performance of polyclonal antibody-enriched functions prepared by various methods, *J. Immunol. Methods,* in press.
55. **Howell, E. E., Nasser, J., and Schray, K. J.,** Coated tube immunoassay: Factors affecting sensitivity and effects of reversible protein binding to polystyrene, *J. Immunoassay,* 2, 205, 1981.
56. **Peterman, J. H.,** Factors which influence the binding of antibody to solid-phase antigens: Theoretical and experimental investigations, Ph.D. Thesis, University of Iowa, Ames, 1986.
57. **Nieto, A., Gaya, A., Moreno, C., Jansa, M., and Vives, J. J.,** Adsorption-desorption of antigen to polystyrene plates used in ELISA, *Ann. Instit. Pasteur (Immunol),* 137C, 161, 1986.
58. **Kemeny, M. and Challacombe, S. J.,** Microtiter plates and other solid-phase supports, in *ELISA and Other Solid-phase Immunoassay,* Kemeny, D. M. and Challacombe, S. J., Eds., John Wiley & Sons, London, 1987, 31.
59. **Peterman, J. H., Tarcha, P. J., Chu, V. P., and Butler, J. E.,** The immunochemistry of sandwich ELISAs. IV. A comparison of the antigen capture capacity of antibodies adsorbed or covalently attached to polystyrene, *J. Immunol. Meth.,* 11, 271, 1988.
60. **Yalow, R. W. and Berson, S. A.,** Immunoassay of endogeneous plasma insulin in man, *J. Clin. Invest.,* 39, 1157, 1960.
61. Regulatory Guide 8.20. U.S. Nuclear Regulatory Commission, April 1978.
62. **Greenwood, F. C., Hunter, W. M., and Glover, J. S.,** The preparation of ^{131}I-labelled human growth hormone of high specific radioactivity, *Biochem. J.,* 89, 114, 1963.
63. **Avrameas, S.,** Enzyme markers: their linkage with proteins and use in immune histochemistry, *Histochem. J.,* 4, 321, 1972.
64. **Sternberg, L. A., Hardy, P. H., Jr., Cuculis, J. J., and Meyer, H. G.,** The unlabeled antibody enzyme method of immunohistochemistry. Preparation and properties of soluble antigen-antibody complex (horseradish peroxidase-anti-horseradish peroxidase) and its use in identification of spirochetes, *J. Histochem. Cytochem.,* 18, 315, 1970.
65. **Kraehenbuhl, J. P., Galardy, R. E., and Jamieson, J. D.,** Preparation and characterization of an immunoelectron microscope tracer consisting of a heme-octapeptide copuled to Fab., *J. Exp. Med.,* 139, 208, 1974.
66. **Avrameas, S. and Guilbert, B.,** Enzyme-immunoassay for the measurement of antigens using peroxidase conjugates, *Biochimie,* 54, 837, 1972.
67. **Rohr, T. E., Cotton, T., Fan, N., and Tarcha, P. J.,** Immunoassay employing surface-enhanced Ramon spectroscopy, *Analyt. Biochem.,* 182, 383, 1989.
68. **Pecht, I., Givol, D., and Sela, M.,** Dynamics of hapten-antibody interaction. Studies on a myeloma protein with anti-2,4-dinitrophenyl specificity, *J. Mol. Biol.,* 68, 241, 1972.
69. **Smith, T. W. and Skubitz, K. M.,** Kinetics in interactions between antibodies and haptens, *Biochemistry,* 14, 1496, 1975.
70. **Day, L. A., Sturtevant, J. M., and Singer, S. J.,** The kinetics of the reactions between antibodies to the 2,4-dinitrophenyl group and specific haptens, *Ann. N. Y. Acad. Sci.,* 103, 611, 1963.
71. **Herron, J. N.,** Equilibrium and kinetic methodology for the measurement of binding properties in monoclonal and polyclonal populations of anti-fluorescyl-IgG antibodies, in *Fluorescein Hapten: An Immunological Probe,* Voss, E. W., Jr., Ed., CRC Press, Boca Raton, FL, 1984, 50.
72. **Friquet, B., Djavadi-Ohaniance, L., and Goldberg, M. E.,** Polypeptide-antibody binding mechanism: Conformational adaptation investigated by equilibrium and kinetic analysis, *Res. Immunol.,* 140, 355, 1989.
73. **Noble, R. M., Reichlin, M., and Schrieber, R. D.,** Studies on antibodies directed toward single antigenic sites on globular proteins, *Biochemistry,* 11, 3326, 1972.
74. **Fersht, A.,** *Enzyme Structure and Mechanisms,* W. H. Freeman & Co., New York, 1985, 150.
75. **Nygren, H., Kaartinen, M., and Stenberg, M.,** Determination by ellipsometry of the affinity of monoclonal antibodies, *J. Immunol. Meth.,* 92, 219, 1986.
76. **Li, C. K. N.,** ELISA-based determination of immunological binding constants, *Mol. Immunol.,* 22, 321, 1985.

77. **Scatchard, G.,** The attraction of protein for small molecules and ions, *Ann. N.Y. Acad. Sci.,* 51, 660, 1949.
78. **Langmuir, I.,** The absorption of gases on plane surfaces of glass, mica and platinum, *J. Am. Chem. Soc.,* 40, 1361, 1918.
79. **Steward, M. W. and Petty, R. E.,** The use of ammonium sulphate globulin precipitation for determination of affinity of anti-protein antibodies in mouse serum, *Immunology,* 22, 747, 1972.
80. **Nygren, H. and Stenberg, M.,** Immunochemistry at interfaces, *Immunology,* 66, 321, 1989.
81. **Steward, N. W.,** Affinity of the antibody-antigen reaction and its biological significance, in *Immunochemistry: An Advanced Textbook,* Glynn, L. E. and Steward, M. W., Eds., John Wiley & Sons, London, 1977, 233.
82. **Butler, J. E., Spradling, J. E., Dierks, S. E., Heyermann, H., Peterman, J. H., and Suter, M.,** The immunochemistry of sandwich ELISAs. I. The binding characteristics of immunoglobulins to monoclonal and polyclonal capture antibodies adsorbed on plastic and their detection by symmetrical and asymmetrical antibody-enzyme conjugates, *Mol. Immunol.,* 237, 971, 1986.
83. **Koertge, T. E. and Butler, J. E.,** The relationship between the binding of primary antibody in solid-phase microtiter assays and its detection using ELISA technology, *J. Immunol. Meth.,* 83, 283, 1985.
84. **Butler,. J. E., Spradling, J. E., Peterman, J. H., Joshi, K. S., and Challacombe, S. J.,** Humoral immunity in root carries in an adult population. I. Establishment of a reliable approach for measuring specific antibodies in serum and saliva, their specific activities, rate of salivary secretion and relationship to other, salivary proteins, *Oral Microbial Immunol.,* 5, 98, 1990.
85. **Steiner, L. A. and Eisen, H. N.,** The nature of antigen-antibody interactions, in *Immunological Diseases,* Samter, M. and Alexander, H. L., Eds., Little Brown, Boston, 1965, 122.
86. **Butler, J. E., Feldbush, T. L., McGivern, P. L., and Stewart, N.,** The enzyme-linked immunoasorbent assay (ELISA): A measure of antibody concentration or affinity? *Immunochemistry,* 15, 131, 1978.
87. **Lew, A. M.,** The effect of epitope density and antibody affinity on the ELISA as analyzed by monoclonal antibodies, *J. Immunol. Method.,* 72, 171, 1984.

Chapter 2

THE CHEMICAL PROPERTIES OF SOLID-PHASES AND THEIR INTERACTION WITH PROTEINS

Peter J. Tarcha

TABLE OF CONTENTS

I. INTRODUCTION

The development of solid phase immunoassay (SPIA) techniques and technology has taken place primarily in the laboratories of the biological scientist. Close collaboration with the materials science laboratory on the design, surface chemisty, and bulk chemical properties of the solid support is often lacking. As a result, many solid phase support materials used in SPIA were originally designed for an entirely different purpose. For example, polystyrene latex made to serve as size calibrations standards for microscopy or other instrumentation, is sometimes used to develop agglutination tests.

Often, no attempt is made to purify or physically characterize the surface chemistry of solid support reagents. Variation in performance from lot to lot is adjusted by varying thc amount of antibody (Ab) or antigen (Ag) in the coating solution, pH, or addition of other components which help stabilize and maintain the activity of the biological molecule, but may detrimentally block some of the surface available for immobilization.

With injection-molded supports, such as microtiter trays, beads, or tubes, thc resins used are usually formulated to assist the mold operator in making the part. This is done by adding mold release agents and lubricants, many of which are quite surface active, and can effect the coatability of the support, and the performance and shelf-life stability of the immunoreagent immobilized on the support.

Failure to understand and control the surface chemistry of the solid phase can be costly, when time is wasted on empirical approaches for optimizing each lot of immunoreagent, without specifically understanding the effects of the variable(s) used to achieve a given level of performance. As a result, basic improvements in thc solid phase material often proceed slower than necessary.

This chapter will outline the chemistry and methods of preparation, purification, and analysis of solid supports currently used in the field, as well as those which are likely to become common in the future. Chemical aspects will be emphasized, since thc biological treatment will be well covered in other chapters in this monograph. This background information is designed to help the biological scientist developing new immunoassay methods to understand the chemical properties of various solid phases. This can only help to inspire creativity and foster closer interaction between the immunochemist and the materials chemist.

II. PREPARATION OF SOLID PHASES

A. RESINS

Polymer resins commonly used for the injection molding of tubes, beads, trays, dipsticks, and other configurations are usually thermoplastic homopolymers made by the chain growth polymerization of vinylic monomers. Thermoplastics can be differentiated from thermosetting materials in that they can be softened by heating to a given level of fluidity and cooled repeatedly without appreciable changes in physical properties. Thermosetting materials change apprecia-bly under the influence of heat due to chemical cross-linking and/or increase in chain length. Since the material hardens and becomes intractable, it cannot be resoftened by heating. Examples of the latter are epoxies and polyimides. The most common thermoplastic for SPIA is polystyrene, because of its low cost, ease of processability, and optical clarity. Poly(vinylchloride), poly(ethylene), and poly(propylene) are also used as supports in the development of some test formats.

The homopolymer of styrene is usually produced by free radical initiation in bulk or in suspension. In bulk polymerization, monomer, polymer, and initiator are the only components. The amount of oxygen present is usually controlled, because its diradical nature can cause it to function as both a free-radical inhibitor and an initiator. Styrene is unique in that it is one of the few monomers that can polymerize thermally in the absence of any added initiator. At 127°C, styrene shows a thermal sclf-initiation rate of 14% per hour.[1]

During the bulk polymerization process, including thermal polymerization, the polymer formed is soluble in the monomer, and hence the solution becomes quite viscous, even at low conversion. Technology has been developed to control the reaction so that it does not polymerize violently due to inefficient transfer of the heat released as a consequence of the polymerization reaction. Unfortunately, details of these processes are rarely made public. A general description of a typical industrial process can be found in a textbook by Rodriguez.[2]

Suspension polymerization is a second common method for producing polystyrene. In this process the monomer is dispersed into droplets in an aqueous phase with the aid of a surfactant. A hydrophobic peroxide initiator, such as benzoyl peroxide or dilauroyl peroxide, is dissolved in the monomer prior to dispersing. Initiation takes place predominantly in the droplets (10 to 100 μm in diameter), and the heat of polymerization is transferred out of the reactor via the water phase. A small percentage of latex (0.1 to 1 μm) usually also forms and is generally treated as a waste by-product. Upon completion of the reaction, the particlcs, except for the latex, settle and are recovered. This process is also well suited for producing cross-linked resin beads for ion-exchange applications, or ligand immobilization, and will be discussed more fully later. It is not the best method for producing resins where optical clarity is required.

Ethylene, propylene, and vinyl chloride are all gases at room tcmperature, hence reactors must fulfill the additional requirement of withstanding higher pressure. Polyvinyl chloride (boiling temperature, –13°C) can be made by suspension polymerization in water at elevated temperatures below the boiling point of water. Gas-liquid polymerization is also possible[3] and can be used to produce materials where optical clarity is important. Ethylene (T_b = –104°C) and propylene (T_b = –48°C) can be polymerized at extremely high pressure (1000 to 3000 atm) using peroxides or oxygen to initiate polymerization. Under these conditions, the monomer is a liquid and the polymer remains dissolved in the monomer.

Alternatively, heterogeneous catalysts can be used, such as the Ziegler-Natta type. These catalysts usually consist of the reaction products of organometallic compounds and transition metal compounds. One example is the triethylaluminum/titanium tetrachloride system. The polymerizations are carried out in an oxygen-free system in a hydrocarbon solvent. Pressures and temperatures are significantly lower (5 to 30 atm and 170°C or less), and it is possible to produce stereoregular polymeric structures. For optical clarity, and to avoid the catalysis of unwanted reactions during an immunoassay, the catalyst should be removed. Descriptions of these processes can be found in Miles and Briston.[4]

A general description of many of the commercially available resins (not limited to those discussed) is found in an annual publication, *Modern Plastics Encyclopedia*.[5]

The first chapter of this reference book devotes, on the average, $1^1/_2$ pages to each of approximately 70 common commercial resins. A description of the chemistry and chemical properties, processing conditions, grades available, and applications are given for most resins mentioned.

B. INJECTION MOLDING

Conventional tubes, trays, beads, and other configurations are fabricated by the process of injection molding. The basics of injection molding are explained by referring to Figure 1, which is a drawing of a simple molding machine. In parts with an internal void, such as microtiter trays or tubes, a cavity, which defines the outer dimensions of the part, is machined into a metal block and serves as one half of the mold. The second half is machined to form the core, which defines the internal dimensions of the void. With solid parts, such as spheres or sticks, both halves of the mold contain cavities. The grade of metal used to make the mold is determined by the number of injection cycles required from the mold and parting line flash tolerances. For example, some prototype molds are made of brass, since it is easy to machine. Production molds are made of stainless steel or other wear-resistant grades of steel. The two halves of the mold are clamped between two platens and held there mechanically by a toggle device, or hydraulically via a

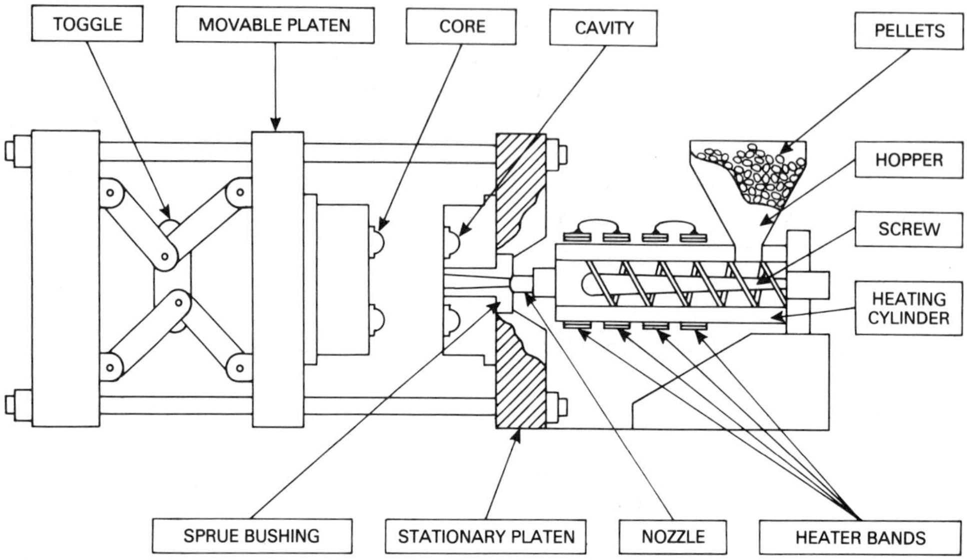

FIGURE 1. Diagram of a typical injection molding machine. (Courtesy of D. V. Rosato, Rhode Island School of Design.)

piston. Through the hopper, resin pellets enter the cylinder of the molding machine where they are plasticized by a combination of external and mechanical heating. After uniform temperature and viscosity are achieved, the molten plastic is forced through the nozzle into the cavity of the mold. After solidification, the mold is opened and the part is ejected.

From a surface chemistry standpoint, the medical diagnostic scientist should realize that additives can be added at several points during processing, from compounding the resin pellets, to molding the finished parts. The additives most likely to be found in molded diagnostic parts are flow agents, plasticizers, and mold release agents. Flow agents serve to interrupt intermolecular forces among the polymer chains, thus reducing melt viscosity. They often possess some incompatibility with the polymer itself and can bloom to the surface of the plastic part during molding. The most commonly used flow agent is calcium stearate.

From a manufacturing standpoint, flow agents can be good because they aid in the release of the part from the mold and prevent parts from sticking, if they are packaged in close contact to each other. Plasticizers also serve to interrupt interchain forces in the material and impart to the plastic a variety of properties, such as reduction in processing temperatures, reduction of the plastic's hardness, and improved impact resistance. These additives are usually completely compatible in the resin in that they have solubility parameters very similar to the resin itself. Phthalate, adipate, and stearate esters are commonly used as plasticizers.

Mold release agents are substances that are applied to the surfaces of molds to prevent wetting of the surface by the plastic, and consequently, reduce sticking of the part in the mold. They can be waxes, fluoropolymers, soaps, or a variety of proprietary compounds or blends.

Lists of commonly used flow agents, plasticizers, and mold release agents can be found in many plastics handbooks, a few of which are given.[6-8] The *Modern Plastics Encyclopedia* gives short descriptive sections on various types of processing aids.[5]

Since the amount and type of additive(s) can vary depending on resin type, resin manufacturer, and molding shop, it is important for the diagnostic scientist to be aware of these components in his specific solid phases. The vendor and molder can provide this information or it can be determined by standard analytical means. It is possible to reduce or totally eliminate additives through the proper selection of resin type and molding conditions. This is important in two respects: (1) the procedure for immobilizing a particular Ab or protein can be optimized especially for a given solid phase and its immunoassay. If vendors and molders are not aware that changes in the amounts or types of processing aids can effect the consistency of performance of the SPIA product, subtle changes might be inadvertently made at the materials or fabrication stage, which cause assay lot failures during end usage. (2) Many of the additives are surface active and can bloom to the surface during storage. This can effect the shelf life of the product, since these compounds can compete with the coated Ab or Ag for hydrophobic anchor sites and/ or disrupt the three-dimensional structure of the immobilized biomolecule.

C. BULK ANALYSIS OF MOLDED SOLID PHASES

There are many thorough textbooks and review articles on polymer analysis and characterization.[9-11] Briefly outlined herein are two main bulk analysis techniques that may be useful to the medical diagnostic scientist. These techniques measure molecular weight and its distribution and the amount and type of residuals and additives.

Molecular weight and distribution are important properties for high-performance plastics, and plastic parts which must perform under mechanical stress or in harsh environments. They can also be important parameters in diagnostic solid phases, but for other reasons. For example, the number of polymer chain end groups is related to the number average molecular weight (M_n). End groups usually originate from the polymerization initiator and can also result if polymer degradation occurs. If charged, they are surface active and can migrate to the surface of the material. Perhaps more importantly, molecular weight and distribution are handles on the lot-to-lot consistency of the resin used to mold the solid phase.

Molecular weight analysis is usually done by size-exclusion chromatography.[12] This is usually performed as a high-performance liquid chromatography (HPLC) technique, and the mechanism of separation is based solely on the size of the molecule; affinity interactions with the porous column packing should, in theory, be nonexistent. The size-exclusion chromatography system is calibrated with known molecular-weight standards of different types of polymers to generate a universal calibration curve.[13]

The chromatogram yields data that allows calculation of M_n, weight average molecular weight, M_w, and polydispersity index M_w/M_n. M_n is the arithmetic mean molecular weight, $\Sigma n_i M_i / \Sigma n_i$ and M_w is a weighted average, where M_i is multiplied by the weight of that class, i, i.e., $\Sigma w_i M_i / : \Sigma w_i$. Note that M_w will always be greater than or equal to M_n, since larger molecules are weighted more heavily than smaller ones in the M_w average. The ratio of the two averages M_w/M_n is related to the heterogeneity in the molecular weight of the sample. A totally monodisperse polymer will have a polydispersity index of 1; fairly monodisperse distributions will exhibit values from 1 to 2.5, while very broad distributions can show values of 6 or greater.

A limitation of this technique is that the data are generated on the basis of molecular size, not neccesarily true molecular weight. Ideally, if linear standards are available, measured by a technique which yields true molecular weight, e.g., by osmometry, more accurate estimates of molecular weights can be obtained, if the linear standards are of the same polymer being analyzed.

Analysis of additives is generally done as a separate procedure. The material is usually ground up (a clean coffee grinder is sometimes adequate) into fine pieces and extracted with a Soxlet extractor for several days. Methanol is a good solvent for the extraction, since it will not dissolve most plastics and will effectively extract most residual monomers and additives. Analysis of the solvent is done by HPLC, coupled with spectroscopic methods, if standards for each extractable componcnt are not available.

The potential exists for using a size exclusion instrument which can give information about both molecular wcight and additives as well. Krishen and Tucker[14] describe the use of size exclusion columns capable of separations in the molecular weight range of 100 to 2000.

D. NANOPARTICLES (LATEXES) AND MICROPARTICLES

1. Polystyrene and Carboxylated Polystyrene

The preparation of a polymer colloid involves a basic recipe consisting of a continuous phase (usually water), monomer(s), an initiator, and sometimes a colloidal stabilizer. The process involves the emulsification of monomer having limited solubility, into a continuous phase solvent, usually with the aid of a surfactant. The reaction is initiated by a free-radical initiator, which is usually dissolved in the continuous phase. A basic qualitative theory was given by Harkins for emusion polymerization in the presence of surfactants.[15] Such reactions will yield particles that are covered with adsorbed emulsifier and the particle size distribution can be broad, especially if the concentration of emulsifier is above the critical micelle concentration and the rate of initiation is low.

For medical diagnostic products it is desirable, but not always essential, to have monodispersed particles. They should be free of adsorbed emulsifiers, since amphiphiles may compete for hydrophobic binding sites on the particles' surfaces. It is possible, however, to clean a preformed latex, removing most of the adsorbed species. A brief working outline of such a procedure is given by Van den Hul and Vanderhoff,[16] which involves the use of precleaned mixed-bed ion-exchange resins.

The emulsifier in these systems serves mainly to stabilize the newly formed latex particles against aggregration. There are simpler systems employing only monomer(s), water, and ionic initiator. Conditions are chosen such that sufficient charge stabilization is achieved via the polymer endgroups, which originate from an ionogenic peroxide initiator, such as potassium persulfate. The mechanism of such self-stabilization in an emulsifier-free environment is outlined by Fitch.[17] Fitch subsequently published a practical guide for designing emulsion polymerization recipes, the principles of which can be applied both to systems which contain emulsifiers and those that are emulsifier free.[18]

The most widely used diagnostic application of "clean" polystyrene latex is in the design of latex agglutination tests. Here, the Ab or Ag, possessing epitopes against the analyte of interest, is passively adsorbed to the particle surface. In another format, Brown[19] describes a SPIA for human chorionic gonadotropin (hCG) which employs immobilized latex particles possessing bound anti-β-hCG Ab. The test is not a slide agglutination, since the particles are fixed on the surfaces of glass fibers. Here, the particles serve the same purpose as the surface of an Ab-coated microtiter well, however, their tremendous surface relative to their volume, and greatly reduced diffusion distances for the aqueous reactants, allows a significant reduction in the time necessary to complete the required immunoreactions and obtain an assay result.

It is increasingly more important to develop polymer colloid solid phases that possess chemically reactive groups, through which covalent linkage of the capture molecule is possible. The synthesis of latexes which contain chemically reactive surface copolymer units can be accomplished by simply adding the desired functional monomer to the monomer feed in amounts from 0.5 to 5%. The monomer possessing the polar group will be incorporated into polymer at the particle surface, buried in the interior, and also in the water-soluble polymer dissolved in the continuous phase. For instance, Greene found that when acrylic acid was copolymerized with butadiene and styrene in a latex reaction, the distribution of the ionogenic polymer subunits was 2, 3, and 1 corresponding to water phase, polymer/water interface, and interior, respectively.[20] Adjusting the pH so as to reduce the amount of ionized carboxylic acid increases the amount of particle-bound polyacrylic acid units.

The synthesis of microparticles (larger diameter than typical latexes) of sparingly water-soluble vinylic monomers such as styrene, methyl methylacrylate, or vinyl chloride can be accomplishcd by a process of suspension polymerization. This process is a true "emulsion" polymerization in that the monomer dropets are the main locus of polymerization. In contrast, in an emulsifier-free latex polymerization, particle nucleation is predominantly in the continuous phase (homogeneous nucleation) and the droplets serve mainly as monomer reservoirs, feeding the growing particles.

In the suspension polymerization process, oil-soluble initiators are the most effective, and they are usually dissolved in the monomer phase prior to initiation of polymerization. Benzoyl peroxide is commonly used. The monomer is dispersed into the water phase by stirring with the aid of nonionic emulsifiers and protective colloids, e.g., bentonite or ionic soaps have also been used. Polymerization is initiated by raising the temperature while stirring. A good starting point for reviewing general procedures is an article by Hohenstern and Mark[21] which reviews the literature from 1910 to 1945.

Since this early review, many more advances have been made with regard to understanding the mechanisms of particle formation and controlling particle size and distribution. One of the most significant studies in this regard was reported by Ugelstad and Hansen,[22] who outline the kinetics, thermodynamics, and mechanism of suspension polymerization in great detail. In this work, they describe the principles for the synthesis of monodisperse microparticles with sizes greater than 2 μm. Though the exact recipes are not clearly given, they involve a seeded polymerization of a monodisperse polymer colloid in the presence of additives. This seed is dispersed into water and swollen with monomer and low molecular weight oligomers of the polymer. Additives such as long-chain alkanes and fatty alcohols and/or ionic emulsifiers may be included. These methods allow the production of very monodisperse microparticles in size ranges not previously possible without subsequent size fractionation.

The authors attribute the effect of the added oligomer in the droplets *before* initiation of polymerization by considering the monomer/water interface to be a semipermeable membrane — permeable to the transport of monomer, but not to the transport of oligomer. The oil-in-water emulsion, formed before initiation, is fairly monodispersed; however, some transport by diffusion of monomer from the droplets of slightly smaller diameter to those of slightly larger diameter will occur, as dictated by the Laplace equation. A metastable state will soon be reached, since the oligomer cannot diffuse with the monomer and an osmotic pressure counterbalances the difference in partial free energy of the monomer in different-sized droplets, due to the differences in curvature. Such a mechanism should, therefore, stabilize the emulsion before and during polymerization, resulting in a monodisperse polymer microparticle dispersion, provided the droplet/polymer particles are stabilized against coalescence.

A later review by Ugelstad et al. suggests medical applications of such monodisperse spheres.[23]

2. Polyacrylate Latexes and Microparticles

Acrylate latexes and microparticles can be synthesized employing the same principles discussed previously. These materials have an added feature. If a small amount of cross-linking monomer, usually ethylene diacrylate or ethylene glycol dimethacylate, is used, the surface obtains a respectable level of bound surface carboxylic acid, due to low-level hydrolysis of the polyacrylate or methacrylate ester during polymerization. If initiators which produce phosphate,[24] sulfate, and/or sulfonate[25] end groups are used, such as potassium peroxydiphosphate, potassium persulfate, or sodium bisulfite, the particles will possess covalently bound, strong acid groups, if put into the acid form with a strong acid ion-exchange resin. Warming these preparations, as shown by Fitch et al.,[26] produces particles with defined levels of carboxylic acid. A quantitative model which describes the hydrolysis of such heterogeneous systems has been developed by Tarcha.[27]

An advantage of carboxylated materials for medical diagnostic solid supports is the existence of well-developed methods for covalent coupling of Ags and Abs to these surfaces. In addition, at high levels of surface carboxylation, the polymer/water interface is quite hydrophilic and less prone to capture unwanted proteinaceous species by non-immunospecific means. A hydrophilic environment also minimizes the possible distortion of the three-dimensional structure of the biomolecules due to strong interaction with the surface, which can result in decreased activity (see a review of this topic by Sato et al.[28]). Additional potential exists for the development of a latex-based immunoreagent which can be redispersed after drying, since the surface contains loops and tails of hydrophilic polymer known to be effective steric stabilizers.[29]

Polyglycidyl-co-styrene latexes have been prepared by similar methods.[30,31] The surface epoxide groups are reactive to attack by protein-bound amine or thiols, permitting covalent linkage of immunoreactive capture molecules to the particles' surfaces.[32]

3. Polyacrolein Particles

Polymer colloids and microparticles made by the polymerization of acrolein have been the subject of several investigations. Most papers stress their applicability as a carrier for Ab sor Ags, since the polymer structure contains reactive pendant aldehyde groups which form Schiff bases with primary amino groups. The monomer is most commonly polymerized by a free-radical mechanism, initiated by γ-irradiation of aqueous solutions using a Colbalt 60 source, or anionically under strongly basic conditions.[33] The methods developed permit the production of monodisperse particles, the diameter of which can vary from 0.04 to 8 μm.

The chemical structure of polyacrolein has been studied by solid state ^{13}C NMR and FTIR spectroscopy.[34] It was concluded that radical polymerization produces mainly polymerization via the vinyl group. Under basic conditions, Michael addition and carbonyl addition probably occurs simultaneously. Further studies compared the loading level between particles prepared by the two methods, as a function of the pH of the coating solution, using mouse IgG.[35] The immobilization reactions were carried out in 0.1 *M* sodium acetate buffer which contained 1% Tween-20, a nonionic detergent, to minimize hydrophobic binding. It was found that pH = 5.3 was optimal for both types and the radiation polymerized particles had the highest loading level. Murgita and Vas studied mouse immunoglobulins by electrophoretic separation and reported that normal mouse immunoglobulins generally have isoelectric points in the range of 6.3 to 8.0.[36] The polyacrolein particle may obtain a net negative charge as a result of limited surface oxidation of pendant aldehyde groups forming bound carboxylic acid. This should facilitate chemical immobilization of the IgG by the localization of it near the surface by charge attraction.

The synthesis of both magnetizable and fluorescent polyacrolein particles has been reported,[37] as well as copolymers of acrolein with hydroxyethyl methacrylate and other monomers.[38] Applications to cell labeling[39] and cell separations have been described.[37]

4. Polyacrylamide Latexes and Microparticles

Cross-linked polyacrylamide and its derivatives are well known in the liquid chromatography area as a solid support for protein desalting, separations, and for enzyme immobilization.[40,41] It is widely used in SDS gel electrophoresis as the separation matrix. In our laboratory experience, we have found that protein tends not to bind to polyacrylamide via nonspecific interactions, since it is very hydrophilic. This is important for high-sensitivity, "sandwich-type" enzyme immunoassays, where free enzymes or enzyme conjugates must be completely washed from the support before adding the substrate.

Over the last 7 to 8 years, inverse emulsion polymerization has been a technique of active research. The method involves a latex polymerization in nonaqueous media employing water-soluble monomers dissolved in water, which are subsequently dispersed into an oil phase. If a cross-linking comonomer is used, the production of water-swellable nanoparticles results. Early examples are latexes of poly-*N*-vinylpyrrolidone[42] and polyacrylamide.[43] Recent studies discuss the effects of experimental variables on the polymerization and stability of inverse emulsion latexes[44] and some of the kinetic and mechanistic aspects.[45-48] Microparticles of acrylamide and other hydrophobic particles are made by suspension polymerization techniques in organic media. A textbook edited by Barrett[49] provides an extensive treatise on the subject, including a thorough listing of the patent literature.

In order to utilize polyacrylamide for solid supports, provision must be made for covalent linkage of the capture molecules. Methods to produce carboxylic acid groups are limited to hydrolysis of polyacrylamide or copolymerizing with acrylic acid. Kraemer et al.[40] described methods for producing polyacrylamide beads that possessed active esters for the covalent linkage of proteinaceous materials. This involved the incorporation of acryloyl *N*-hydroxysuccinimide as a comonomer. Tarcha et al.[50] formed the same active groups on polyacrylamide particles, which possessed carboxylic groups, by esterifying them with *N*-hydroxysuccinimide. In this work, the particles were designed to possess capture Ab only at the particles' surfaces. The interior of the particle was cross-linked to retard entry of the analyte of interest, but retained a high swelling capacity for water. In practice, the solid-phase served as an absorbent to remove water during the immunoassay, thus concentrating the reactants. This reduced by almost 50% the time necessary for sufficient immunobinding to occur, compared to particles that were preswollen before addition of Ag.

5. Polysaccharide Particles

Polysaccharide microparticulates have found wide applications in the gel filtration and affinity chromatography field. Particles of various pore sizes and surface functional groups are commercially available. One of the most common is agarose, a complex polysaccharide, believed to have alternating β-D-galactopyranosyl and 3,6,anhydro-α-1-galactopyranosyl units coupled 1 → 3. The material can be formulated into beads which lend themselves not only as a column-packing, but as a water-insoluble hydrophilic support for antibodies. Dextrans, polysaccharides composed mainly of 1,6,α-D-glucopyranosyl linkages, have also been used as a macromolecular separation media. They are water-soluble polymers, but can be cross-linked into a gel with epichlorohydrin. Murry and Whistler[51] review dextrans and their applications, including a treatise[52] on polysaccharides and their derivatives.

6. Nylon Particles

Nylon has been used as a solid phase for the covalent linkage of Abs. It is a polyimide, and pendant groups (carboxyl and amine) can be generated by the partial hydrolysis of the polymer chain. The preparation of microparticles can be accomplished by dissolving the nylon in concentrated hydrochloric acid and reprecipitating into distilled water.[53] During the dissolution into the acid, the chain becomes partially hydrolyzed, resulting in an increase in surface endgroups upon particle reformation.

7. Nitrocellulose

Cellulose nitrate has demonstrated good properties as supports for antibodies.[54,55] It is prepared by the direct nitration of cellulose, usually to 10 to 15% nitrogen by weight. To process it, plasticizers and solvents must be present; their addition results in a viscous material. The solution can then be processed into filters or membranes.[56,57]

8. Metal Colloids and Surfaces

Gold colloids, coated with Ab, have been used for some time as an electron rich "dye" for labeling and contrasting biological materials under electron microscopic examination. Gold and silver have also been used as colloidal labels for Abs in sandwich immunoassays.[58,59] Typically, a microtiter plate is coated with Ab and successively incubated with sample, and the metal colloid-containing Ab, bound at the particle surface. Afterwards, the wells are washed to remove unbound materials, and the bound metal conjugate is dissociated chemically and analyzed colorimetrically or by atomic absorption.

More recently, metal colloids and various metal surfaces have taken on new importance in the area of analytical chemistry and biochemistry. It was discovered that molecules near these surfaces exhibit tremendously enhanced Raman scattering signals,[60] and this was recognized to be a unique phenomenon.[61] This has applications for the detection of trace components in various environments and the topic has been recently reviewed.[62] The first utilization of the surface-enhanced Raman scattering effect for immunoassay applications was described by Rohr et al.[63,64]

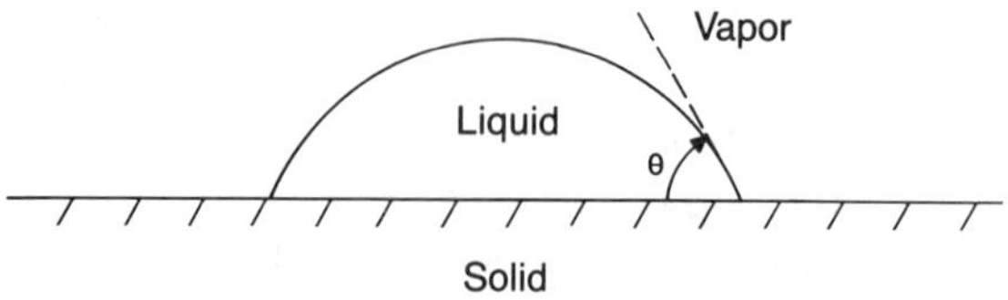

FIGURE 2. Contact angle between liquid and solid phase.

III. SURFACE CHEMISTRY OF SUPPORT SURFACES

A. SURFACE ANALYSIS

Thc coatability of a surface with biomolecules, either by adsorption or covalent linkage, is controlled to a large extent by the nature of the surface during the binding process. Once a particular solid phase has been selected, the major concerns are (1) the reproducibility or the surface coating, (2) the biological activity of the immobilized biomolecules, and (3) the shelf life of the product. Sources of variability in molded plastic surfaces can arise from blooming to the surface of surface-active additives used to aid in processing. The problem usually manifests itself after a change has been made in the type or amount of additive(s) used, or the molding parameters. Storage or transport of the solid supports can introduce surface contaminants from the storage environment or the packaging materials. Change in surface charge can affect coatability.[65]

Surfaces modified for covalent linkage should be analyzed quantitatively for levels of modification, or at minimum, subjected to a quality control method which identifies lots outside the working specifications. Proteins which are attached by chemically stable linkages will not change in surface concentration; however, their biological activity may change. For example, this can result from the further reaction of the protein with unquenched surface groups, which may distort the proteins' three-dimensional structure in an unfavorable manner.

There are a myriad of classical methods for measuring interfacial energetics.[66] In SPIA development, one is usually concerned with surface wettability and contamination; consequently, the methods of analysis are restricted to those applicable to solid-liquid interfaces. In this section, three techniques will be discussed: contact angle, surface infrared and fluorescence spectroscopy, and X-ray photoelectron spectroscopy.

1. Contact Angle

Thc contact angle (Θ) is defined as the angle which is formed at the junction of three phases (liquid, solid, gas) and is measured through the liquid as shown in Figure 2. It is related to the interfacial tensions of the three interfaces as

$$\text{Cos } \Theta = \gamma_{sv} - \gamma_{sl}/\gamma_{lv}$$

where γ_{sv}, γ_{sl}, and γ_{lv} represent the interfacial tension between the solid and vapor, solid and liquid, and liquid and vapor, respectively. Hence, if a solid surface becomes more wettable by pure water due to contamination by surface active material or changes in the surface chemistry, the contact angle will decrease. Table-top devices are commonly used in the industry to measure and record contact angles, and an example schematic of such a device is shown in Figure 3. Newman and Good[67] have reviewed the subject.

2. Surface Infrared and Fluorescence

Attachments are available for infrared spectrometers which permit spectral sampling of the surfaces of solid materials, provided they are fairly flat. The technique is known as attenuated total reflection and the principles are illustrated by Figure 4. This method samples depths much

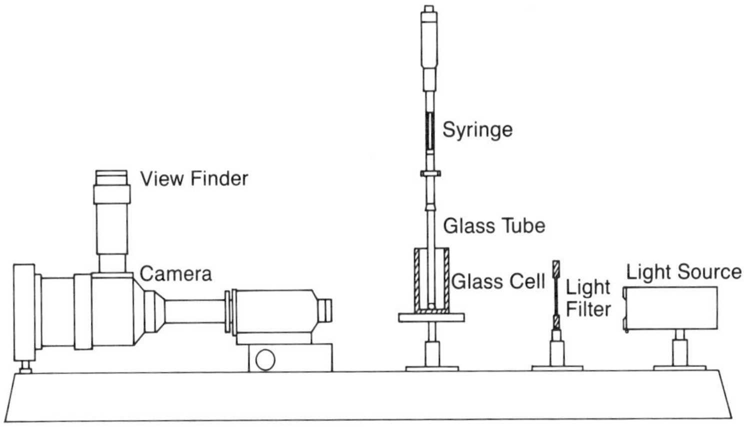

FIGURE 3. The captive bubble method apparatus (Courtesy of R. H. Ottewill).

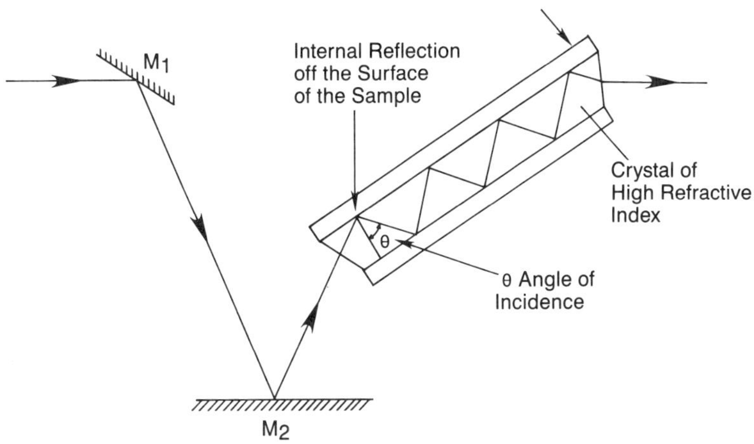

FIGURE 4. Principle of the attenuated total reflectance accessory. (Reprinted from *Am. Lab.*, 10(3), 111, 1978, ©1978 by International Scientific Communications, Inc., with permission.)

greater than the first tens of monolayers and hence is not an extremely surface-sensitive method. The advent of Fourier transform infrared, coupled with the minicomputer, provides the ability to subtract the bulk spectrum from the surface signals. This permits the identification and measurement of the surface species in many cases. Two recent papers apply this technique to surface-adsorbed proteins, one employs infrared spectroscopy[68] and the second, fluorescence spectroscopy.[69] Basic principles of the infrared technique, complete with examples of typical analyses from the polymer industry has been reviewed by Osland, Price, and Devenish.[70]

3. X-Ray Photoelectron Spectroscopy

X-ray photoelectron spectroscopy (XPS) often referred to as electron spectroscopy for chemical analysis, or ESCA, is a surface spectroscopy which is truly sensitive to the first 100 Å of a surface. The top 25 Å is the region where approximately 70% of the spectral information is obtained.

Shown in Figure 5 is the basic XPS technique for an atom located 0 to 100 Å beneath a surface. X-ray photons from a nearly monoenergetic beam are directed onto the sample and absorbed by the surface atoms. A photoelectron is emitted for every X-ray photon absorbed. The kinetic energy of the photoelectron, measured by the spectrometer, is equal to the energy of the

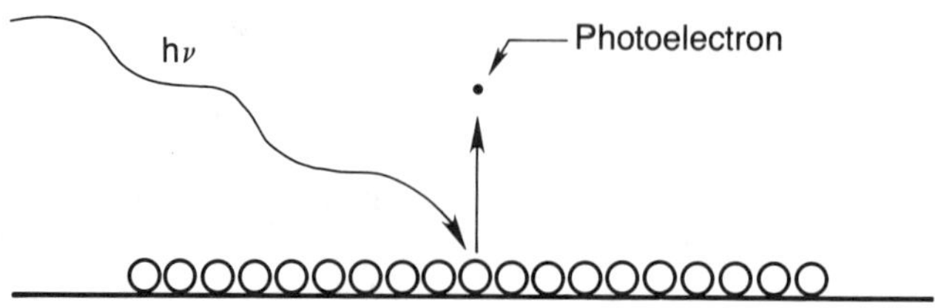

FIGURE 5. Principle of X-ray photoelectron spectroscopy.

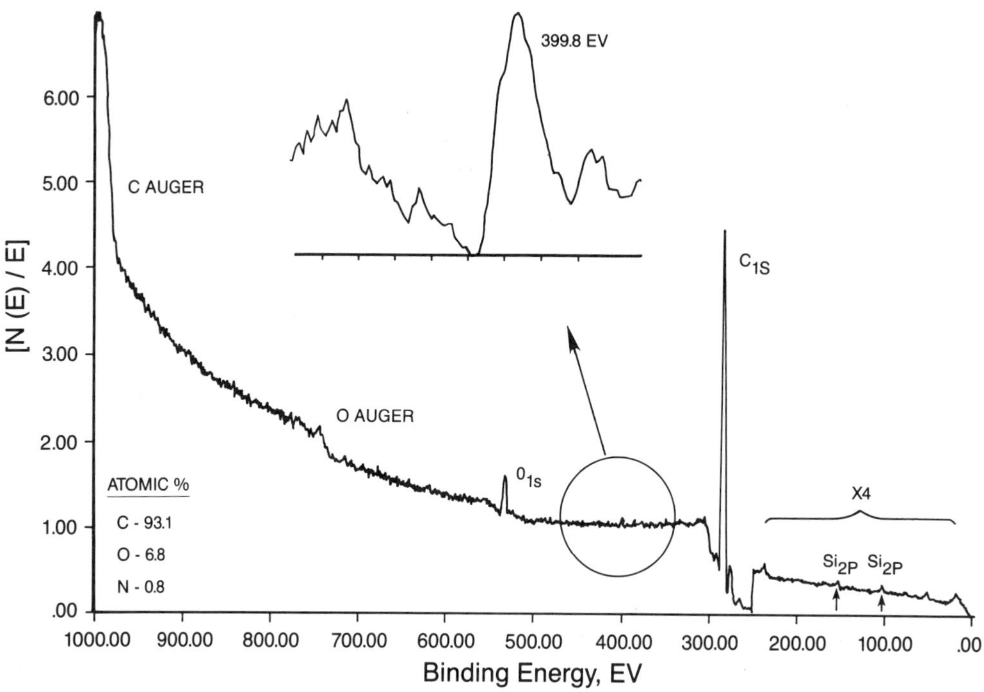

FIGURE 6. X-ray photoelectron spectrum from an injection-molded polystyrene bead.

impinging X-rays (hυ)) minus its photoelectron binding energy in the original atom. Although electrons in an atom of binding energy less than hυ are excited, the probability of emission depends on the photoelectron cross section of the atomic level. For atoms of greatest interest in organic materials (carbon, nitrogen, and oxygen), the 1S orbital has the largest cross section. Its value is at least 20 times greater than all others. This results in one observable peak for each type of atom in a unique binding energy region. For a given element, the peak can be shifted within its region depending on its chemical environment. Therefore, not only are the surface elements identified, but the nature of the chemical bonding or functional groups are determined as well. The spectrum is plotted as the number of counts at any given energy divided by that energy, plotted in arbitrary units, as a function of the binding energy of the electrons.

The technique has been used to study the adsorption and morphology of protein organization on the surface of several plastic surfaces.[71,72,73] By using a variable-angle XPS technique, Ratner et al.[74] concluded that hemoglobin adsorbed to the surface of teflon in islands, but formed a continuous film on platinum.

XPS was used by Tarcha[75] to characterize the surface chemistry of $^1/_4''$ polystyrene beads. Figure 6 shows the complete XPS spectrum of a bead showing the typical carbon and oxygen peaks. The oxygen peak is most likely due to adsorbed water since the carbon region did not show any carbons directly bound to oxygen. Silicon peaks are present, but are most likely due to

a)

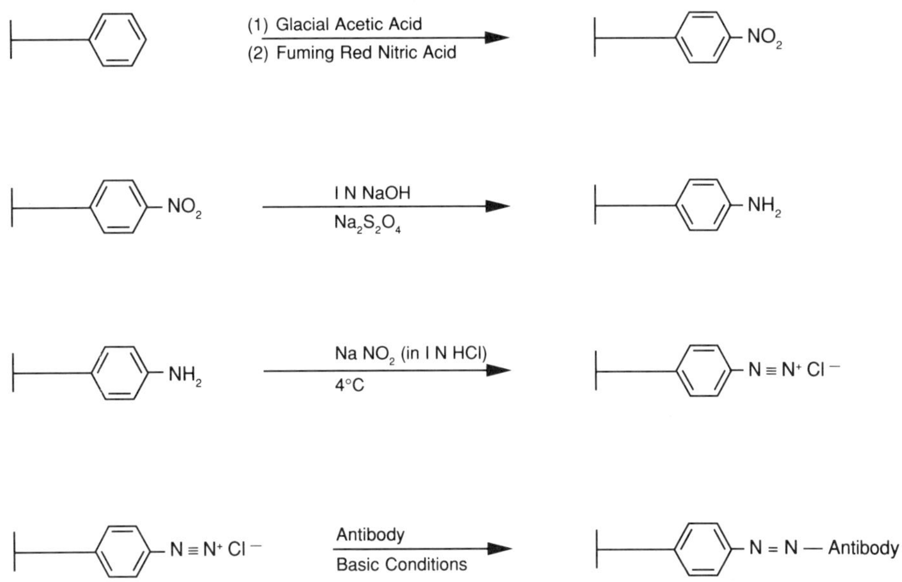

FIGURE 7. Reaction scheme for producing diazotized polystyrene surfaces and subsequent binding of antibody. (From Tenoso, H.J. and Smith, D.G., NASA Tech. Brief Report #MCS-13906, 1972. With permission.)

contamination by ubiquitous surface-active silicones, which is a common occurrence in chemistry labs. Most interesting is the peak in the nitrogen region (400 EU) shown in the insert. Integration of this peak indicated that there was approximately 1% nitrogen on the bead surface. It was concluded that the nitrogen was coming from bisstearamide, a flow agent which is commonly used in very small amounts with this type of resin. Calculations indicate that the beads contain approximately 19% bisstearamide in the surface monolayers. This surface concentration is several thousand times higher than would be expected based simply on the amount originally added to the bulk resin.

XPS is only one of many surface-sensitive spectroscopies. The reader is referred to a textbook by Czanderna[76] which describes XPS along with many other surface sensitive spectroscopies currently available.

B. SURFACE MODIFICATIONS AND COUPLING METHODS

The literature concerning the chemistry of surface-functionalization of organic polymers and inorganic solid supports is vast. These include reviews on the chemistry of affinity adsorbants for chromatography,[77,78] on chemical cross-linking reagents,[79] and on specific immobilization procedures used in SPIA.[80] This section will not attempt to duplicate the reviews in this area, but rather focus on select methods used in SPIA which are of concern to the diagnostic scientist. In addition, several future technologies will be discussed.

1. Polystyrene

There are almost as many ways to modify polystyrene as there are derivatives of benzene or toluene. Generally, the methods which lend themselves to SPIA are (1) latex copolymerizations of styrene with a functionalized monomer (discussed earlier), (2) reactions in organic media of cross-linked polystyrene surfaces, and (3) surface reactions in an organic medium which do not dissolve the polystyrene or its functionalized surface. Nitration and subsequent reduction has been reported as a method for producing a poly(aminostyrene) which can be diazotized and used to couple antibodies[81] (Figure 7). Reproducibility in these types of reactions can be difficult to control, and publications often fail to describe adequately the characterization of these surfaces.

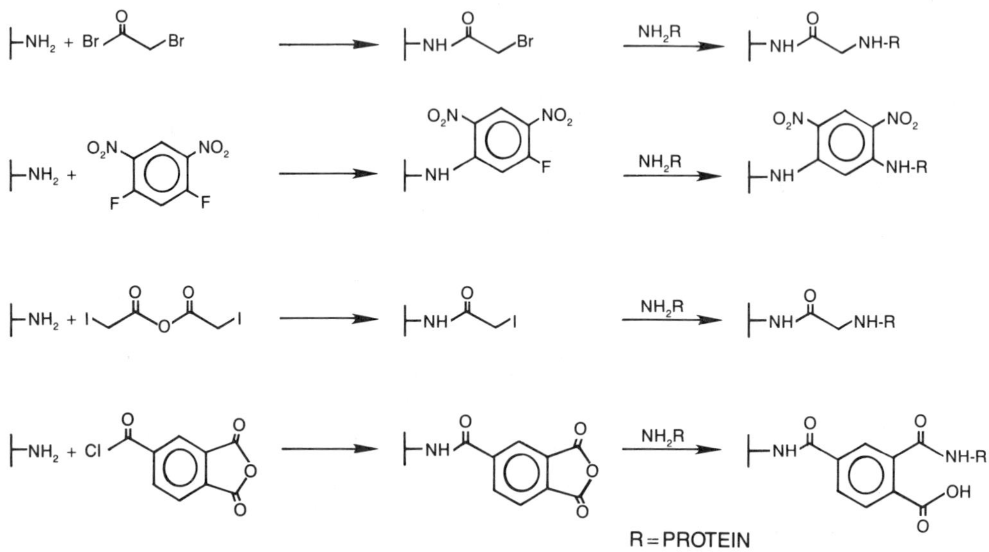

FIGURE 8. Surface immobilization schemes via bromoacetyl, dinitrofluoro, iodoacetyl, and trimellitic anhydride, groups respectively. (From Chu, V. P. and Tarcha, P. J., *J. Appl. Polym. Sci.*, 34, 1917, 1987. With permission.)

Cross-linked polystyrene resins can be brominated and converted to the litheated derivative. The poly(*p*-lithiostyrene) can be converted to a variety of derivatives[82] in which thiol, aldehyde, carboxylic acid, and boronic acid should be of most interest to the diagnostic scientist.

Often, it is necessary to derivatize polystyrene which has been molded into a bead, stick, or plate. Invariably, these parts are not cross-linked, since the resin would be too viscous for molding applications. Chu and Tarcha describe methods for producing activated surfaces on uncross-linked polystyrene, namely those displaying bromoacetyl, dinitrofluoro, iodoacetyl, or trimellitic anhydride groups.[83] Nitromethane, a nonsolvent for polystyrene, is used. In this solvent, the resin is chlorosulfonated, which-cross links the surface and simultaneously produces reactive sulfonyl chloride groups. Subsequently, the surface is reacted with an excess of di- or triamine to form an amino surface. Proteins can be coupled directly to this surface, or the amino groups can be further reacted to form the aforementioned protein-reactive groups (Figure 8).

2. Hydroxylated Surfaces

Hydroxylated surfaces can be activated for coupling with proteins by cyanogen bromide. March et al. describe the method and apply it to the activation of agarose for coupling with alanine and albumin.[84] In another example, Tabata et al. used the method to immobilize collagen onto cellulose and polyvinylalcohol.[85] Cyanogen bromide-activated microparticles are used to formulate chromatography supports and are readily available from speciality supply houses. 1,1′ carboxyldiimidazole has also been described as a coupling agent which can be used with hydroxylated surfaces.[86]

3. Glutaraldehyde

One of the most common coupling reagents found in the laboratories of diagnostic scientists are solutions of glutaraldehyde. Since amino groups form Schiff bases easily with aldehydes, this reagent is often seen as a convenient means by which antibodies can be attached to solid phases via a 5-carbon spacer. The reagent is frequently used to couple protein to surfaces possessing amines or simply used as a coating on nonaminated surfaces to which protein is subsequently coated. Monsan et al.[87] studied the mechanism of the glutaraldehyde-coupling reaction and found that the assumption that this reagent is a homobifunctional coupling agent

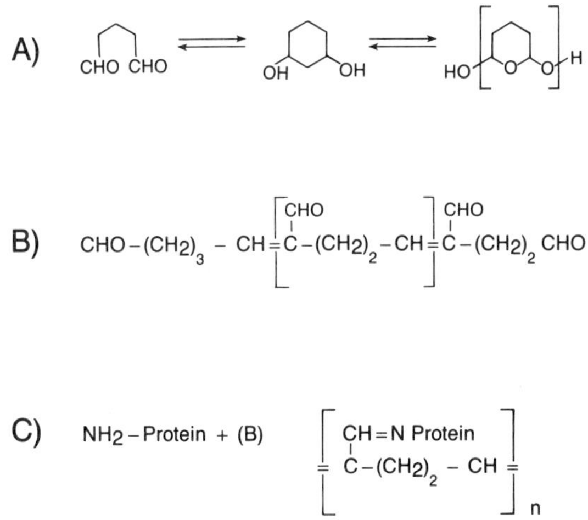

FIGURE 9. Reaction of glutaraldehyde under (A) acidic conditions, (B) basic conditions, and (C) reaction of glutaraldehyde polymer with protein.

is oversimplified. Under acidic conditions, where glutaraldehyde solutions are usually stored, the molecule is in equilibrium with its cyclic hemiacetyl and polymers of the cyclic hemiacetyl. Upon raising the pH to 7 or above, conditions where coupling reactions are usually run, the dialdehyde undergoes an aldol condensation with itself followed by dehydration, forming α,β unsaturated polymers. When protein is added to surface-immobilized glularaldehyde polymer, pendant aldehyde groups form Schiff bases with the lysine residues of the protein, and these linkages are quite stable toward hydrolysis, because they are conjugated with the chain olefinic group. Consequently, it is not necessary to reduce these Schiff bases to secondary amines for stability. Because of the polymeric nature of this reagent, removal of the unreacted reagent (subunits) is not usually feasible. This is due to multipoint attachment of the chain via some of its subunits to both the solid phase and the biomolecules being immobilized. Care must be taken to consume any unreacted aldehyde groups to avoid further cross linking of the immobilized protein during storage. Figure 9 illustrates the glutaraldehyde chemistry discussed.

4. Carbodiimide

Carbodiimides are coupling promoters which have seen wide use in peptide synthesis. The most widely used carbodiimide is dicyclohexylcarbodiimide. It has not seen much use as a modification or coupling reagent for proteins or antibodies because it is not water soluble and must be used in conjunction with organic solvents. Another disadvantage is that the by-product of the coupling reaction is insoluble in most solvents and can produce a separation problem if a particulate solid phase is used. Water-soluble carbodiimides can be used to modify carboxyl groups on proteins and can potentially be used to couple antibodies to aminated surfaces. Hoare and Koshland[88] used *N*-benzyl-*N*^{1}-3-dimethylaminopropyl carbodiimide to modify the enzymes lysozyme, chymotrypsin, and trypsin with glycine methyl cestin. Another analogue, 1-ethyl-3(3-dimethylamino-propyl) carbodiimide (EDC) has been used for the modification of lysozymes.[89-91]

Figure 10 shows the structure of EDC and outlines the chemistry and side reactions that can occur during a coupling reaction. The carboxyl group can be activated for reaction with protein-bound amides or thiols in two ways. Coupling can occur by nucleophilic attack on species B1 or attack by a neighboring carboxylate anion can form the anhydride; the anhydride is also very reactive towards amines or thiols. It is important to have the protein present either concurrently or immediately after the carboxyl groups are activated, because several nonproductive side

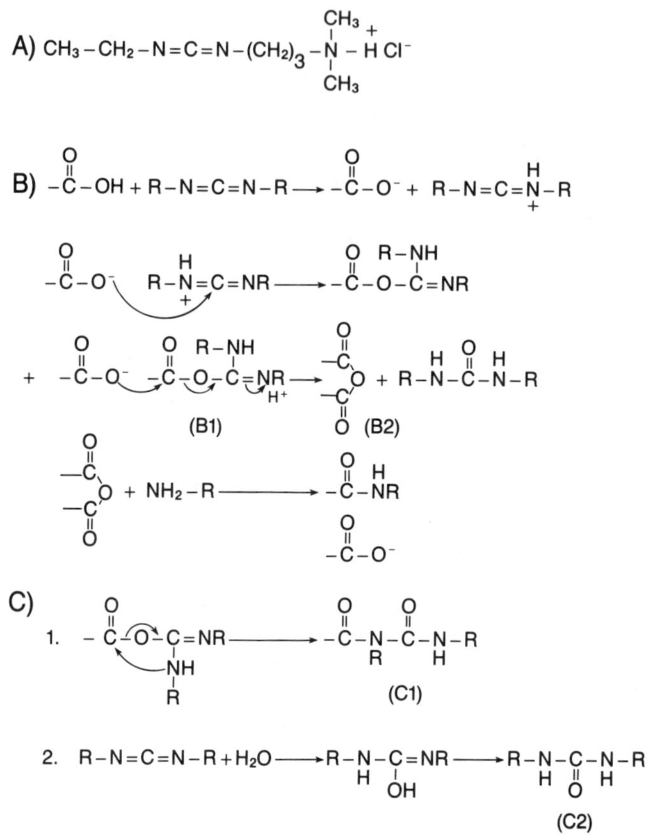

FIGURE 10. (A) Structure of 1-ethyl-3-(3-dimethylamino) carbodiimide, (B) mechanism of carboxylic acid activation and reaction of protein nucleophiles, and (C) side reactions possible with carbodiimides.

reactions can occur. First, the anhydride is very unstable in water and will hydrolyze back to carboxylic acid, if no nucleophiles stronger than water are present. Second, the activated species B1 can rearrange forming an unreactive urea, (C1). In addition, carbodiimides are good dehydrating agents and can form ureas (C2) easily by reaction with water. In any case, an advantage of this agent is that the free urea is water soluble and can be easily removed by washing with water or buffer or by dialysis.

IV. IMMOBILIZED ANTIBODY ACTIVITY

The direct conjugation of an Ab or Ag to a solid support or to another functional molecule such as an enzyme or toxin is often not the most optimal situation for maximum biological activity. In examining the structural features that influence efficacy in holotoxin-Ab conjugates, Marsh and Neville[92] found that incorporation of a flexible spacer between ricin and the targeting Ab increased tenfold the potency of inhibiting protein synthesis in target cells. The conjugate did not become more toxic to normal cells as a result of the modification. The spacer was derived from the insulin B chain which possessed two amino groups, the N-terminal and the ε-amino group of the lysine at position 29. Both amino groups were converted to sulfhydryl by reaction with 2-iminothiolane. Both ricin and the Ab were separately activated with *m*-maleimidobenzoyl-*N*-hydroxysuccinimide ester. The conjugate was made by first reacting the activated ricin with the spacer, purifying this product, and then adding the activated Ab.

Positive effects on Ab performance in SPIA have been observed with spacers. Dierks et al.[93]

reported that Ab adsorbed directly to the solid phase was recognized less efficiently in an ELISA assay than Ab bound to previously immobilized Ag. Suter and Butler[94] showed that the Ag capture capacity of a biotinylated antifluorescein Ab was 5- to 400-fold higher when bound via an adsorbed avidin spacer, as opposed to direct adsorption to plastic. Peterman et al.[95] went on to study covalent linkage and reported that binding a biotinylated Ab covalently to a bromoacetylated surface reduced its Ag capture capacity, relative to Ab bound to avidin, where the avidin had been previously covalently linked to the same surface.

The same type of observations were made in related cases where haptens are immobilized and an antibody's ability to recognize and bind to them is measured. Kennel[96] showed that fragment D of human fibrinogen (FgD), bound directly to polystryene plates, was least efficient at binding labeled monoclonal Ab. FgD bound to plates through poly(L-lysine) was slightly more effective, while binding FgD to Sepharose® beads by cyanogen bromide activation was three to ten times more effective at maintaining antigenic recognizability. Steric factors and selective destruction or shielding of the epitope from the Ab due to nonrandom attachment of the Ag to the solid support may play a role. This work suggests that a hydrophilic surface environment, where the protein is attached to loops and tails of bound soluble polymer, may provide more protein mobility and hence greater opportunities for favorable interaction of the capture moiety with its complimentary partner. Custrecasas and Anfinsen[78] cite many examples in their review of affinity chromatography, where spacing of small ligands away from the support results in an effective adsorbent for enzymes, whereas direct attachment often produces a totally ineffective adsorbent.

In view of these findings, the greatest potential for improving the effectiveness of immunosorbants with respect to activity and nonspecific interactions lies in the engineering of surfaces which can immobilize antibodies or antigens via reactive spacer arms. A base material which is very hydrophilic and minimizes nonspecific interactions should be utilized. Numerous crosslinking reagents for proteins are now commercially available which permit attachment of the reagent to the support via one specific end of the spacer and leaving the other end free for reaction with the capture protein to be immobilized. One of many schemes is suggested as a possible example: *N*-succinimidyl-3(2-pyridinylthio)propionate could be uscd in conjunction with hydrazido polyacrylamide particles. The cross linker could be reacted with the hydrazido group on the particle, leaving the potential 4-atom spacer available for reaction with sulfhydryl-containing Ab or fragments. The polyacrylamide surface, being hydrophilic, should exhibit low nonspecific adsorption.

Novel heterobifunctional coupling agents have been recently reported[97] along with methods for attaching Abs and Ags to surfaces.

ACKNOWLEDGMENT

I sincerely appreciate the support of Abbott Laboratories in the preparation and Sandi Johnson for her kind assistance in the typing and formatting of this chapter.

REFERENCES

1. **Walling, C.,** *Free Radicals in Solution,* John Wiley & Sons, New York, 1957, 180.
2. **Rodriguez, F.,** *Principles of Polymer Systems,* McGraw-Hill, New York, 1970, 387.
3. **Nauman, E. B.,** Synthesis and reactor design, *Plastics Polymer Science & Technology,* Baijal, M. D., Ed., John Wiley & Sons, 1982, 483.
4. **Miles, D. C. and Briston, J. H.,** *Polymer Technology,* Chemical Publishing, New York, 1965, 134.
5. **Joan Agfanoff, Ed.,** *Modern Plastics Encyclopedia,* (An annual publication by McGraw-Hill , New York.)
6. **Dick, J. S., Ed.,** *Compounding Materials for the Polymer Industries,* Noyes Publishing, Park Ridge, NJ, 1987.
7. **Kresta, J. E.,** *Polymer Additives,* Plenum Press, New York, 1984.
8. **Flick, E. W.,** *Plastics Additives — An Industrial Guide,* Noyes Publishing, Park Ridge, NJ, 1984.

9. **Haslam, J., Willis, H. A., and Squirrel, D. C. M.,** *Identification and Analysis of Plastics,* John Wiley & Sons, New York, 1979.
10. **Cobler, J. C. and Chow, C. D.,** Analysis of high polymers, *Anal. Chem.,* 49, 159R, 1977.
11. **Coe, G. R.,** Systematic analysis of polymer compounds, *Austr. Plastic Rubber,* August, 36, 1977.
12. **Abbott, S. D.,** Size exclusion chromatography in the characterization of polymers, *Am. Lab.,* August, 1977.
13. **Grubisic, Z., Rempp, P., Benoit, H.,** A universal calibration for gel permeation chromatography, *J. Polym. Sci.,* B5, 753, 1967.
14. **Krishen, A. and Tucker, R. G.,** Gel permeation chromatography of low molecular weight materials with high efficiency columns, *Anal. Chem.,* 49, 898, 1977.
15. **Harkins, W. D.,** A general theory of the mechanism of emulsion polymerization, *J. Am. Chem. Soc.,* 69, 1428, 1947.
16. **Van den Hul, H. J. and Vanderhoff, J. W.,** Clean monodisperse latexes as model colloids, in *Polymer Colloids,* Fitch, R. M., Ed., Plenum Press, New York, 1971, chap. 1.
17. **Fitch, R. M.,** Preparation and characterization of charge-stabilized polymer colloids, in *Polyelectrolytes and Their Applications,* Rembaum, A. and Selegny, E., Eds., Reidel, 1975, 51.
18. **Fitch, R. M.,** A practical guide to the control of latex particle size and its distribution in emulsion polymerization, in *Science and Technology of Polymer Colloids,* Vol. 1, Polhlein, G. W., Ottewell, R. H., and Goodwin, J. W., Eds., Martinus Nijhoff, Amsterdam, 1983, 100.
19. **Brown, W. E., III,** Microparticle-capture membranes: applications to TestPack hCG-urine, *Clin. Chem.,* 33, 1567, 1987.
20. **Greene, B. W.,** Quantitative determination of surface carboxyl groups in vinyl acid modified styrene/butadene copolymer latexes. I. Latexes prepared with acrylic acid, *J. Colloid Interface Sci.,* 43, 449, 1973.
21. **Hohenstein, W. P. and Mark, H.,** Polymerization of olefins and diolefins in suspension and emulsion. Part I, *J. Polymer Sci.,* 4, 127, 1946.
22. **Ugelstad, J. and Hansen, F. K.,** Kinetics and mechanism of emulsion polymerization, *Rubber Chem. Technol.,* 49, 536, 1976.
23. **Ugelstad, J., Mork, P. C., Mfutakamba, H. R., Soleimany, E., Nordhuus, I., Schmid, R., Berge, A., Ellingsen, T., and Aune, O.,** Thermodynamics of swelling of polymer, oligomer and polymer-oligomer particles. Preparation and application of monodisperse polymer particles, in *Science and Technology of Polymer Colloids,* Vol. 1, Poehlein, G. W., Ottewill, R. H., and Goodwin, J. W., Eds., Martinus Nijhoff, Amsterdam, 1983.
24. **Lenka, S. and Nayak, P. L.,** Vinyl polymerization initiated by peroxydiphosphate, *Colloid Polym. Sci.,* 260, 771, 1982.
25. **McCarvill, W. T. and Fitch, R. M.,** The surface chemistry of polystyrene latexes initiated by the persulfate/bisulfite/iron system, *J. Colloid Interface Sci.,* 67, 204, 1978.
26. **Fitch, R. M., Gajria, C., and Tarcha, P. J.,** Acrylic polymer colloids: kinctics of autocatalyzed hydrolysis, *J. Colloid Interface Sci.,* 71, 107, 1979.
27. **Tarcha, P. J.,** (1) Autocatalyzed Hydrolysis of Acrylic Polymer Colloids: A Kinetic and ^{13}C NMR study, (2) Synthesis of Monodisperse Spin-Labeled Amphiphilic Polymers of Stearic Acid and Poly(ethylene glycol), Ph.D. thesis, University of Connecticut, Storrs, 1983.
28. **Sato, H., Tomiyama, T., Morimoto, H., and Nakajima, A.,** Structure and activity changes of proteins caused by adsorption on material surfaces, in *Proteins at Interfaces, Physiochemical and Biochemical Studies,* Brash, J.L. and Horbett, T.A., Eds., ACS Symp. Ser, Vol. 343, American Chemical Society, Washington, D.C., 1987, 76.
29. **Napper, D. H.,** Steric stabilization, *J. Colloid Interface Sci.,* 58, 390, 1977.
30. **Janca, J., Pribylova, D., Bouchal, K., Tyrackova, V., and Zurkova, F.,** Characterization of polystyrene and poly(styrene-glycidyl methacrylate) copolymer latexes by sedimentation field flow fractionation, *J. Liquid Chromatogr.,* 9, 2059, 1986.
31. **Zurkova, E., Bouchal, K., Zdenkova, D., Pelzbauer, Z., Svec, F., Kalal, J., and Batz, H.G.,** Preparation of monodisperse reactive styrene-glycidyl methacrylate latexes by the emulsifier-free dispersion copolymerization technique, *J. Polym. Sci., Polym. Chem. Ed.,* 21, 2949, 1983.
32. **Dorman, L. C. and Mani, I.,** Method of coupling a protein to an epoxylated latex, U.S. patent 4,210,723, 1980.
33. **Margel, S. and Wiesel, E.,** Acrolein polymerization: monodisperse, homo and hybrido microspheres synthesis, mechanism, and reactions, *J. Polym. Sci., Polym. Chem. Ed.,* 22, 145, 1984.
34. **Chang, M. and Rembaum, A.,** Microstructure of polyacrolein, *J. Polym. Sci. Polym. Lett. Ed.,* 22, 279, 1984.
35. **Rembaum, A., Chang, M., Richards, G., and Li, M.,** Structure and immunological propertics of polyacrolein formed by means of ionizing radiation and base catalysis, *J. Polym. Sci. Polym. Chem. Ed.,* 22, 609, 1984.
36. **Murgita, R. A. and Vas, S. I.,** Isoelectric separation of mouse immunoglobulins, *J. Immunol.,* 104, 514, 1970.
37. **Margel, S., Beitler, U., and Ofarim, M.,** Polyacrolein microspheres as a new tool in cell biology, *J. Cell. Sci.,* 56, 157, 1982.
38. **Kumakura, M. and Kaetsu, I.,** Polymeric microspheres by radiation copolymerization of acrolein and various monomers at low temperatures, *Colloid Polym. Sci.,* 262, 450, 1984.

39. **Kumakura, M., Suzuki, M., Adachi, S., and Kaltsu, I.,** Polyacrolein microspheres as immunoreagents, *J. Immunol. Methods,* 63, 115, 1983.
40. **Kraemer, D., Lehmann, H., Pennewiss, H., Plainer, H., and Schweder, R.,** U.S. patent 4,070,348, 1978.
41. **Pollak, A., Blumenfeld, H., Wax, M., Baughn, R. L., and Whitesides, G. M.,** Enzyme immobilization by condensation copolymerization into cross-linked polyacrylamide gels, *J. Am. Chem. Soc.,* 102, 6324, 1980.
42. **Kim, K.U., Lee, H. Y., and Song, D. M.,** Inverse emulsion polymerization of *N*-vinyl pyrrolidone, *Pollimo,* 6, 305, 1982.
43. **Kim, K. U., Lee, H. Y., and Park, B. H.,** Inverse emulsion polymerization of acrylamide, *Pollimo,* 6, 188, 1982.
44. **Visioli, D.,** Formation and Stabilization of Inverse Emulsion Polymers, Ph.D. thesis, Lehigh University, Bethlehem, PA, 1984.
45. **Vanderhoff, J. W., DiStefano, F.V., El-Aasser, M. S., O'Leary, R., Shaffer, D. L., and Visioli, D. L.,** Inverse emulsion polymerization of acylamide: polymerization kinetics and process development, *J. Dispersion Sci. Technol.,* 5, 323, 1984.
46. **Visioli, D. L., El-Aasser, M. S., and Vanderhoff, J. W.,** Kinetics and mechanism of inverse emulsion polymerization, *Polym. Mater. Sci. Eng.,* 51, 258, 1984.
47. **Carver, M. T., Dreyer, U., Knoesel, R., Candau, F., and Fitch, R. M.,** Kinetics of photopolymerization of acrylamide in AOT reverse micelles *J. Polym. Sci., Polym. Chem. Ed.,* 27, 2161, 1989.
48. **Carver, M. T., Candau, F., and Fitch, R. M.,** Effect of solution components on the termination mechanism in acrylamide microemulsion polymerizations, *J. Polym. Sci., Polym. Chem. Ed.,* 27, 2179, 1989.
49. **Barrett, K. E .J., Ed.,** *Dispersion Polymerization in Organic Media,* John Wiley & Sons, New York, 1975.
50. **Tarcha, P. J., Bindseil, W., and Chu, V. P.,** Absorption enhanced solid-phase immunoassay method via water-swellable poly(acrylamide) microparticles, *J. Immunol. Methods,* 125, 241, 1989.
51. **Murphy, P. T. and Whistler, P. L.,** Dextrans, in *Industrial Gums,* Whistler, RL. and BeMiller, J. N., Eds., Academic Press, New York, 1973, chap. 23.
52. **Whistler, R. L. and BeMiller, J. N., Eds.,** *Industrial Gums,* Academic Press, New York, 1973.
53. **Hendry, R. M. and Herrman, J. E.,** Immobilization of antibodies on nylon for use in enzyme-linked immunoassay, *J. Immunol. Methods,* 35, 285, 1980.
54. **Palfree, R. G. E. and Elliott, B. E.,** An enzyme-linked immunosorbent assay (ELISA) for detergent solubilized Ia glycoproteins using nitrocellulose membrane discs, *J. Immunol. Methods,* 52, 395, 1982.
55. **Hawkes, R., Niday, E., and Gordon, J.,** A dot-immunobinding assay for monoclonal and other antibodies, *Anal. Biochem.,* 119, 42, 1982.
56. **Maier, K. H.,** Method of continuous manufacture of membrane filters, U.S. patent 3,026,571, 1962.
57. **Kesting, R. E.,** *Synthetic Polymer Membranes,* John Wiley & Sons, New York, 1985, 251.
58. **Leuvering, J. H. W., Thal, P. J. H. M., van der Waart, M., and Schurrs, A. H. W. M.,** Sol particle immunoassay (SPIA), *J. Immunoassay,* 1, 77, 1980.
59. **Wilmott, N. J., Miller, J. N., and Tyson, J. F.,** Metals as labels in immunoassay, *Anal. Proc. (London),* 22, 4, 1985.
60. **Fleischmann, M., Hender, P. H., and McQuillan, A. J.,** Raman spectra from electrode surfaces, *J. Chem. Soc., Chem. Commun.,* 80, 1973.
61. **Jeanmarie, D. L. and Van Duyne, R. P.,** Surface Raman spectroelectrochemistry. Part I. Heterocyclic, aromatic, and aliphatic amines adsorbed on the anodized silver electrode, *J. Electroanal. Chem.,* 84, 1, 1977.
62. **Vo Dinh, T., Alak, A., and Moody, R. L.,** Recent advances in surface-enhanced Raman spectroscopy for chemical analysis, *Spectrochim. Acta,* 43B, 605, 1988.
63. **Rohr, T. E., Cotton, T., Fan, N., and Tarcha, P. J.,** Application of surface-enhanced Raman spectroscopy to immunoassay, *J. Cell Biol.,* 107(3), 214a, 1988.
64. **Rohr, T. E., Cotton, T., Fan, N., and Tarcha, P. J.,** Immunoassay employing surface-enhanced Raman spectroscopy, *Anal. Biochem.,* 182, 388, 1989.
65. **Cosgrove, T., Obey, T. M., and Vincent, B.,** The configuration of sodium poly(styrene sulfonate) at polystyrene/solution interfaces, *J. Colloid Interface Sci.,* 111, 409, 1986.
66. **Gans, D. M.,** Surface Energetics, in *Paint Testing Manual,* 13th ed., Sward, G. G., Ed., The American Society for Testing Materials, 1972, chap. 3.3.
67. **Newman, A. W. and Good, R. J.,** *Surface Colloid Techniques,* Vol. 1, Plenum Press, New York, 1976.
68. **Chattur, K. K., Fink, D. J., Hutson, T. B., Gendreau, R. M., Jalcobsen, R. J., and Leininger, R. I.,** Fourier transform infrared spectroscopic and attenuated total reflectance studics of protein adsorption in flowing systems in *Proteins at Interfaces, Physicochemical and Biochemical Studies,* (ACS Symp. Ser., Vol. 343), American Chemical Society, Washington, DC, 1987, chap. 23.
69. **Anderson, A. B., Darst, S. A., and Robertson, C. R.,** Protein adsorption at polymer surfaces: a study using total internal reflection fluorescence, in *Proteins at Interfaces, Physicochemical and Biochemical Studies,* Brach, J. L. and Horbett, T. A., Eds., (ACS Symp. Ser., Vol. 343), American Chemical Society, Washington, DC, 1987, chap. 20.
70. **Osland, R. C. J., Price, W. J., and Devenish, A. H.,** Infrared techniques in the analysis of rubber, *Am. Lab.,* March, 111, 1978.

71. **Gildins, D. K., Payater, R. W., and Castle, J. E.,** Quantitative evaluation of water structuring and protein adsorption on the surface of hydrophilic polymers by ESCA, *Biomaterials,* 1, 163, 1980.
72. **Golander, C. G., Jonsson, S., Ladkova, T., Stenius, P., and Eriksson, J. C.,** Preparation and protein adsorption properties of photopolymerized hydrophilic films containing n-vinyl pyrrolidone (NVP), acrylic acid (AA), or ethylene oxide (EO) units as studied by ESCA, *Colloid Surf.,* 21, 149, 1986.
73. **Golander, C. and Kiss, E.,** Protein adsorption on functionalized and ESCA-characterized polymer films studied by ellipsometry, *J. Colloid Interface Sci.,* 121, 240, 1988.
74. **Ratner, B. D., Horbett, T. A., Shuttleworth, D., and Thomas, H. R.,** Analysis of the organization of protein films on solid surfaces by ESCA, *J. Colloid Interface Sci.,* 83, 630, 1981.
75. **Tarcha, P. J.,** unpublished data, 1984.
76. **Czanderna, A. W., Ed.,** *Methods of Surface Analysis,* Elsevier Scientific, New York, 1975.
77. **Lowe, C. R. and Dean, P. D. G.,** *Affinity Chromatography,* John Wiley & Sons, New York, 1974.
78. **Cuatrecacas, P. and Anfinsen, C. B.,** Affinity Chromatography, *Annu. Rev. Biochem.,* 40, 259, 1971.
79. **Peters, K. and Richards, F. M.,** Chemical cross-linking: reagents and problems in studies of membrane structure, *Annu. Rev. Biochem.,* 46, 323, 1977.
80. **Tijssen, P.,** in *Laboratory Techniques in Biochemistry and Molecular Biology,* Vol. 15, Elsevier, New York, 1985, chap. 13.
81. **Tenoso, H. J. and Smith, D. G.,** Covalent bonding of antibodies to polystyrene latex beads: a concept, *NASA Tech. Brief Report #MCS-13906,* 1972.
82. **Farrall, M. J. and Frechet, M. J.,** Bromination and lithiation: two important steps in the functionalization of polystyrene resins, *J. Org. Chem.,* 41, 3877, 1976.
83. **Chu, V. P. and Tarcha, P. J.,** Protein-reactive, molded polystyrene surfaces having applications to immunoassay formats, *J. Appl. Polym. Sci.,* 34, 1917, 1987.
84. **March, S. C., Parikh, I., and Cuatrecasas, P.,** A simplified method for cyanogen bromide activation of agarose for affinity chromatography, *Anal. Biochem.,* 60, 149, 1974.
85. **Tabata, Y., Conikar, S. V., Horii, F., and Ikada, Y.,** Immobilization of collagen onto polymer surfaces having hydroxyl groups, *Biomaterials,* 7, 234, 1986.
86. **Bethell, G. S., Ayers, J. S., Hancock, W. S., and Hearn, M. T. W.,** A novel method of activation of cross-linked agarose with 1,1′-carbonyldiimidazole which gives a matrix for chromatography devoid of additional charged groups, *J. Biol. Chem.,* 254, 2372, 1979.
87. **Monsan, P., Puzo, G., and Mazarquil, H.,** Study of the mechanism of glutaraldehyde-protein bond formation, *Biochemie,* 57, 1281, 1975.
88. **Hoare, D. G. and Koshland, D. E.,** A procedure for the selective modification of carboxyl group proteins, *J. Am. Chem. Soc.,* 88, 2057, 1966.
89. **Atassi, M. Z., Rosemblatt, M. C., and Habeeb, A. F. S. A.,** Enzymatic and immunochemical properties of lysozyme. VIII. Modification of two carboxyl groups and their role in the antigenic reactivity, *Immunochemistry,* 11, 495, 1974.
90. **Lin, T.-Y. and Koshland, D. E.,** Carboxyl group modification and the activity of lysozyme, *J. Biol. Chem.,* 244, 505, 1969.
91. **Yamada, H., Imoto, T., Fujita, K., Olcazaki, K., and Motomura, M.,** Selective modification of aspartic acid-101 in lysozyme by carbodiimide reaction, *Biochemistry,* 20, 4836, 1981.
92. **Marsh, J. W. and Neville, D. M.,** A flexible peptide spacer increases the efficacy of holoricin and anti-t cell immunotoxins, *J. Immunol.,* 140, 3674, 1988.
93. **Dierks, S. E., Butler, J. E., and Richerson, H. B.,** Altered recognition of surface-adsorbed compared to antigen bound antibodies in the ELISA, *Mol. Immunol.,* 23, 403, 1986.
94. **Suter, M. and Butler, J. E.,** The immunochemistry of sandwich ELISAs. II. A novel system prevents the denaturation of capture antibodies, *Immunol. Lett.,* 13, 313, 1986.
95. **Peterman, J. H., Tarcha, P. J., Chu, V. P., and Butler, J. E.,** The immunochemistry of sandwich-ELISAs. IV. The antigen capture capacity of antibody covalently attached to bromoacetyl surface-functionalized polystyrene, *J. Immunol. Methods,* 111, 271, 1988.
96. **Kennel, S. J.,** Binding of monoclonal antibody to protein antigen in fluid phase or bound to solid supports, *J. Immunol. Methods,* 55, 1, 1982.
97. **Bieniarz, C., Welch, C., Barnes, G., and Schlesinger, C. A.,** Covalent attachment of antibodies and antigens to solid phases using extended length heterobifunctionall coupling agents, European patent 314, 127, 1989.

Chapter 3

IMMUNOCHEMICAL CONSIDERATIONS IN THE ANALYSIS OF DATA FROM NON-COMPETITIVE SOLID-PHASE IMMUNOASSAYS

Jeffrey H. Peterman

TABLE OF CONTENTS

I. INTRODUCTION

Since their initial development in the early 1970s,[1,2] noncompetitive solid-phase immunoassays (SPIAs) have gained widespread use in the measurement of both antigens and antibodies. While SPIAs can be used qualitatively merely to determine the presence or absence of a component in a sample, their use as quantitative tests, through comparison with a known standard (the reference standard method; RS method), is widespread. Such quantitation may be performed manually by visually comparing data obtained from samples to a plot of a standard curve, the latter being generated with concentration and signal (e.g., absorbance) values obtained from a standard preparation. The manual procedure is tedious and of limited accuracy. With the advent of programmable calculators, and the recent availability of powerful personal computers, it has become practical to automate aspects of sample quantitation, with greatly improved accuracy and simplicity. However, accurate quantitation with a calculator or computer requires the use of an appropriate curve-fitting routine, such as one derived from a mathematical model; the latter should approximate the underlying immunochemistry of the system. While there have been extensive investigations concerning the theoretical basis of both fluid-phase[3-9] and solid-phase[10-14] immunoassays, the results of these studies have not generally been applied to data analysis.

This chapter will present a simple mathematical model for SPIA-binding curves and show the implications of this model in terms of assay performance, curve-fitting routines, and sample quantitation. Theoretical curves will be compared with curves from actual SPIA data. Finally, the influence of the method of plotting and the curve-fit routine on sample quantitation will be examined.

A thorough understanding of the expected SPIA titration curve (i.e., dose-response curve) can provide information useful in the establishment of an optimal assay. Knowledge of the effect of the method of analysis on sample estimation is important for determining the accuracy of the estimation. Without such information, SPIA data may be misinterpreted and result in erroneous conclusions about the biological phenomena under investigation.

For the purposes of this chapter, the reagent initially attached to the solid phase is referred to as the receptor, while the component to be measured is referred to as the ligand.*

II. DERIVATION OF A MATHEMATICAL MODEL FOR SPIA TITRATION DATA

Establishment of a simple mathematical model for SPIAs may begin with the assumption that the interaction is noncooperative, not influenced by binding-site heterogeneity, and at equilibrium.[13] Under such conditions, the interaction between the ligand and the solid-phase receptor would obey the Law of Mass Action:

$$\text{Association Constant} = \text{AB}/(\text{A} \times \text{B})$$

where A and B are the molar concentrations of the reactants and AB is the molar concentration of the product.

A. APPLICATION OF THE LAW OF MASS ACTION

With the assumption of noncooperative binding, the concentration of the product is equal to the molar concentration of bound ligand sites (L_b). Thus, if L_f is the molar concentration

* In order for the equations to be applicable to both assays for antigen and assays for antibody, the terms ligand and receptor are used. In assays for antigen, the receptors would be antigen-binding sites exposed on antibodies attached to the solid phase while ligands would be epitopes on the antigen. In assays for antibody, the receptors would be epitopes exposed on the antigen attached to the solid phase while the antibody antigen-binding sites would be ligands.

of available (free) ligand sites, R_f is the molar concentration of available receptor sites, and K is the association constant for the interaction (the affinity):

$$K = L_b/L_fR_f$$

Multiplying by R_f:

$$L_b/L_f = KR_f$$

If L_t = total concentration of ligand epitopes, then

$$L_f = L_t - L_b$$

Thus substituting for Lf,

$$L_b/(L_t - L_b) = KR_f$$

Hence,

$$L_b = KR_fL_t - KR_fL_b$$

Adding KR_fL

$$KR_fL_b + L_b = KR_fL_t$$

Dividing the latter equation by KR_fL_b:

$$KR_f/KR_f + 1/KR_f = L_t/L_b$$

Sirnplifying:

$$L_t/L_b = (KR_f + 1)/KR_f$$

Inverting:

$$L_b/L_t = KR_f/(1 + KR_f) \tag{1}$$

Equation 1 provides information on the proportion of ligand bound (L_b/L_t), which depends on the term $KR_f/(1+KR_f)$. Under conditions of receptor excess, R_f approximates the total receptor-binding site concentration (R_t). Therefore, Equation 1 shows that the maximum proportion of ligand bound is independent of the ligand concentration when the receptor is in excess, but dependent on the absolute values of K and R_t. Hence, the maximum proportion of ligand bound will be less in a low-affinity ligand-receptor system than in a higher affinity system. It can be seen that $KR_f/(1+KR_f)$ approaches unity whenever the product of K and R_f is greater than 10, under which conditions more than 90% of the ligand will be bound. Thus, assuming that the minimum affinity of a specific antibody-antigen reaction is $10^5\ M^{-1}$, any assay system will be affinity independent whenever the concentration of receptor binding sites available on the solid phase (R_f) is above $10^{-4}\ M$. The accuracy of sample estimation will depend on sample affinity at receptor levels below $10^{-4}\ M$; the assay will be affinity dependent, with the amount of ligand bound depending on both concentration and affinity. The actual reaction volume cannot be easily determined in solid-phase systems (see Chapter 12) precluding the accurate estimation of R_t and the subsequent calculation of the affinity dependency.

However, the relatively low levels of reactive receptor which may be attached to the surface of a solid phase make it likely that SPIAs will be affinity dependent, except in systems with very large reactive surface areas. The affinity dependency of SPIAs will therefore depend on the amount of receptor attached to the solid phase.[11,12,15-18]

The influence of affinity on ligand binding, as defined by Equation 1, can be useful in examining the immunochemistry of an assay in terms of the functional binding capacity of the solid-phase receptor. For example, if it is known that the same amount of receptor has been attached to a solid phase by various methods, and the reaction volume is a constant for the systems, then differences in the maximum proportion of ligand bound reflect differences in the functional affinity of the receptor.[17,18] Such differences may be a reflection of variations in the conformation of the receptor, variations in availability of its binding sites, or variations in levels of denaturation resulting from the method of attachment[19] (see also Chapters 10, 11, 13, and 17).

It should be emphasized that the above equation shows that the proportion of ligand bound is independent of ligand concentration while the receptor is in excess. Thus, contrary to popular belief, merely lowering the ligand concentration, such that the concentration of the receptor is in large excess, is not sufficient to achieve 100% binding of the ligand, but will only result in its maximal binding. For example, if $K = 10^5\,M^{-1}$ and $R_t = 10^{-6}\,M$, a maximum of 9% of the ligand will be bound, no matter how low the value of L_t.

B. GENERATION OF THEORETICAL PLOTS FROM THE MATHEMATICAL MODEL

Equation 1 can be useful in estimating the performance of a system, but it is of limited use in interpreting routine SPIA data. However, a simple rearrangement of the equation can predict titration curves for receptor-excess systems.

1. A Model of SPIA Titrations in Receptor-Excess Systems

SPIA titration data are commonly plotted in the format of ligand added compared to ligand bound, thus, from Equation 1:

$$L_b = KR_f/(1 + KR_f) \times L_t \tag{2}$$

With a logarithmic transformation:

$$\log(L_b) = \log(L_t) + \log[KR_f/(1 + KR_f)] \tag{3}$$

For a given ligand, K will generally be a constant in a system. Under the conditions of receptor excess, R_f approximates the constant R_t. With R_f and K as constants, Equations 2 and 3 fit the format of an equation for a straight line ($Y = AX + B$). It can therefore be predicted that a simple (nonlog) plot of ligand bound against ligand added will give a straight line with an intercept of zero and a slope dependent on $KR_f/(1 + KR_f)$. Similarly, a logarithmic (log-log) plot will give a straight line with a slope of unity (i.e., independent of L_t, K, and R_t) and an intercept dependent on $\log(KR_f/(1 + KR_f))$. Therefore, the influence of affinity on the titration curve depends on the type of plot used. This dependency can be important in sample estimation by the RS method, which requires parallelism between the plot generated by titrating the RS and the plots produced from the samples. Such parallelism should always occur with logarithmic plots (while the receptor is in excess) but will only occur in the nonlog plot if the RS and samples have the same affinity for the receptor, or if R_f is sufficiently large for the assay to be affinity independent.*

* While affinity would affect a log-log plot, the effect is on the position of the curve relative to the x-axis rather than its shape. Such an effect could affect sample estimation, resulting in overestimation of high-affinity samples and underestimation of low-affinity samples.[12,16]

2. A Model for SPIA Titrations without Receptor Excess

Equations 2 and 3 are useful when the receptor is in excess, but as L_t is raised and the receptor excess diminishes, R_f will drop and cease to be a constant. Most ligand titrations are not restricted to the region of receptor excess and so a more complex model must be used. As R_f represents the concentration of free receptor binding sites, and the concentration of bound receptor sites is equal to the concentration of bound ligand sites,

$$R_f = R_t - L_b$$

Rearranging Equation 2 and substituting for R_f:

$$L_b = L_tKR_t - L_tKL_b/(1 + KR_t - KL_b)$$

Multiplying by $1+KR_t - KL_b$,

$$-KL_b^2 + (1 + KR_t)L_b = L_tKR_t - L_tKL_b$$

Thus,

$$-KL_b^2 + (1 + KR_t + KL_t)L_b - (L_tKR_t) = 0$$

Dividing by –K,

$$L_b^2 - (1/K + R_t + L_t)L_b + (L_tR_t) = 0 \qquad (4)$$

Equation 4 is in a form solvable as a quadratic equation, thus, either

$$L_b = \{(1/K + R_t + L_t) - [(1/K + R_t + L_t)_2 - 4(L_tR_t)]^{1/2}\}/2 \qquad (5)$$

or

$$L_b = \{(1/K + R_t + L_t) + [(1/K + R_t + L_t)^2 - 4(L_tR_t)]^{1/2}\}/2 \qquad (6)$$

Equation 6 allows values for L_b greater than L_t; therefore, Equation 5 is the valid root.

III. ANALYSIS OF THEORETICAL CURVES GENERATED WITH THE MATHEMATICAL MODEL

Equation 5 allows data for an entire range of ligand titrations to be calculated if the functional affinity (K) and concentration of the receptor (R_t) are known. The range of values are generated simply by varying the value for L_t (ligand added). These data may be used to produce theoretical titration curves (Figure 1).

A. A COMPARISON OF DATA-PLOTTING TECHNIQUES

SPIA titration data are typically plotted (1) untransformed (nonlog, or linear, plots), (2) with logarithmic transformation of only the x-axis (semilog plots), or (3) with logarithmic transformation of both axes (log-log plots). Figure 1 shows five different plots generated using a set of theoretical data obtained with Equation 5; values used were $R_t = 10^{-7}$ *M*, $K = 10^7$ M^{-1} (x) or 10^5 M^{-1} (+), and L_t ranging from 10^{-4} to 10^{-9} *M*. SPIAs are usually performed using serial dilutions; thus, the dilutions are logarithmic (e.g., twofold dilutions are logarithmic dilutions to the base two). Nonlog plots (Figure 1A), therefore, tend to compress the data

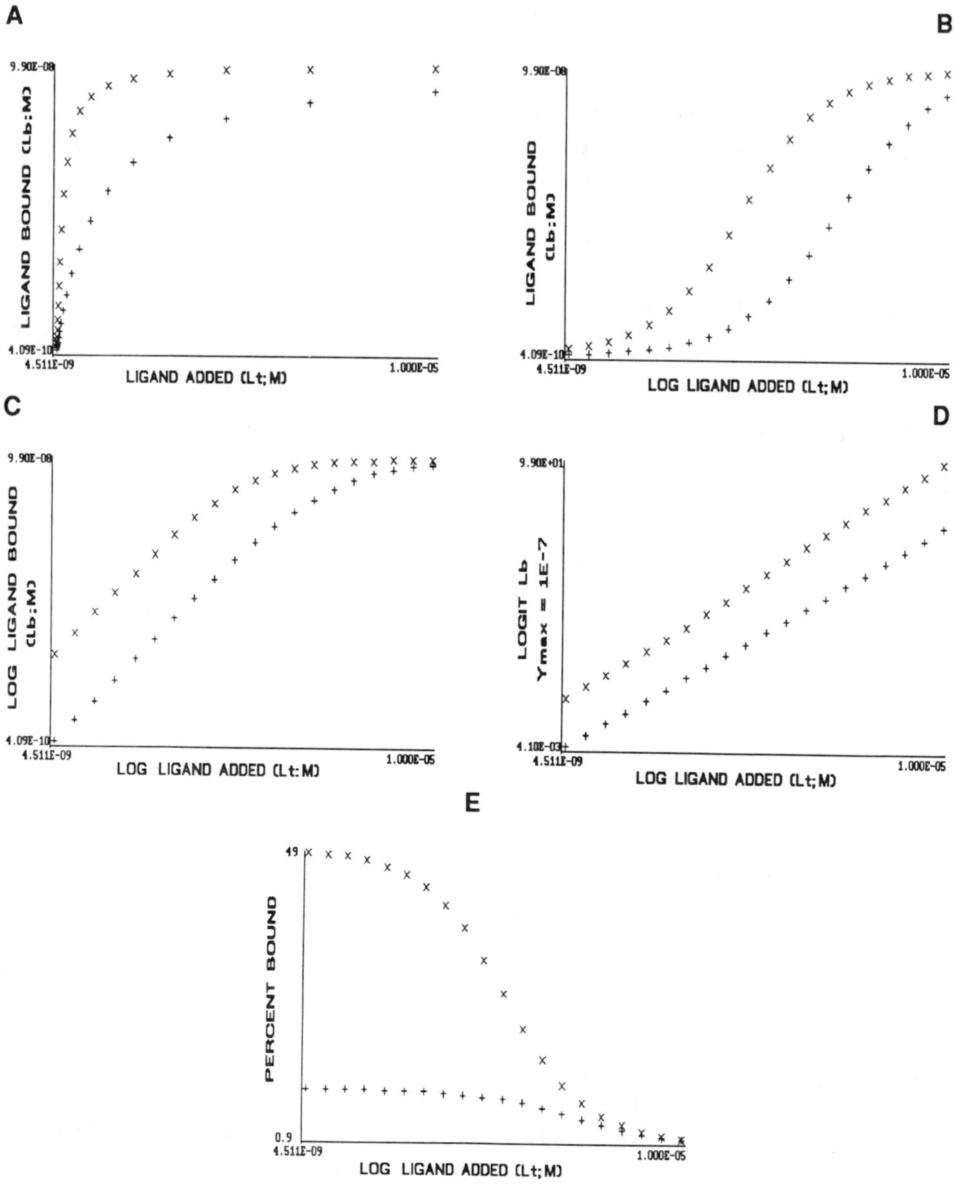

FIGURE 1. Plots of the theortical data generated using Equation 5. Data were generated according to Equation 5 using 10^7 M^{-1} (x) and 10^5 M^{-1} (+) for K. A value of 10^{-7} M was used for R_t, and L_t ranged from 4.5×10^{-9} to 10^{-4} M. The corresponding values of L_b were calculated. The data were analyzed with software used for routine analysis of SPIA data (ELISANALYSIS®[20]) and plotted in nonlog (A), semilog (B), log-log (C), logit-log (D), and percent bound (E) formats. Scientific notation is used for the axes labels, e.g., 1.2E2 corresponds to 1.2×10^2.

points from the highest dilutions, the latter being least affected by steric hindrance, competition, and inhibitory factors.[20] The commonly used semilog plot (Figure 1B) only partially overcomes this problem, as the data from the bound ligand (e.g., the final absorbance values) are still compressed, resulting in the typical, sigmoidal curve. Therefore, the most useful type of plot is the log-log plot (Figure 1C). The log-log plot has the following advantages: (1) it is mathematically valid based on the Law of Mass Action (as seen in Equation 3), (2) it gives equal dispersion to data derived from serial dilutions, (3) on the basis of Equation 3, the region of receptor excess can be readily identified as the linear region (which should have a slope of one), and (4) it should have a true linear region allowing a simple "curve-fit" by linear regression analysis. Finally, one other type of plot, the logit-log plot[6] is common in sample quantitation. The logit-log transformation uses log $L_b/(L_m - L_b)$ in place of log L_b, where L_m is the amount of ligand bound at saturation; the magnitude of the transformation is therefore proportional to the value of L_b. This transformation linearizes the hyperbolic log-log plot, extending the linear region of the curve into the region of saturation (Figure 1D). Logit-log transformation requires an accurate estimate of L_m; overestimates of L_m will not completely straighten the curve whereas underestimates will result in an upward-curving line.

The data generated with Equation 5 can be used to predict curves showing the proportion of ligand bound compared to the amount added (Figure 1E). As predicted with the simple equation (Equation 1), the curve approaches a maximum in the region of receptor excess, and the value of the maximum is dependent on the affinity of the interaction and the value of R_t (the latter is a constant in Figure 1). Additionally, the plot shows how the binding cannot be increased to 100% simply by lowering L_t until the receptor is in large excess; at the lowest dilution, the ratio of R_t to L_t is >20:1 but only approximately 1% of the low-affinity ligand is bound.

B. A COMPARISON OF CURVE-FITTING TECHNIQUES

The purpose of the mathematical model discussed above was to derive the theoretical shape of the ligand titration curve in order to find the best method of fitting a line (curve fitting) to the data points. The resultant line would be the standard curve used in estimating the ligand concentration of the unknown samples. The most popular method of plotting SPIA data is semilog, to which the most popular curve fits are based on simple linear regression analysis, and the so-called four-parameter log-logistic (4-PLL) curve fit proposed by Rodbard.[7,13]

Linear regression analysis determines the straight line which best fits a set of points. As described above, and as shown in Figure 1B, semilog plots will always produce sigmoidal curves. Therefore, a linear regression fit of such data is not appropriate except in the region of the pseudolinear inflection point; this combination of plot and curve fit greatly restricts the usable range of the standard curve. The 4-PLL curve fit is a reliable fit for a sigmoidal plot (assuming a linear central region) and is an extended form of the logit-log transformation. The 4-PLL curve fit requires an estimate of the background (L_0) in addition to the estimate of saturation (L_m). Many other types of curve fits are used with the sigmoidal semilog plot, including those based on polynomials[21] (e.g., the third degree polynomial $L_b = aL_t^3 + bL_t^2 + cL_t + d$, where a, b, c, and d are constants), spline curves,[22] and simply determining the best line between each pair of points and using those lines to produce the fit (linear interpolation). Polynomial-, spline-, and interpolation-based methods can adequately fit the theoretical sigmoidal curve; however, they are not restricted to a sigmoidal curve and so may be greatly affected by outlying data points, producing multimodal curves. Such multimodal curves do not fit the theoretical data and are of limited use in estimating the concentration of ligand in test samples, as they allow multiple values of L_t to produce a single value of L_b.

The complexity of curve fitting is greatly reduced if a log-log plot is used and the analysis restricted to the region of receptor excess. As described above, log-log plots are more valid immunochemically than are semilog plots; furthermore, there is a greater chance of parallelism between binding curves and more equal distribution of data from serial dilutions. Log-log plots have the advantage of placing the emphasis on the values obtained in receptor excess — the region in which the assay is least affected by inhibitory factors. Such data are therefore more reliable and should produce a straight line which can be readily fitted by linear regression analysis.

If it is necessary to obtain data from values above the limit of receptor excess, the logit-log transformation can extend the linear region and a linear regression fit may still be used. The range of usable data may also be extended by the application of the more complex curve-fitting routines. As discussed above, these routines (with the exception of the 4-PLL) have the inherent disadvantage of allowing multimodal curves. The 4-PLL fit is valid for the log-log plot, but requires sufficient data to accurately estimate both the maximum (L_m) and minimum (L_0) for the curve;[7,13] the logit-log transformation requires only an estimate of the maximum of the curve, while neither estimate is required when the linear region of a log-log plot is selected. While data from within the region approaching saturation can be used for sample analysis, it is of limited value; there are many factors which can distort the upper portion of the curve (see Section III.D, below) and a small error in signal measurement will cause a large error in sample estimation (as a large change in concentration will only have a small change in the signal).

C. IS THE LAW OF MASS ACTION REALLY APPLICABLE TO SPIAS?

The model derived above is based on the Law of Mass Action, as applied to fluid-phase systems, and makes the assumption of noncooperativity. It could therefore be argued that such a model is not appropriate for solid-phase systems. Lehtonen[23] has claimed that the Mass Law cannot apply to SPIAs with an enzymatic detection system (EIAs), because of observed discrepancies between affinities estimated for a monoclonal antibody in fluid-phase and solid-phase systems.* However, a wide range of data derived from solid-phase systems, both radioimmunoassays (RIAs) and EIAs, have been found to fit the above model,[10,16-20,24] thus supporting the validity of applying the model described here to SPIAs. Therefore, while ligand may bind cooperatively to the solid-phase receptor, and the assay may not actually reach equilibrium under normal conditions (see Chapter 18), these deviations from the conditions assumed for the model discussed here do not appear to significantly affect the validity of its application.

D. THEORETICAL EFFECTS OF BACKGROUND, SATURATION, STERIC HINDRANCE, BINDING-SITE COMPETITION, AND WASHING

While the mathernatical model for ligand-receptor interactions does appear to apply to routine SPIAs, there are factors which distort the titration curve. These factors are illustrated and summarized in Figure 2. The portion of the titration plot corresponding to lower concentrations may be distorted by direct binding of the detection system to the receptor or the solid-phase (background binding).[20] Background binding would have a minimal effect on a nonlog or semilog plot, but would distort the lower end of a log-log or logit-log plot and result in a curve with a slope which decreases at higher dilutions. A log-log plot of such data could resemble a sigmoidal curve; however, unlike a semilog plot, it would have a true linear region. The background binding can be estimated based on the knowledge that *the plot should form a straight line with a slope of one while in receptor excess.* This procedure can therefore be used to estimate a theoretical background by subtracting progressively larger values until a value is found which, in a log-log plot, results in straight line with a slope of one. Overcorrection for background (by subtracting too large a value) results in a curve with a maximum slope greater than one (the slope increasing at higher dilutions). It could be argued that the distortion which may occur in log-log plots due to inaccurate background estimation is a good reason to avoid such plots. However, a log-log plot, unlike a semilog plot, allows the background to be readily identified and thus signals potential nonspecificity in an assay. Once the background has been identified, the assay conditions can be optimized to reduce it, removing the need for mathematical manipulation. The modification in the assay to reduce background should also reduce nonspecific binding and thus potentially improve the assay specificity and sensitivity.

* The discrepancies observed by Lehtonen may be explained by inaccurate estimates of the actual solid-phase antigen concentration due to a lack of knowledge of the true reaction volume and/or an actual difference in the functional affinity of the antibody in the two systems.[16]

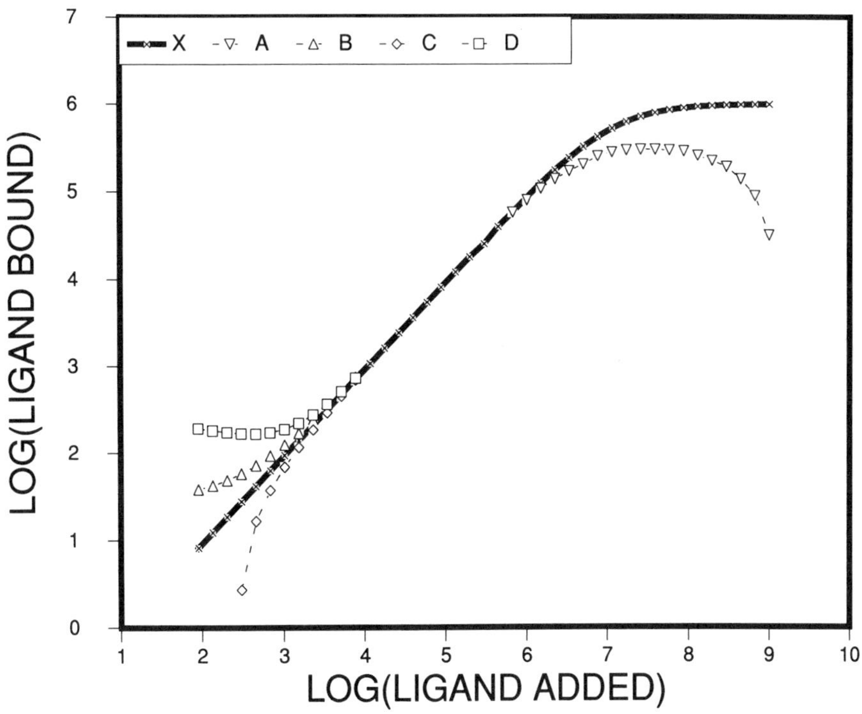

FIGURE 2. A diagrammatic representation of the effect of various factors on the ELISA titration curve. The theoretically predicted log-log titration curve (x; see Figure 1C) can be distorted by (A) steric hindrance,[24] ligand competition,[25,29,30] receptor heterogeneity,[14] or inadequate washing;[14] (B) undercorrection for the system background;[20] (C) overcorrection for the system background[20] or binding of ligand to desorbed receptor;[16] and (D) nonspecific binding of the detection system to the solid phase.[10]

Distortion from linearity in the upper region of the titration curve can be due to a wide range of factors. The simplest factor, saturation of the receptor, is included in the model. Similar effects may result from progressively increasing steric hindrance of the detection system,[24] competition between detected and nondetected ligands (e.g., when measuring specific IgA in the presence of specific IgG[25]), receptor heterogeneity,[14] inadequate washing of reagents on the solid phase,[14] or nonspecific inhibition by factors in the sample.[10] The model still applies if these effects mimic saturation; under such circumstances, either the logit-log transformation plus linear regression analysis or a 4-PLL curve fit are appropriate for the inclusion of the upper region. However, the factors (excluding saturation) can create the "high-dose hook effect",[14] in which a maximum is followed by a decreasing amount of ligand binding as higher amounts of L_t are added. Logit-log and 4-PLL analyses do not apply to such curves, but they can be applied if the values after the point of maximum binding are excluded from the analysis.

IV. ANALYSIS OF DATA FROM A SAMPLE ASSAY

The various plots shown in Figures 1 and 2 are all theoretical. Two questions which cannot be answered from the theoretical data are (1) how well do the various types of plots and curve-fit routines work with real SPIA data and (2) to what extent does the method of curve fitting affect sample analysis? To answer these questions, data from a routine assay currently in use by the author were analyzed. The data concern the measurement of total human IgA2 and polymeric IgA1 by sandwich enzyme-linked immunosorbent assay (ELISA).[26]

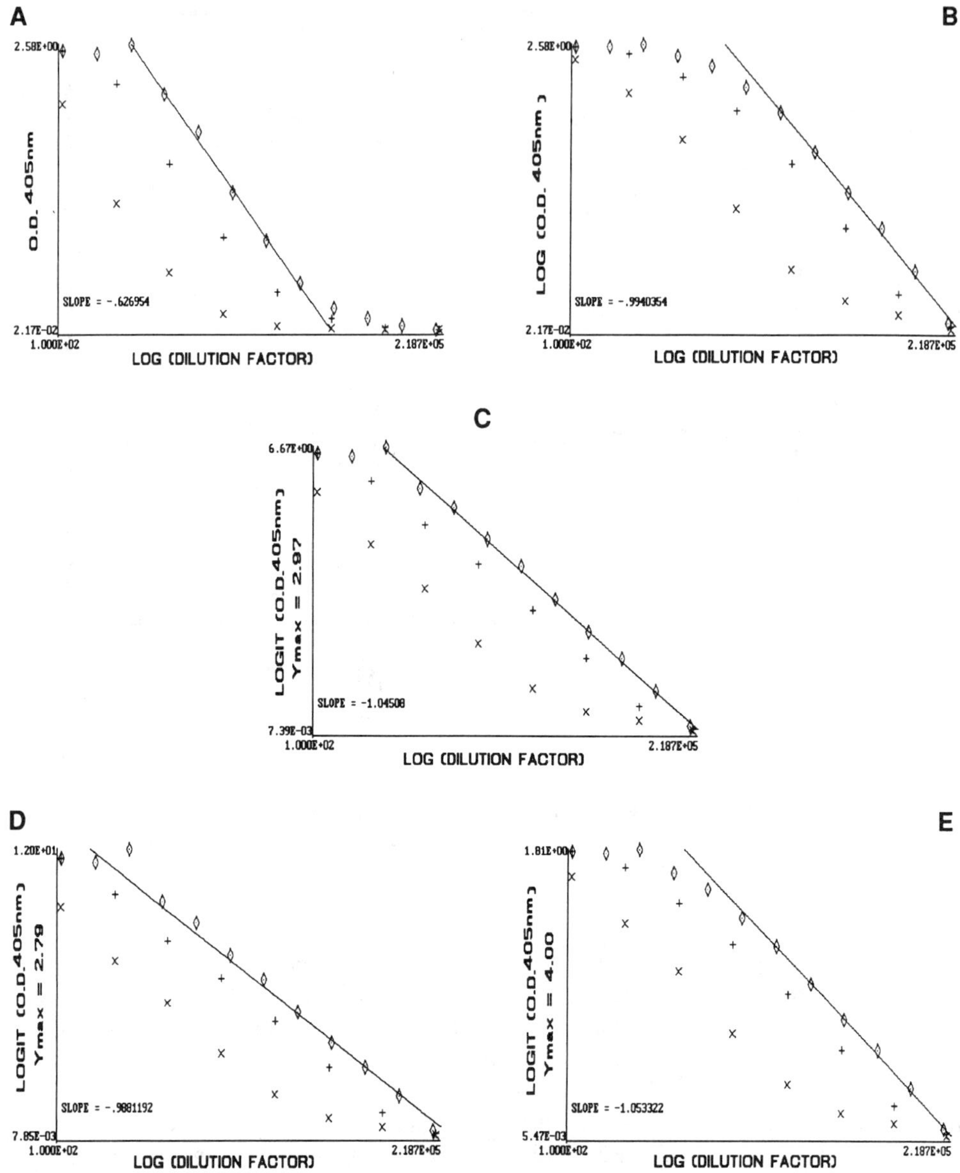

FIGURE 3. Plots of data from a sandwich ELISA for the quantitation of human IgA2.[26] Data are plotted as log dilution factor vs. the log of the absorbance value, for the RS (◇) and two test samples (x and +). The standard curve obtained by linear regression analysis is shown as a solid line overlaying the data points obtained from the dilutions of the standard. The data are plotted in various formats: semilog (A), log-log (B), and logit-log (C, D, and E). Three different logit-log plots are shown to illustrate the effects of under- (D) or over- (E) estimation of the value of L_m.

A. EXAMINATION OF ACTUAL SPIA TITRATION CURVES

The plots shown in Figure 3 were generated using data from three serum samples (an RS and two "unknown" test samples) in a sandwich ELISA.[26] It can be seen that semilog, log-log, and logit-log plots of the data have the same shapes as predicted theoretically (Figure 1). It is apparent from the upper portion of the RS curve that there is inhibition of binding at high concentrations (low dilutions). Analyses of various plots of the RS to determine their linear regions (using ELISANALYSIS;[20] Figure 3, solid lines) show that the linear regions of the semilog and log-log plots each cover a range of six twofold dilutions (from $^1/_{400}$ to $^1/_{12,800}$, and

$^1/_{6400}$ to $^1/_{20,4800}$, respectively). The linear region of the semilog plot consists of data from the lower dilutions (higher concentrations) while the linear region of the log-log plot involves the higher dilutions. Thus, if the inhibition is due to a component of the sample, and therefore likely to vary from sample to sample, the semilog plot will be less reliable for sample analysis than the log-log plot. The data used for analysis in the log-log plot cover the dilutions least affected by the inhibitory factors. The slope of the linear region of the log-log plot is approximately one, as predicted from Equation 3, and therefore must fall into the region of antibody (receptor) excess. Using the log-log plot, it can be determined that competition between components of the sample is not likely to occur below a dilution of $^1/_{6400}$. Furthermore, the form of plot used for the data affects the apparent sensitivity of the assay since the linear region of the curve in the log-log plot covers higher dilutions than those in the semilog plot; the assay is more sensitive if data analysis is based on a log-log plot.

The linear region of the logit-log plot of data from the RS covers a wider range of dilutions than either the semilog or log-log plots: 10 2-fold dilutions, ranging from $^1/_{400}$ to $^1/_{20,4800}$. The logit-log plot also provides the same sensitivity as the log-log plot. However, the logit-log plot does not provide information for the determination of the region of antibody excess, and its upper portion is subject to the same limitations as a semilog plot, i.e., the influence of inhibitory factors can prevent a good curve fit. Additionally, the logit-log plot requires an accurate estimation of the theoretical saturation value for the curve (L_m). The value of L_m cannot generally be measured in an SPIA. This problem may occur due to limitations in instrumentation (e.g., if L_m in an ELISA corresponds to an absorbance above 2.0 and the spectrophotometer is only linear up to 2.0) or due to distortion of the upper portion of the curve resulting from steric hindrance of the detection system (such that the observed curve does not parallel the sample binding curve[24]). Therefore, the value of L_m must be estimated. On the basis that a logit-log curve should be linear when a correct value for L_m is employed, L_m may be estimated by an iteration routine which varies L_m until the best linearity (determined using linear regression analysis) is achieved.[20] However, such a routine requires that the data are first analysed in a log-log format to be sure that they do not include values affected by inhibitory effects. For example, with the data obtained for the RS shown in Figure 3, exclusion of the upper 2 dilution points results in an L_m value of 2.97 (the maximum absorbance value observed in the assay was 2.75) and the resultant curve is linear over 10 dilutions (Figure 3C). However, inclusion of those data points results in (1) an underestimation of L_m (2.79), (2) overcorrection of the curve, and (3) production of a curve with a linear region composed of only nine dilutions (Figure 3D). Overestimation of L_m (e.g., 4.00) results in an undercorrected curve which closely resembles the untransformed log-log plot with a linear region including only 7 dilutions (Figure 3E).

B. EFFECTS OF INHIBITORY FACTORS AND BACKGROUND BINDING

The data shown in Figure 3 are from a well-established assay for serum IgA2 without known immunochemical problems. However, some systems measure samples containing components which can compete for the solid-phase receptor or require extreme sensitivity such that background binding of the detection system becomes a problem. An assay for human polymeric IgA1 in serum[26] is one such system in which the captured polymeric IgA is denatured and then detected with an anti-J chain reagent. This assay requires an anti-IgA1 capture antibody as the solid-phase receptor; J-chain is not adequately exposed on the captured IgA until it has been denatured. Only a small proportion of the IgA in human serum is polymeric,[27] and there is potential competition between the monomeric and polymeric forms of IgA for the solid-phase anti-IgA.[28] The degree of competition can vary between samples, depending on the proportions of monomer to polymer. When there is a high amount of monomer and a low amount of polymer in a sample, the polymer may be undetectable if the sample is sufficiently dilute for the anti-IgA to be in excess compared to the total IgA

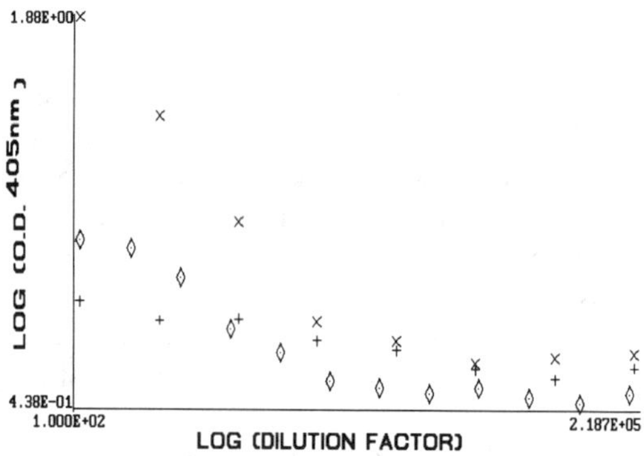

FIGURE 4. Plot of data from a sandwich ELISA for the quantitation of human polymeric IgA1.[26] Data are plotted as log dilution factor vs. log absorbance value; values for the RS (◇) and two test samples (x and +) are shown. The plot suggests that the RS reaches saturation at a lower absorbance value than one of the test samples. This presumably results from competition between monomeric and polymeric IgA1 for the solid-phase anti-IgA1 capture antibody. Furthermore, background absorbance value appears to vary from sample to sample.

concentration. If the anti-IgA is not in excess, competitive inhibition results in distortion at the upper portion of the curve, with the amount of distortion varying from sample to sample. An example of such data, for one serum sample and two intestinal wash samples, is shown in Figure 4. Similar ligand competition has been reported in ELISA systems for measuring specific IgA[25] or IgE[29,30] in the presence of competing IgG.

The problem of ligand competition can be avoided by removing the nondetected ligand (e.g., IgG removed in an assay for specific IgA[25]), reducing the inhibition and lengthening the linear regions. Ligand competition should not occur under conditions of receptor excess. However, it can be difficult to use an assay under such conditions if the ligand to be detected is a very low proportion of the total ligand captured by the receptor; such a situation requires an extremely sensitive assay. When high sensitivity is required, the background binding of the detection system can become significant. If the background is a constant, it may be determined from measuring a sample without ligand and subtracted from all the data. However, the background may vary from sample to sample, depending on variations in the levels of blocking components in the samples, or even vary due to factors which encourage the background binding.[16] Such variations in background are more likely to occur with heterogeneous samples such as secretions.

The data presented from the polymeric IgA1 assay, shown in Figure 4, illustrate the problems described above, with variations in L_m and L_0 from sample to sample, and very short (or nonexistent) linear regions. Therefore, data obtained from such an assay cannot be reliably used to estimate the concentrations of samples. Fortunately, such data are not common. Nevertheless, a reliable sandwich assay might be configured for this system by (1) using a detection system with higher affinity and specificity (increasing the sensitivity) so that samples may be measured at lower concentration (in higher receptor excess for reduced competition) and (2) using more effective blocking conditions (to reduce background interactions).

C. APPLICATION OF VARIOUS CURVE-FITTING FUNCTIONS TO THE SAMPLE DATA

The purpose of analyzing the RS is to determine the equation of the best curve fit for use when estimating the concentrations of test samples. As shown in Figure 3, the simplest curve fit is obtained by linear regression analysis of semilog or log-log plots of the data. The linear region of the log-log plot may be extended by performing the logit-log transformation on the data. Alternatively, the nonlinear curve fits may be used, such as the 4-PLL equation,[6,7,13,14] polynomials of various complexities,[21] or interpolation. The principle problem with nonlinear curve fitting is the difficulty in determining how well the derived curve fits real data. It can be difficult to differentiate between a poor fit and high variability, and it is much easier to visually confirm that a straight line, rather than a complex curve, is the best fit to a set of points. The nonlinear polynomial and interpolation-fitting methods do not provide any immunochemical information on the assay and can produce multiphasic (nonmonotonic) curves resulting from outliers or high assay variability. The 4-PLL curve fit is based on more sound immunochemistry, however, it requires careful selection of the data to exclude values from the region of inhibitory factors and a sufficient range of data points for extrapolation to find L_m and L_0.

Thus, a linear regression-derived fit applied to a log-log or a logit-log plot is the easiest type of curve fit to interpret and provides the most immunochemical information about the assay. As a log-log plot of data from a well-functioning assay should have a substantial linear region, analysis of a log-log plot can be useful in optimizing an assay. Extension of the linear region of an optimized assay with a logit-log transformation further removes the need for a nonlinear curve fit.

The lower region of log-log plots can be distorted by inaccurate measurements of the background (Figure 2); even a small inaccuracy can substantially distort the lower region of the titration curve.[20] The distortion is not a problem in most systems because the value measured for a ligand-free sample can be used as the background. However, where the distortion does occur, the "theoretical background" can be estimated by finding the value which results in the least distortion of the lower portion of the curve *and* produces a slope close to unity.* A significant difference between the measured background and the theoretical background is an indication that the background binding in the assay is variable and thus suggests that the specificity of the assay is questionable.

Applications of the various types of curve fits to data (obtained from the RS shown in Figure 3) are shown in Figures 5 and 6. The data were analyzed using two software packages: ELISANALYSIS® (developed by the author) and ELISA+® (from Meddata, Inc., New York, NY). Data from the dilutions which did not conform to the shape required for the various fits were excluded before performing the analyses, e.g., the two highest dilutions were excluded from all fits except for interpolation. The simple linear regression fit to a semilog (Figure 5E) or log-log (Figure 6E) plot covers the smallest range of dilutions (only six), while interpolation (Figures 5C and 6C) covers the widest range (all 12). However, as mentioned above, the log-log plot with a linear regression fit provides the most information about the immunochemistry of the assay, whereas the interpolation fit provides the least. It can also be seen that it is much easier to visually confirm that a straight line fits a log-log plot of the data set than it is to confirm any of the more complex curve fits for the nonlinear plots.

D. EFFECT OF VARIOUS CURVE-FITTING FUNCTIONS ON SAMPLE ESTIMATION

There are two common methods for estimating sample concentrations by comparison to an RS of known or assigned ligand concentration. One method is to compare entire sample curves with the curve produced from the RS and then determine the dilution at which each reaches either 100% or 50% saturation;[31] this describes the so-called "end-point titer" method. Such a method can be reliable when all samples contain the same components in the same ratios. However, the method becomes extremely unreliable if there are variations in the apparent

* Such data manipulation can be disturbing to those unfamiliar with the immunochemitsry of antigen-antibody reactions. This is justified by the demonstrated fact that a log-log plot of actual primary binding data has a linear region with a slope of one10,16,24 and is valid so long as the same manipulation is performed on all data points.

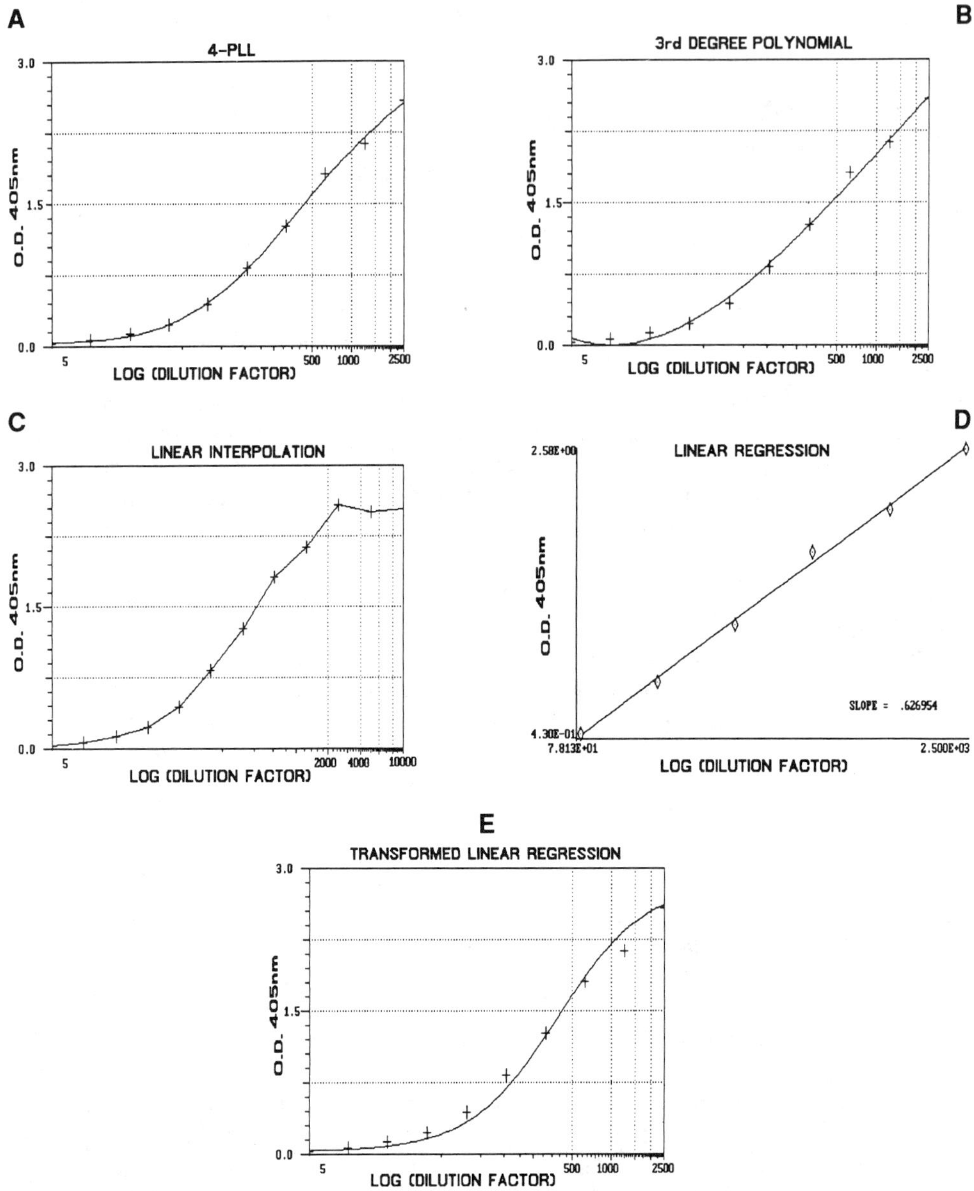

FIGURE 5. Application of various curve fits to a semilog plot of a referrence standard. The semilog plot of data from the IgA2 reference standard shown in Figure 3 was analyzed using the ELISA+® and ELISANALYSIS®[20] software packages. The best range of data points for each type of curve fit was selected and the calculated curve fit was overlaid on those points. The points were fitted with applications of the 4-PLL equation (A, ELISA+®, a third-degree polynomial (B, ELISA+®), linear interpolation (C, ELISA+®), linear regression (D, ELISANALYSIS®), or a semilog transformation of a linear regression fit to logit-log transformed data (E, ELISA+®). Note that the logit-log fit produced by ELISA+® uses linear regression of a logit-log plot to derive the best fit and then plots the data and the fit in a semilog manner.

point of saturation due to competition or inhibitory factors[25,29,30] (Figure 4 and Chapter 9). Additionally, the method requires detailed analysis of complete titration curves for all samples in order to estimate the saturation points for each curve. It thus requires a substantial amount of resources. An alternative, and more reliable, form of the end-point titer method is to determine the dilution necessary for a selected binding value (e.g., the dilution giving an absorbance value of 0.500); this variation of the method also requires extensive analysis of the titration curves of all samples.

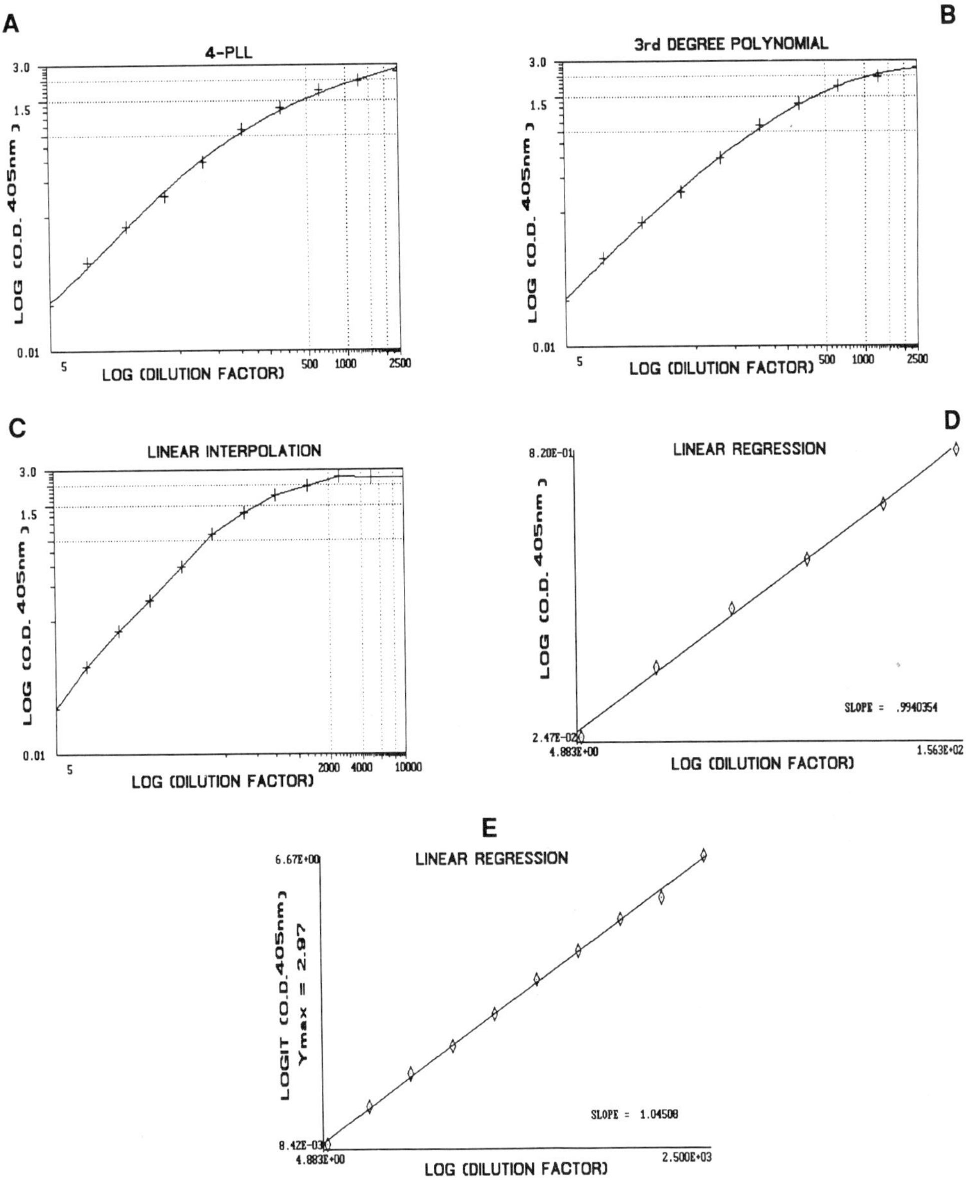

FIGURE 6. Application of various curve fits to a log-log plot of a reference standard. The log-log plot of data from the IgA2 reference standard shown in Figure 3 was analyzed using the ELISA+® and ELISANALYSIS®[20] software packages. The best range of data points for each type of curve fit were selected, and the calculated curve fit was overlaid on those points. The points were fitted with applications of the 4-PLL equation (A, ELISA+®), a third-order polynomial (B, ELISA+®), linear interpolation (C, ELISA+®), or the equations derived by linear regression analysis of either the log-log plot (D, ELISANALYSIS®), or a logit-log plot (E, ELISANALYSIS®).

With the limitations of the end-point titer method, many (including the author) prefer a second method; the so-called "reference equation" method. In this method, an equation is first obtained for the plot of an RS. The concentrations of the samples are then estimated using that equation together with the binding data from individual sample dilutions. This method requires fewer dilutions of each sample and does not require direct analysis of sample curves.

Both the end-point titer method and the reference equation method are only valid with sample data taken from the region where the sample curve parallels the standard curve. This "parallel" region may be identified by analyzing the individual curves, and it may be readily

TABLE 1
Effect of the Type of Plot on Sample Estimation by the End-Point Titer Method

Plot format	Sample A	Sample B
LOGIT-LOG	87	441
LOG-LOG	86	521
SEMI-LOG	75	412

Note: The data from the reference standard and the two samples were each plotted individually and the equations for their linear regions determined with linear regression analysis (using ELISANALYSIS®[20]). The absorbance value at the midpoint of the linear region of the reference standard curve was estimated for each type of plot and entered into the equations for the sample curves. The dilution values obtained from the equations were then used to calculate the sample concentration; e.g., an absorbance value of 1.00 in a logit-log plot would be given by a $^1/_{5715}$ dilution of the RS and a $^1/_{495}$ dilution of Sample A. Therefore, Sample A is equal to $^1/_{11.5}$ of the RS (which is 1000 E.U./ml), i.e., 87 E.U./ml. This method provides no statistical information on the final data.

identified in the reference equation method by simply determining whether three or more successive dilutions produce similar estimated concentrations.[20] For the latter to occur, such values must come from the parallel region.

Estimates of the sample data shown in Figure 3, derived by the end-point titer and reference equation methods, are show in Tables 1 and 2, respectively. With the end-point titer method, Sample A has a concentration of between 75 and 87 ELISA units (E.U.)/ml, while Sample B has a concentration of 412 to 521 E.U./ml, dependent on the plot format (Table 1). These data show variations of approximately 14 and 20%, respectively, simply based on the type of plot used. The variation is even greater with the wider range of plots used for the reference equation method. Depend;ing on both the type of plot and the type of curve fit utilized, Sample A varies from 79 to 113 (data not shown) and Sample B varies from 412 to 658 (shown in Table 2) E.U./ml (this represents 30 and 37% variation, respectively). These examples indicate that the method used to estimate sample concentration can significantly effect the estimated value. However, this should not be a problem if the same method, with one type of plot and one type of curve fit, is used for all sample estimations. Ideally, with perfect fits, the sample estimations should be independent of the types of plot and fit used. The observed variations (shown in Tables 1 and 2) probably result from the influence of the type of plot on weighting of the data. For example, a logarithmic plot automatically gives more weight to the lower values on both axes, whereas a semilog plot only weights the dilution factor values.

Many researchers who use the reference equation method for sample estimation only measure their samples at three or four dilutions (or less). Furthermore, they assume that estimates of the sample are accurate as long as a value falls within the linear range of the RS; sometimes this procedure means that estimations are made on the basis of a single dilution point! The latter estimations can be valid *only* if the titration curve of the sample parallels the standard curve over its entire range. The theoretical effects of ligand competition and receptor heterogeneity, and the examples in Figure 4, show that this assumption is not always true. Thus, it is dangerous to estimate the concentration of a test sample based on data from a single dilution point.

V. SUMMARY

This chapter has discussed various aspects of SPIA data analysis, using a mathematical model derived from the Law of Mass Action and using data obtained from routine sandwich assays performed by the author. The similarity between the theoretical and empirical binding

TABLE 2
Comparison of Sample Quantitation with the Reference Equation Method and Various Curve-Fit Routines

Sample dilution factor	4-PLL[a,b]	Semilog 3P[b]	Fit LI[b]	Logit[b,c]	LR[d]
100	240[e]	226	1,000	250	229
300	396[f]	420[f]	435[f]	309	419[f]
900	394[f]	416[f]	385[f]	376	397[f]
2,700	449[f]	415[f]	440[f]	530	419[f]
8,100	493	440[f]	494[f]	668[f]	574
24,300	551	634	474[f]	732[f]	1,185
72,900	472	350	483[f]	575[f]	3,130
218,700	1,050	1,050	1,050	1,050	9,140
Mean	413	423	452	658	412
SD	31	12	40	79	13
CV	8	3	9	12	3

Sample dilution factor	4-PLL[b]	Log-Log[2] 3p	Fit LI[b]	LOGIT[c,g]	LR[d]
100	196	250	1,000	233	46
300	392	352	440[f]	368	122
900	443[f]	397	397[f]	396	246
2,700	477[f]	478[f]	444[f]	482[f]	415
8,100	472[f]	493[f]	520[f]	514[f]	518[f]
24,300	490[f]	487[f]	474[f]	504[f]	523[f]
72,900	544	525	520[f]	512[f]	517[f]
218,700	1,050	1,050	1,050	896	885
Mean	471	486	467	503	519
SD	20	8	49	15	3
CV	4	2	11	3	1

Note: Mean = mean of the selected values, SD = standard deviation of selected values, and CV = coefficient of variation (%) for selected values. The concentrations estimated for each dilution of Sample B are shown, based on the plots shown in Figures 5 and 6. Each value is from a single well in an ELISA. From these values, the regions of apparent parallelism were selected (indicated by f, where at least three successive dilutions gave similar values) and the mean of these values used as the estimated concentration for the sample. Note that the existence of such a region is questionable for the semilog plot analyzed with the ELISA+® version of logit-log transformation.

[a] Curve-fitting equation: LR = Linear regression analysis, 3P = 3rd degree polynomial function, LI = linear interpolation, and 4-PLL = 4-parameter log-logistic.
[b] Determined using ELISA+® software.
[c] While the ELISA+® software determines a linear regression fit to logit-log transformed data, it uses a semilog plot.
[d] Determined using ELISANALYSIS® software.
[e] Estimated concentration for one dilution of the sample.
[f] Values selected for estimate of sample concentrations.
[g] Actually a linear regression fit to a logit-log plot.

curves shows the validity of the theoretical model. Analysis of theoretical and empirical binding curves shows how the method of plotting the data can influence the interpretation of the titration curves and the superiority of the log-log plot over the more common semilog plot. Analysis of SPIA titration plots, to derive the equation of the curve, can be performed by a range of curve-fitting methods. The concentration values estimated for samples can depend on both the method of plotting and the type of curve fit employed. Selection of the appropriate data points is important for the accurate application of the curve fits since most methods require that values from within the region of inhibition (at high concentration) be excluded. For most SPIA systems, the best method of analysis involves a simple log-log plot to examine the assay immunochemistry and estimate the appropriate system background followed by linear regression analysis of logit-log transformed data for the derivation of the reference equation. Finally, estimations from multiple dilutions of the sample should be used routinely to ensure that the concentration value obtained for the sample is based on the region of the sample titration curve which parallels the standard titration curve.

Application of the principles described herein to routine SPIAs, such as the ELISA, may result in more reliable SPIAs and a higher level of confidence in the data obtained from those assays. Such data may provide better insight into the biological phenomenon which is the primary interest of the SPIA user.

ACKNOWLEDGMENTS

The author would like to thank Dr. H. Gadler, from Meddata Inc., for supplying a copy of ELISA+ for use in the sample analyses. Meddata may be contacted at 212-860-7320. The author would also like to thank Drs. S. Jackson and J. Mestecky, from the University of Alabama at Birmingham, for their support during the initial preparation of this chapter and Dr. R. Massey, from IGEN Inc., for his support during its completion.

REFERENCES

1. **Catt, K. J., Niall, H. D., and Tregear, G. W.,** Solid phase radioimmunoassay, *Nature,* 213, 825, 1967.
2. **Catt, K. and Tregear, G. W.,** Solid-phase radioimmunoassay in antibody-coated tubes, *Science,* 158, 1570, 1967.
3. **Griswold, W. R.,** Theoretical analysis of the reaction of multivalent antigen with heterogeneous antibody: a model for soluble phase antibody assays, *J. Immunoassay,* 8, 145, 1987.
4. **Griswold, W. R. and Nelson, D. P.,** Theoretical analysis of the Farr antibody assay with a computer model: importance of antigen concentration and antibody affinity, *J. Immunoassay,* 5, 71, 1984.
5. **Griswold, W. R. and Nelson, D. P.,** Calculation of monoclonal antibody affinity constants directly from antibody dilution curves, *Immunol. Lett.,* 9, 15, 1985.
6. **Rodbard, D. and Lewald, J. E.,** Computer analysis of radio-ligand assay and radioimmunoassay data. Karolinska symposia, research methods reproductive endocrinology, Geneva, March, 1970, *Acta Endocrinol.,* 64(Suppl. 147), 79, 1970.
7. **Rodbard, D. and Hutt, D. M.,** Statistical analysis of radioimmunoassays and immunoradiometric (labeled antibody) assays: a generalized, iterative least squares method for logistic curve fitting, in *Proceedings, Symposium on Radioimmunoassay and Related Procedures in Medicine,* International Atomic Energy Agency, Vienna, Unipub, New York, 1974, 165.
8. **Rodbard, D., Lenox, R. H., Wray, H. L., and Ramseth, D.,** Statistical characterization of the random errors in the radioimmunoassay dose-response variable, *Clin. Chem.,* 22, 350, 1976.
9. **Shaw, W., Smith, J., Spierto, F. W., and Agnese, S. T.,** Linearization of data for saturation-type competitive protein binding assay and radioimmunoassay, *Clin. Chim. Acta,* 76, 15, 1977.
10. **Butler, J. E., Peterman, J. H., and Koertge, T. E.,** The amplified enzyme-linked immunosorbent assay (a-ELISA), in *Enzyme-Mediated Immunoassay,* Ngo, T. T. and Lenhoff, H. M., Eds., Plenum Press, New York, 1985, 241.
11. **Griswold, W. R.,** Theoretical analysis of the sensitivity of the solid phase antibody assay (ELISA), *Mol. Immunol.,* 24, 1291, 1987.
12. **Griswold, W. R., Lucas, A. H., Bastian, J. F., and Garcia, G.,** Functional affinity of antibody to the *Haemophilus influenzae* type b polysaccharide, *J. Infect. Dis.,* 159, 1083, 1989.
13. **Rodbard, D. and Feldman, Y.,** Kinetics of two-site immunoradiometric ("sandwich") assays. I. Mathematical models for simulation, optimization, and curve fitting, *Immunochemistry,* 15, 71, 1978.
14. **Rodbard, D., Feldman, Y., Jaffe, M. L., and Miles, L. E. M.,** Kinetics of two-site immunoradiometric ("sandwich") assays. II. Studies on the nature of the "high-dose" hook effect, *Immunochemistry,* 15, 77, 1978.

15. **Devey, M. E. and Steward, M. W.,** The role of antibody affinity in the performance of solid phase assays, in *ELISA and Other Solid Phase Immunoassays,* Kemeny, D. M. and Challacombe, S. J., Eds., John Wiley & Sons, New York, 1988, 135.
16. **Peterman, J. H.,** Factors which Influence the Binding of Antibody to Solid-Phase Antigens: Theoretical and Experimental Investigations, Ph.D. thesis, University of Iowa, Iowa City, 1986.
17. **Peterman, J. H., Tarcha, P. J., Chu, V. P., and Butler, J. E.,** The immunochemistry of sandwich-ELISAs. IV. The antigen capture capacity of antibody covalently attached to bromoacetyl surface-functionalized polystyrene, *J. Immunol. Methods,* 111, 271, 1988.
18. **Suter, M., Butler, J. E., and Peterman, J. H.,** The immunochemistry of sandwich ELISAs. III. The stoichiometry and efficacy of the protein-avidin-biotin capture (PABC) system, *Mol. Immunol.,* 26, 221, 1989.
19. **Dierks, S. E., Butler, J. E., and Richerson, H. B.,** Altered recognition of surface-adsorbed compared to antigen-bound antibodies in the ELISA, *Mol. Immunol.,* 23, 403, 1986.
20. **Peterman, J. H. and Butler, J. E.,** Application of theoretical considerations to the analysis of ELISA data, *Bio. Tech.,* 7, 608, 1989.
21. **Gripenberg, M. and Gripenberg, G.,** Expression of antibody activity measured by ELISA. Anti-ssDNA antibody activity characterized by the shape of the dose-response curve, *J. Immunol. Methods,* 62, 315, 1983.
22. **Guardabasso, V., Rodbard, D., and Munson, P. J.,** A model-free approach to estimation of relative potency in dose-response curve analysis, *Am. J. Physiol.,* 252 (*Endocrinol. Metab.* 15), E357, 1987.
23. **Lehtonen, O.-P.,** Immunoreactivity of solid phase hapten measured by a hapten binding plasmacytoma protein (ABPC 24), *Mol. Immunol.,* 18, 323, 1981.
24. **Koertge, T. E. and Butler, J. E.,** The relationship between the binding of primary antibody to solid-phase antigen in microtiter plates and its detection by the ELISA, *J. Immunol. Methods,* 83, 283, 1985.
25. **Butler, J. E., Heo, Y., Adams, P., and Richarson, H. B.,** The antigen limited nature of microtiter ELISAs results in underestimation of serum IgA antibody activity, *Mol. Immunol.,* in press.
26. **Peterman, J. H., Julian, B. A., Kirk, K. A., and Jackson, S.,** Monomeric IgA1 is selectively elevated in IgA nephropathy patients with normal renal function, submitted.
27. **Mestecky, J., Russell, M. W., Jackson, S., and Brown, T. A.,** The human IgA system: a reassessment, *Clin. Immunol. Immunopathol.,* 40, 105, 1986.
28. **van den Wall Bake, A. W. L., Daha, M. R., Haaijman, J. J., Radl, J., van der Ark, A., and van Es, L. A.,** Elevated production of polymeric and monomeric IgA1 by the bone marrow in IgA nephropathy, *Kidney Int.,* 35, 1400, 1989.
29. **Haba, S. and Nisonoff, A.,** Quantitation of IgE antibodies by radioimmunoassay in the presence of high concentrations of non-IgE antibodies of the same specificity, *J. Immunol. Methods,* 85, 39, 1985.
30. **Metzger, W. J., Butler, J. E., Peterman, J. H., and McGivem, P. L.,** Immunoglobulin E. Determination of isotype antibody, in *Methods of Enzymatic Analysis,* Vol. 10, Bergmeyer, H. U., Bergmeyer, J., and Grassl, M., Eds., VCH Publishers, Weinheim, West Germany, 1986, 127.
31. **Harlow, E. and Lane, D.,** *Antibodies, a Laboratory Manual,* Cold Spring Harbor Laboratory, Cold Spring Harbor, NY, 1988.

Chapter 4

PHOTOPHYSICS OF HETEROGENEOUS IMMUNOASSAYS

Omar S. Khalil

TABLE OF CONTENTS

I. INTRODUCTION

The purpose of this chapter is to review the various photophysical principles involved in the detection of a signal generated in a heterogeneous immunoassay (HIA) reaction, which determines the presence or the concentration of an analyte. The different detection methodologies include absorbance, fluorescence, evanscent wave total-internal-reflection fluorescence, time-resolved fluorescence (TRF), and chemiluminescence (CL). The choice of the detection method depends on the assay configuration and the limits of detection and quantitation required in the assay.

HIAs can be broadly divided into two types: cut-off assays and quantitative wide-dynamic-range assays. Cut-off assays are performed to detect analytes that are extraneous to the human body such as infectious agents, illicit drugs, and disease markers. Positive samples contain the analyte at a concentration higher than a certain cut-off value that is arbitrarily determined.

Cut-off assays require a low limit of detection and a high sensitivity. The sensitivity S of an analytical measurement is defined as

$$S = dR/dC \qquad (1)$$

where R is the detector response and C is the concentration. The higher the rate of change of signal with respect to concentration, the easier is the ability to differentiate between low concentrations that are close to the cut off of the assay. Sensitivity of an assay may also be expressed as the limit of detection (LOD) or the limit of quantitation (LOQ). The LOD is defined as the concentration at which[1]

$$\text{Mean of Sample Signal} \geq \text{Mean of Blank Signal} + 3\ \text{SD} \qquad (2)$$

The mean of the blank is the mean of the signal generated by the negative controls or the mean of the signal generated from a negative population. The negative samples should be confirmed by an independent method. SD is the standard deviation of the mean of the negative control replicates or the mean signal of a negative population. The limit of quantitation LOQ is the concentration at which

$$\text{Mean of Sample Signal} \geq \text{Mean of Blank Signal} + 10\ \text{SD} \qquad (3)$$

Thus the relative error of the blank signal (the zero dose response) determines the LOD and LOQ of an analytical measurement.

Quantitative, wide-dynamic-range HIAs are used to measure physiological markers such as thyroid hormones and disease-state-related antigens. Examples of this type of HIA are those for thyroid-stimulating hormone (TSH) carceno-embryonic antigen (CEA), α-fetoprotein (AFP) and β-HCG. For these substances it is important to precisely quantify the analyte at low as well as at high concentrations. Thus, the immunochemical reaction, the signal generating reaction, and the detection method should have sufficient precision over the whole dynamic range to allow measurement of clinically significant concentration changes without experimenting with different sample dilutions. The precision and accuracy at the high end of the dynamic range of the assay is determined by (1) The limitations of the binding reaction such as saturation and hook effects, (2) the factors that affect the signal-generating reaction such as substrate depletion or enzyme inhibition, and (3) the perturbations of the photophysical measurement such as quenching or stray light.

Ekins[2,3] has shown that the sensitivity of an immunometric assay is better expressed as an SD of the zero dose response rather than a slope of the response curve.[4] Thus, the LOD in Equation 2 should be considered as the best criterion for assay sensitivity. The number of SDs that should be used depends on the degree of confidence needed in the assay.

The experimental variables that can contribute to the relative error (SD) of a HIA include several factors. These affect cut off and quantitative assays to different extents. The type of assay dictates the method of optimization and determines the sources of errors to avoid. Some of these factors are

- The accuracy and precision of samples and reagents pipetting
- The reproducibility by which the capturing agent, i.e., antigen or antibodies, or solid-phase receptors can be immobilized on the solid phase. This will depend on the nature of the solid phase and manufacturing variations in different lots of materials
- If the solid phase is in the form of suspended microparticles, the precision of dispensing these particles contributes to the precision of the assay, especially in the case of a competitive binding assay
- The efficiency of separating the bound immune complex from the remainder of the reaction mixture, especially the labeled reagent
- The sources of nonspecific binding, including those which are sample related or associated with the solid phase
- The temperature equilibration during the assay and its effect on the primary binding reaction
- Factors which affect the signal-generating reaction such as temperature, substrate depletion, nonspecific lysis of liposomes, quenching of fluorescence, or CL
- The precision of the optical detection method.

The photophysical measurement is performed either on the clear supernatant solution above the solid phase or directly on the immune complex entrapped on the solid phase. In the former case, the precision and accuracy of dispensing the final solution or diluent and the shape and optical properties of the vessel contribute to the final accuracy and precision of the assay. In the latter case, the interaction of the solid phase with the reaction products and its interference in the optical measurement will contribute to the accuracy and precision of the assay. In the following sections we will discuss the application of absorbance, fluorescence, TRF, and CL as detection systems for HIAs.

II. ABSORPTION

A. SINGLE-LABEL ABSORPTION MEASUREMENTS

Absorption of light in solution is quantitated using Beer's law:

$$I = I_0 e^{-\epsilon C l} \tag{4}$$

where I is the transmitted light intensity, I_0 is the incident light intensity, ϵ is the molar extinction coefficient, C is the molar concentration, and l is the path length in centimeters. The concentration of the chromophore can be deduced from this equation where

$$C = A/(\epsilon l) \tag{5}$$

The following factors contribute to the precision and accuracy of this type of photometric measurement.

- Stray light impinging on the detector, either directly from the source or as reentry light through reflections of the principal light beam on the walls of the photometer compartment
- The optical quality of the solution, such as the presence of particulate material or bubbles that cause light scattering

- The optical quality of the vessel such as its dimensional tolerance, its optical clarity, and the presence of scratches or digs on its surfaces
- Flicker noise in the light source and drift in light source intensity during the measurement
- Variations in the spectral bandwidth of interference filters with changes in temperature
- Detector noise such as scale and quantization error, Johnson noise, and shot noise in photon counting[5]
- The linearity of the response of the detector and detection electronics
- Autooxidation, aggregation and dissociation, and chemical effects that can cause spectral shifts of the absorption maxima of the product.

Experimental conditions affect the limits of an absorbance measurement. The measured absorbance is related to the light intensity at the detector by

$$A = -\log (I \text{ at the detector})/(I_0 \text{ at the detector}) \quad (6)$$

In order to have a measurable absorbance, $(I_0 - I)$ must be larger than the sum of all noise contributions from the detector, electronics, cuvette variations, positioning errors, differences in optical properties of blank and test solutions, fluctuations in light source, and bandwidth variations. At $I \approx 0.999\ I_0$, and with a noise level of $\pm 0.1\%\ I_0$, the uncertainty in absorbance measurement can be calculated as $\Delta A = \pm\ 0.00043$. Thus, in direct label assays, tracers with absorbances higher than $3 \times \Delta A$ (=3 SD) will be detectable. This detectable absorbance in the example photometer is 0.0013 absorbance units. For a typical tracer with an extinction coefficient of 3×10^4 l/mol^{-1}/cm^{-1}, the lowest calculated LOD will be 4.3×10^{-7} mol/l. Thus, direct label absorption measurement does not yield the sensitivity needed for most HIAs. Signal enhancement can be achieved by the use of metal or dye sols,[6,7] where either a higher extinction coefficient or a higher effective concentration of label is achieved. Leuvering et al. report the use of 0.6 μm colloidal gold particles as a label in an assay for human placental lactogen (HPL) and measure the absorbance at 540 nm.[6] The LOD in the assay is 5.4×10^{-12} mol/l, using the blank mean plus 2 SDs. Dispersions of water-insoluble dyes such as Palanil® or Luminous Red (BASF Corporation) containing dye sol particles of 0.3 μm median diameter are used to label anti-βHCG and anti-AFP antibodies, which are utilized in HIAs for these analytes. After separation and washing, the immune complex labeled with the dye particle is dissolved in propanol and the absorbance of the solution is measured at 540 nm. The detection limit of AFP is reported to be 6×10^{-11} mol/l.[7] Disperse dyes suitable for conjugation can be found only by screening dye dispersions against different antibody preparations. The particular dye dispersion selected appears to depend on the antibody preparation.

At high absorbances, the transmitted light intensity I approaches zero and the measured absorbance at the detector can be expressed as

$$A = -\log (I+I_s)/(I_0 + I_s) = -\log I_s/(I_0 + I_s) \quad (7)$$

where I_S is the stray light intensity reaching the detector. I_S can be expressed as a fraction of the incident light intensity I_0. For a good photometer, $I_S = 0.1\%\ I_0$. Thus, the calculated maximum absorbance is

$$A = -\log (0.001/1.001) = 3.0004 \quad (8)$$

Precautions to eliminate stray light must be taken in order to achieve this maximum absorbance. These include

- The use of narrow acceptance angles and proper baffling of the detector

- The use of filters or dispersive elements with narrow spectral bandwidth.
- Placement of the light filter or the exit slit of the dispersive element in close proximity to the detector.
- The use of chopped illumination and synchronous detection

Some new computerized spectrophotometers claim upper absorbance limits as high as 6. This is achieved by measuring the absorbance at the wavelength maximum of the chromophore's absorption band up to A = 3.0. The detection wavelength is then shifted to a side of the absorption band and the absorption measurement is continued to the upper limit of the dynamic range of the photometer. The absorbance at the peak wavelength is calculated from that measured at the side band wavelength and the ratio of the values of the extinction coefficients at the wavelength of maximum absorption and the side band wavelength.

B. ENZYME AMPLIFIED ABSORPTION MEASUREMENTS

Enzyme amplification is widely used in order to improve the LOD in an absorption measurernent. An enzyme is used in a HIA as a label to detect the immune complex on the solid phase which has been separated and washed. An enzyme substrate is then added to generate a colored product. The color generated in a HIA can be measured in terms of rate of formation or as color developed at a predetermined end point. The enzyme reaction can be stopped which can itself increase the extinction coefficient of the chromophore as is the case for some protonated oxidized dyes. A review of the different chromogens used in HIA measurements is given by Porstman and Porstman.[8] A variety of chromogenic substrates are described that have absorption maxima in the 400- to 700-nm range. The LOD of horseradish peroxidase (HRPO) can be calculated from the reported data 1×10^{-10} mol/l, using 2,2′-azino-di(3-ethylbenzthiazoline sulfonic acid-6) (ABTS) as a substrate, and 1.3×10^{-11} mol/l using *o*-phenylene diamine (OPD) as a substrate (assuming a molecular weight of the HRPO of 40,000 Da). The LOD of alkaline phosphatase (AP) using *p*-nitro phenyl phosphate (PNPP) can be calculated as 2×10^{-10} mol/l from the reported LOD in ng/mL and assuming a molecular weight of the enzyme of 100,000 Da. Since it is possible to label a macromolecule with more than one enzyme molecule, it is expected that the LOD in a macromolecular assay to be lower than those reported for the enzymes. Thus, an EIA thyrotropin (TSH) assay from Abbott Laboratories that uses HRPO as a label and OPD as a substrate has a reported LOD of 0.05 mIU/l,[9] which is equivalent to 2.5×10^{-13} mol/l.

The visible absorption of these substrates allows the use of tungsten-halogen lamps as light sources and silicon photodiodes as detectors. Tungsten-halogen lamps with various geometries, power output, and color temperatures are available from several manufacturers. They are inexpensive, easy to regulate, and have broad spectral output. Silicon photodiodes are inexpensive detectors for the visible range of the spectrum. They have a flat response over the visible spectrum, low noise, wide dynamic range, and it is easy to amplify their signals.[5] Several commercial instruments are available for use with microtitration plates, coated tubes, and coated polystyrene beads.

III. FLUORESCENCE

A. SINGLE-LABEL FLUORESCENCE MEASUREMENTS

The intensity of a fluorescent signal generated when light of appropriate wavelength impinges on a sample is given by

$$I_F = I_0.\text{Ø}.k.g.(1 - e^{-A}) \tag{9}$$

where I_0 = intensity of exciting light, Ø = quantum yield of fluorescence, k = collection efficiency of optics, g = detector response, and A = absorbance of the fluorophore at the excitation wavelength. At low absorbances, A < 0.05, using the expansion

$$e^{-x} = 1 - x + x^2/2! - x^3/3! + ... \tag{10}$$

$$I_F = I_0.Ø.k.g.\ A. \tag{11}$$

$$I_F = (I_0.Ø.\epsilon.k.g.)\ C.\ l \tag{12}$$

Thus, at low absorbances the fluorescence intensity of a fluorophore is linearly proportional to concentration. Fluorescence intensity can be increased by (1) the use of higher excitation intensity, (2) the use of optics with higher collection efficiency, (3) the use of a longer optical path or multiple reflections within the sample, (4) the use of fluorophores with high extinction coefficients and/or high quantum yields, and (5) increasing the effective concentration of the fluorophore.

At low signal levels the intensity of light impinging on the detector is

$$I\ \text{(at the detector)} = I_F + I_s \tag{13}$$

where I_S is the stray light intensity or the ambient light intensity at the detector in absence of sample. I_S depends on the mutual rejection of the excitation and emission monochromators at the wavelength of the measurement. For filter fluorometers, I_S depends on the mutual rejection of the excitation and emission filters outside their transmission bands. For any fluorescence signal to be detectable,

$$I_F > I_S \tag{14}$$

It is possible to estimate the detectable concentration of a fluorophore when a filter fluorometer is used. Commercially available multicavity interference filters can achieve a 10^4 attenuation of light intensity outside the filter's transmission band. Thus, a mutual attenuation (rejection) of the excitation light intensity transmitted through the emission filter is $\approx 10^8$. This yields a direct stray light component $I_S \approx 10^{-8}\ I_0$. Substituting $I_F = I_S$ in Equation 12, $I_F = 10^{-8}\ I_0$, Ø = 0.8 for protein bound fluorescein, $\epsilon = 7 \times 10^4$ l/mol^{-1}/cm^{-1} for protein bound fluorescein, k = 0.2, and l = 1 cm. Thus, a the direct stray light component in this hypothetical fluorometer is equivalent to $\approx 10^{-12}$ mol/l of protein-bound fluorescein. Other stray light components that are not included in this calculation are residual fluorescence of solvents and optics, Raman and Rayleigh scattering of solution, and Tyndall scattering of any particulate material in solution. In front-surface fluorescence from a solid surface, the reflected excitation beam and the residual fluorescence of the matrix will significantly add to the level of stray light reaching the detector.

The LOD in a fluorometric measurement is given by Equation 2.

$$\text{Mean } I_F \text{ (sample)} = \text{Mean } I_F \text{ (blank)} + 3\ \text{SD} \tag{15}$$

where SD (I_F) is the SD of the fluorescent measurement at zero dose response which depends on

- The stability of the excitation light source
- The stability of the detection system
- The reproducibility of sample positioning
- Variability in the dimensions and optical quality of cuvettes
- Time-dependent photobleaching of the label
- Variations in solid-phase contributions to the fluorescence background in direct front-surface fluorescence measurements

Although an LOD of 10^{-12} mol/l of fluorescein is theoretically attainable, in practice Curry et al. report 3×10^{-11} mol/l using a commercial fluorometer, 8×10^{-12} mol/l using a photon counting filter fluorometer,[10] and $\approx 10 \times 10^{-12}$ mol/l using a photon counting clinical fluorometer.[11] These determinations are made using buffered solutions in a cuvette and 2 SDs above the blank reading. The LOD of a fluorophore can be improved by the use of polarized excitation or fluorescence.[12] Single, horizontally oriented polarizers are efficient at removing first-order Rayleigh scattered signals and crossed polarizers substantially remove second-order signals.[12] Popelka et al. report an LOD of fluorescein in a phosphate buffer = 5×10^{-12} mol/l, using crossed polarizers in a breadboard clinical fluorometer.[13] An LOD of 1×10^{-12} mol/l can be achieved using the commercial fluorescence polarization clinical analyzer, the Abbott TDx, that is based on the same principle. Particle concentration fluorescence immunoassay (PCFIA)[16,17] is an HIA which involves the use of suspended microparticles in solution to improve reaction kinetics, the use of a fluorescent label, concentrating the microparticles by vacuum filteration, and exciting and measuring the label fluorescence using a front surface fluorometer. An increase in sensitivity (lower LODs) is, supposedly, gained by concentrating the labeled microparticles into a volume much smaller than that of the sample. Use of fluorescein labels for direct determinations on a solid phase will lead to higher scattered light levels than is the case for solution fluorescence. Due to its small Stokes shift and accompanying difficulty in rejecting stray light, higher detection limits are expected. Several fluorescein-labeled HIA using double antibody precipitation or PCFIA are reported for analyte concentrations higher than 10^{-11} mol/l.[14-17]

B. MULTIPLE-LABEL FLUORESCENCE MEASUREMENTS

In principle, the sensitivity of the fluorescence immunoassay could be improved by increasing the number of fluorescent groups attached to the reagent probe. This could be directly achieved for macromolecular reagent probes or indirectly through conjugation of the reagent probe to a carrier macromolecule labeled with multiple fluorescent molecules. Increased tagging, however, may lead to concentration quenching.[18,19] One way to avoid concentration quenching is the use of fluorescence photobleaching. Polyethylenimine labeled with a large number of fluorescein molecules can be excited to complete photochemical bleaching with an intense laser pulse and the integrated fluorescence intensity is observed over a period of time. The detection of a single antibody is claimed using this method.[18] Other methods to overcome fluorescence quenching in multifluorophore labels are described. A bulking agent, covalently bound to the antigen, can be used to disperse the fluorescein labels and to minimize concentration quenching.[20] An average of 8.8 molecules of fluorescein isothiocyanate are coupled to the Fab′ fragment of rabbit IgG which is then used to label apoferritin or human IgE. 87% of the sum of the individual fluorescein analog is retained (13% quenching). An LOD for the detection of ferritin of 4.9×10^{-12} mol/l is reported using a double-antibody, competitive-binding assay based on this principle.

In another study, up to 25 molecules of 4-methylumbilliferrone-3-acetic acid are attached to a polylysine chain; the estimated number of umbilliferrones are the same when determined by absorbance or fluorescence measurements. The excitation wavelength (365 nm) and the emission wavelength (455 nm) of this multisubstituted polylysine chain do not shift from those values for the monomer indicating no electronic interaction between the umbilliferrone dipoles and no concentration quenching.[21] LOD data are unfortunately not provided for this system but a steroid assay using a double antibody precipitation technique is reported to achieve the "sensitivity" equivalent to that of a tritium-based radioimmunoassay. Fluorescence measurements are performed on the supernatant solution after separation. A comparison of the detection limits of the monomeric fluorophore and the multiple-fluorophore-substituted polylysine is not reported.

Increasing the extinction coefficient of the label can also be used to improve the sensitivity of a fluoroimmunometric assay. Phycobiliproteins are a class of naturally occurring fluorophores that absorb and emit at longer wavelengths than those of intrinsic fluorescence of serum components and plastics (575 to 650 nm) and have extinction coefficient in the range of 10^5 to 10^6 l/mol^{-1}/cm^{-1} (≈30 times that of fluorescein).[22-25] These have very broad absorption bands and a 0.8 fluorescence quantum yield.[23] Phycobiliproteins can be detected using conventional fluorometers or laser excitation, including the inexpensive stable helium-neon laser. Although the single-label detection limit of phycobiliproteins is reported as 10^{-14} mol/l, their substitution for fluorescein in immunoassays do not yield a comparable improvement in assay detection limit. A latex bead immunoassay for IgG using a conventional fluorometer is reported to have an LOD of ≈10^{-11} mol/l.[24] An improvement in sensitivity by a factor of four over that detected using a fluorescein label is reported using PCFIA.[22] The improvement in sensitivity is less than expected on the basis of extinction coefficient and quantum yield; scattered light detected as fluorescence, nonspecific binding, and steric interferences in the binding reaction are thought to have contributed to this lack of improvement in sensitivity.[22]

C. TIME-RESOLVED FLUORESCENCE MEASUREMENTS

Equation 14 indicates that rejection of stray light I_S can be used to improve the LOD of a fluorometric measurement. This is encompassed in the principle of TRF immunoassay in which the fluorescence of a long-lived label is discriminated from the short-lived endogenous fluorescence light and residual excitation light by use of a pulsed excitation source and time-gated detection.[26-38] The development of new long-lifetime labels, such as lanthanide chelates, with decay time ranging from 10 to 1000 μs, presents a possibility for increasing the sensitivity of a fluorescence immunoassay.[26] This results from

- The high specific activity of the europium chelate which can be increased by using stronger excitation
- Reduction of background fluorescence by using time-resolved fluorescence detection
- The absence of lanthanide labels, presumably in biological samples
- The minimal interference of immunoreactivity resulting from lanthanide labeling

Lanthanide ions have a series of multiplicity forbidden intrashell electronic transitions that are characterized by sharp lines and long life times. The most intense of these are the 545 nm $^5D_4$ — $^7F_5$ transition for terbium and the 613 nm $^5D_0$ — $^7F_2$ transition for europium. Because of the low intrinsic extinction coefficient of these transitions the emitting excited state energy levels are generally populated via energy transfer from a chelate. These transitions are multiplicity forbidden (quintet — septet transitions) and the emission is thus a phosphorescence. The proper designation for this type of HIA is phosphorescence immunoassay. We will continue to use the misnomer TRF because it is generally accepted. An extractive application of this concept is pioneered by LKB-Wallac scientists and is commercialized under the name DELFIA™, "Dissociation Enhanced Lanthanide Fluoroimmunoassay".[26-31] The europium ion is bound to an immunoreactive component through a chelating polycarboxylic acid derivative such as EDTA, EGTA, and DTPA (dimethylenetetramine penta acetic acid). Functional groups such as diazophenyl or isothiocyanophenyl can be grafted onto the polycarboxylic acid backbone to allow attachment to protein molecules. The diazophenyl is coupled mainly to tyrosine residues while the isothiocyanophenyl bonds with free amino groups of the protein. These labeled antibodies are essentially nonfluorescent and can be considered as labeled by metal ions. After the immunochemical reaction is completed and the free and bound reactants are separated and washed, the europium ions are released from the chelate by dissociation at low pH. An enhancement solution is added that contains a β-diketone such as naphthoyel trifluoroaceto-acetonate (NTFA), and a nonionic detergent such as Triton® X-100 to solubilize the sparingly soluble organic component in the micellar phase and to exclude quenching water from the chelated europium ion. The enhancement solution also contains a synergistic agent such as trioctylphosphine oxide which coordinates to the last two coordination sites of europium ion to completely exclude water from the coordination sphere of europium. The solution is excited in the 340 nm absorption band of NTFA using a pulsed xenon flash lamp or a nitrogen laser. The emitted light intensity at 613 nm is measured for a few hundred milliseconds (after 100 μs delay) using gated photon-counting techniques.

A theoretical calculation of sensitivity in a TRF measurement is given by Soini and Kojola as the following. The rate of photon emission n_E from a fluorescent Eu $(\beta\text{-NTA})_3TOPO_{2\text{-}3}$ solution is

$$n_E = 2.3.n_0.f.t.\O.\epsilon.C.l \tag{16}$$

photons per second, where n_0 = number of excitation photons per flash = 7.12 10,[11] f = repetition rate of the flash = 1000 flashes per second, t = factor dependent on the transmission and reflections of the cuvette ≈ 0.74, Ø = quantum yield of the fluorescent probe at 613 nm = 0.06, ϵ = molar extinction coefficient of at 340 nm = $15\text{x}10^4$ l/mol^{-1} l/cm^{-1}, C = concentration of probe in moles per liter, and l = path length = 1 cm. The calculated numerical value of the photon flux n_E at 613 nm for the fluorometer used by Soini and Kojola is 1.1×10^{19} photons per second per mole. The calculated detected signal strength (photon flux n) is 3.36×10^{15} counts per second per mole of europium-NTFA measured in a micellar medium containing the synergistic reagent (see above). Using insulin as an example, and for a detectable signal of 500 counts per second, the LOD is calculated to be 1.49×10^{-13} mol/l.[26]

Hammila et al. report that when an optimized enhancement solution and a TRF fluorometer are used, europium can be quantified over a concentration range from 5×10^{-14} to 10^{-7} mol/l.[27-29] However, the blank used in this measurement is distilled water in a cuvette,[27] which is not a true blank. The SD of the fluorescence background of the enhancement solution should have been reported. The value of the blank reading plus 3 SDs is not used in calculating the limit of detection. Thus, error due to contamination with europium is not included in calculating LOD. Rare-earth oxides are present in the earth crust at 10^{-9} to 10^{-6} weight by weight; this concentration is quite low by standard analytical techniques, but is much higher than the LOD of this extractive-enhancement assay procedure. The estimate by Soini and Kojola seems more realistic.[26] Although the quantum yield and the extinction coefficient values of rare-earth chelates used in the above example are much lower than that of fluorescein, the LOD is lower by one order of magnitude. This improvement in LOD is the result of rejecting the stray excitation light, short-lived impurity fluorescence, and Raman and Rayleigh scattering.

The maximum theoretical sensitivity obtainable in an immunometric assay (which could be equated to LOD) has been defined by Jackson and Ekins.[4]

$$k\ (\Omega)/K = 0.5 \times 10^{-15}\ \text{mol/l} \tag{17}$$

where k = fractional nonspecific binding of the antibody (10^{-14}), K = affinity constant (2×10^{10} l/mol), and (Ω) = the relative error in the response in absence of the analyte = error of the nonspecific bound antibody (10%).

In practice, a sensitivity of 0.02 μIU/ml (zero standard plus two SD) is the package insert claim for the TSH-DELFIA™ assay; this corresponds to an LOD of 1×10^{-13} mol/l. The 100-fold difference between the theoretical sensitivities and verified sensitivities is attributed to the efficiency and precision of separation (including the performance of the solid phase).[30] The extra step in the extractive DELFIA™ i.e., the addition of the enhancement solution, increases sensitivity but is a procedural disadvantage. In addition, care must be taken to use europium-free glassware and to avoid any europium contamination during the various assay steps.

An alternative TRF immunoassay approach that avoids the problem of potential europium contamination is developed by Diamanedis et al.[34-38] In this method, the antibody is labeled with a europium-sensitizing chelate such as 4,7-bis(chlorosulfonyl)-1,10-phenanthroline-2,9-dicarboxylic acid (BCDPA). After incubation, separation, and washing of excess reagents, a micromolar solution of europium salt is added to the solid phase (coated microtitration strips). The strips are washed, shaken, and dried in an air oven and fluorescence is measured in a front surface mode using a pulsed nitrogen laser and gated detection. Because of the very low fluorescence of unchelated europium ion, the intensity measured by a time-resolved fluorometer corresponds to that generated by the bound label. This technique is commercialized by CyberFluor Inc., Toronto, Ontario, Canada. The reported lowest LOD of europium by this method is 10^{-11} mol/l. An assay for cortisol is also reported. An improvement in detection limit (LOD = 5×10^{-13} mol/l) is gained when multiple BCPDA labels are used. Biotinylated antibodies are used for the immunochemical reaction and the signal is generated by adding a reagent containing streptavidin (SA) conjugated to the carrier protein bovine thyroglobulin (TG). The resulting SA-TG conjugate is reported to carry 160 BCPDA residues and retain the ability to bind to biotinylated antibodies. Steric interferences probably do not allow the full amplification potential of this procedure to be achieved. Sensitized europium fluorescence measurements using the phenanthroline chelate achieve limits of detection higher than those obtained with NTFA. This may be due to difference in energy transfer efficiency between the triplet state of the sensitizers and the emitting level of europium or to the presence of water molecules in the coordination sphere of the europium:BCPDA complex. No synergistic reagent is used and micellization may not be sufficient to exclude water. Coordinating water molecules lead to radiationless deactivation of the excited europium ion.

D. ENZYME-MEDIATED IMMUNOFLUOROMETRIC MEASUREMENTS

The sensitivity of a fluorescence immunoassay could be increased by the use of an enzyme label and a fluorescent substrate as a signal generator. As examples, AP can be used with methyl umbelliferone phosphate (MUP), and β-galactosidase can be used with β-galactosyl umbelliferone. Although the LOD of methyl umbelliferone (MU) is approximately 10^{-7} mol/l, the turnover effect of the enzyme leads to immunoassay detection limits close to those reported by TRF measurements. In addition to the advantage of the enzyme amplification, this method has the potential of using rate measurement to overcome variable fluorescent blanks from serum and the solid phase. On the other hand, as a form of kinetic analysis, the fluorescence rate measurement is subject to certain limitations such as temperature effects on the activation energy of the reaction and poor precision due to variation in mixing, particularly at high reaction rates.[39]

Enzyme mediated immunofluorometric measurements are used in combination with radial partition for HIAs.[40] Capture antibodies are immobilized on a fiberglass disk on which sample is dispensed, incubated, and then washed using a radial diffusion method.[40] An AP-labeled antibody is then added, incubated with the bound reactants, and the fiber disk washed. Finally, MUP is added and the rate of change in fluorescence signal vs. time is monitored at 450 nm. Excitation is effected by a mercury lamp. Another method uses microparticles as a capture phase, β-galactosidase as a label and β-galactosyl umbilliferrone as a substrate. The reaction takes place in a well with a membrane filter bottom while separation and washing is effected by vacuum filtration.[41] Fluorescence is read using a front surface fluorometer. The reported LOD for TSH in this system is 0.3 mIU/l (1.5×10^{-12} mol/l). An alternative approach is developed by Abbott Laboratories using derivatized polystyrene microparticles that are covalently coated with antibodies as the capture phase. These are incubated with the sample and subsequently separated from nonbound reactants on a fibrous pad which is in contact with an absorbant material to additionally remove excess reactants.[42] The entrapped microparticles are then washed, AP-labeled antibody is added, the reaction incubated, and microparticles and their bound reactants washed. The rate of hydrolysis of MUP is then measured using a front surface fluorometer. The LOD of AFP determined by this assay is reported for 2 master lots of reagents as 3.2×10^{-12} and 4.7×10^{-12} mol/l, respectively.[42] The package-insert claim for LOD of TSH is 0.05 mIU/l (2.5×10^{-13} mol/l).

E. TOTAL-INTERNAL-REFLECTION FLUORESCENCE MEASUREMENTS

Fluorescence intensity may also be increased by increasing the path length l (see Equation 12). This can be achieved by using multiple path total internal reflection (TIR) fluorescence. Harrick and Loeb show that fluorescence intensity measured from a solution placed on a beveled total internal reflection optical element is higher than that measured in a cuvette in a conventional fluorometer.[43] Kronick and Little use this technique for a competitive binding assay for morphine.[44] Antimorphine antibodies are deposited on a quartz TIR optical element. Morphine and fluorescein-labeled morphine compete for those antibody sites. Fluorescence intensity is measured in the evanescent wave region without interference from the bulk solution. The LOD of morphine is 10^{-8} mol/l.[44] Sutherland et al. use the same technique for a sandwich assay for human IgG.[45,46] Sheep anti-human IgG is immobilized on a quartz TIR optical element which is used as a window for a flow cell. A sample containing human IgG is injected in the cell and incubated for 10 min and followed by a wash solution. The tracer, fluorescein isothiocyanate-labeled rabbit anti-human IgG, is then injected, the reaction incubated, and the resulting complex washed; the fluorescence measurement is then performed. The immune complex is subsequently dissociated and the TIR optical element is reused for up to 20 times.[46] The reported LOD of human IgG is 5.3×10^{-9} mol/l. Although total internal reflection has the potential of increasing fluorescence intensity, a comparable improvement in HIA sensitivity is not achieved.

IV. CHEMILUMINESCENCE

Chemiluminescence (CL), or chemical generation of electronically excited states and subsequent light emission, has been extensively reviewed in the past few years by Seitz,[47] Schuster and Schmidt,[48] McCapra and Perring,[49] and Woodhead and Weeks.[50] Several types of CL reactions are used for HIAs.[51] These include

- Acridinium compounds triggered by alkaline peroxides
- Cyclic hydrazides such as luminol and isoluminol triggered by peroxides in presence of a catalyst
- Energy transfer from a chemically excited oxalate ester to an aromatic fluorophore
- Enhanced luminol CL using phenolic molecules, luciferin, or sterically hindered amines[52]
- Enzyme-catalyzed dioxetane CL[53]

The intensity of a CL signal I_{CL}, expressed as the number of emitted CL photons per second, is given by

$$I_{CL}\,(t) = Ø_{CL}.dC^*/dt \tag{18}$$

$Ø_{CL}$ is the CL quantum efficiency which is photons emitted per reacted molecule:

$$Ø_{CL}(t) = Ø_c.Ø_{ex}.Ø_F \tag{19}$$

where $Ø c$ = the yield of the chemical reaction, the number of intermediate molecules per ground state reacting molecule. $Ø_{ex}$ = the efficiency of excited state production. This measures the number of excited molecules formed per reaction intermediate molecule. $Ø_F$ = the fluorescence quantum yield of the excited product, the number of photons emitted per electronically excited molecule formed.

The variation in dC*/dt with time depends on the rate law for the CL reaction. While the first order case is simple to analyze, most important analytical CL reactions have considerably more complex kinetics. For a first order reaction, Equation 18 reduces to

$$I_{CL}(t) = Ø_{CL}K. \{C(t)\} \tag{20}$$

where [C(t)] is analyte concentration as a function of time.

The intensity-time curve can be analyzed either by measuring intensity at a fixed time after mixing or by integrating intensity over time. The maximum intensity measurement has the advantage of speed, convenience, high sensitivity, and minimizes systematic errors from unwanted side reactions. As a form of kinetic analysis, it is subject to temperature effects on the reaction rate and poor precision due to variation in mixing, particularly for fast reactions.[39]

The intensity-time curve may be completely or partially integrated. If the reaction is initiated in the presence of the light detector, the intensity-time curve can be integrated. The general expression for in rated intensity is

$$I_{CL}\ (\text{integrated}) = Ø_{CL}.(dC^*/dt).dt \tag{21}$$

The time-intensity curve integral will be proportional to concentration and independent of reaction rate. This “equilibrium” or “end-point” analysis is not subject to error resulting from variability in reaction rate.[39]

Light emission in a heterogeneous CL reaction takes place in the bulk of the solution as well as on the surfaces of reaction vessels.[53-55] Theory relating CL intensity to reaction rate and mass transfer efficiency for a heterogeneous CL reaction involving diffusion between two phases has been developed by Freeman and Seitz.[56] The system described is for the measurement of hydrogen peroxide in a bulk phase using a fiber optic CL probe. Convection is the primary means of mass transfer in the bulk analyte phase, reaction kinetics are first order in analyte, and the analyte is assumed to diffuse from solution into the interior of the reactant phase. The steady state CL intensity, I_{CL}, is expressed in terms of

- The first order rate constant for the light generating reaction
- The interface areas in which the bulk reactant could diffuse
- The diffusion coefficient of the analyte in the solution phase
- The diffusion coefficient for the analyte in the reactant phase
- The bulk concentration of analyte
- The thickness of the Nernst diffusion layer

Detection of CL signals offers a potential for sensitivities higher than those achieved by fluorescence measurements in spite of the low quantum yield of the CL process. This is because

- The problems associated with the excitation source such as rejection of scattered excitation light, fluctuation in light source intensity, and heat dissipation from incandescent lamps are alleviated.
- There is no Raleigh and Raman scattering of solvents.
- There is no need for filters or wavelength selection, thus, the full photon flux from the sample is detected.
- Electronic noise generated by arc and flash lamps are absent which allows the use of photon counting and long integration times to encompass the whole emission intensity-time curve.
- The blue CL emissions are maximal in the 420 to 450 nm range that corresponds to the

range in which several photomultiplier tubes with high quantum efficiency and low dark current are available.

Although CL is suggested as a preferred detection technique for high sensitivity immunoassays,[49] it has not been widely used for this purpose. This is due in part to instrumentation and detection problems that are encountered in automating CL batch measurements. Some of these problems are

- The positioning accuracy of the multiplicity of samples with respect to collection optics
- The precision and accuracy of delivering trigger solution
- The mixing efficiency of trigger solution and reactants
- The reliability of repetitive, precise light sealing methods around disposable reaction vessels
- Measurement of low light level CL
- Standardization of count rate against concentration
- The stability of the standard curve and the ability to perform normalization

Several attempts to automate CL measurements include the use of

- Photographic film and a densitometer for recording the signal generated from a reaction in clear microtitration plates[57-58]
- Clear-coated microtitration plates as the solid phase with trigger solution port and detector at opposite sides of the plate well[59]
- Magnetizable microparticles and magnetic separation in a test tube followed by measurement of the signal from suspended particles in a tube luminometer such as is commercially available as the Magic™ Lite system distributed by Ciba-Corning Diagnostics[60,61]
- Enhanced CL reaction in a white microtitration plate followed by measurement of the generated signal in a luminometer having a moving mask and photomultiplier tube such as that described for the Amerlite™ system commercialized by Amersham Inc. In this instrument, the long-lived CL signal is generated by triggering outside the instrument followed by inserting the plate into the reader[50,62]

A comparison of ten commercial luminometers is recently reported by Jago et al.[63] The availability of a large number of commercial instruments indicates the rapid growth in CL as a detection technique. A fully automated heterogeneous CL immunoassay system is not as yet commercially available.

The reported single-label detection of luminol is 10^{-12} mol/l.[64] The limit of detection of acridinium esters has been reported in number of moles in an undefined sample volume rather than the conventional molar concentration.[65] A reported value 8×10^{-15} mol/l is calculated from the specific activity in counts per mole, which assumes a zero background (intercept). Weak background luminescence from solvents and protein solutions is very common; a background of 100 counts per 5 s would raise the LOD in the above example to 10^{-13} mol/l, which is 10 times lower than that of luminol. Enhanced luminol CL yields an LOD of the same order of magnitude as those of acridinium compounds.

The preparation of liposomes containing 2.5×10^5 acridinium molecules per liposome has been reported. Anti-TSH is labeled with these liposomes. A 20-fold increase in the slope of signal vs. TSH concentration (see Equation 1), and a 3-fold improvement in the signal-to-noise ratio at the low end of the response curve compared to a HIA based on direct acridinium labeling of the antibody (10 acridinium residues per antibody molecule) are reported.[61]

The reported LOD of AP using a dioxetane substrate in a Tris buffer containing 0.1% BSA

TABLE 1
Comparison of The Sensitivity of Different Heterogeneous Immunoassay Procedures Expressed as Limit of Detection

Method of assay	Analyte	LOD mol
PCFIA (Fluorescein)[a]	IgG	2.42×10^{-11}
PCF IA (Phycobiliprotein)[b]	IgG	6.06×10^{-12}
EIA (Fluorescence)[c]	TSH	1.5×10^{-12}
EIA (Absorbance)[d]	TSH	2.5×10^{-13}
EIA (Fluorescence)[e]	TSH	2.5×10^{-13}
TRF (DELFIA™)[f]	TSH	1.0×10^{-13}
CL (Acridinium)[g]	TSH	1.4×10^{-13}
CL (Enhanced luminol)[f]	TSH	2.0×10^{-13}
CL (Acridinium)[h]	TSH	5.0×10^{-14}
CL (Dioxetane)[i]	TSH	3.45×10^{-14}

[a] Calculated from the reported sensitivity of 2-4 ng/ml.[16]
[b] B. Francis, D. W. Buck, L. L. Lanier, M. F. LaFoe, D. R. Bethell, and M. E. Jolley, Enhanced sensitivity using B-phycoerythrin conjugates to antibodies in particle concentration fluorescence immunoassay, poster at Conference of Phycobiliproteins in Biology and Medicine, Seattle, September 9-10, 1985; as cited in Reference 22.
[c] Using β-galactosidase label, microparticle capture and vacuum filteration.[41]
[d] Using HRPO as label and OPD substrate.[9]
[e] Abbott Laboratories TSH label claim, using AP label and the microparticle capture method and apparatus.[42]
[f] Using Amerlite™ system.[62]
[g] Using magnetizable microparticles separation.[67]
[h] Using coated tubes, no reported criterion for LOD determination.[51]
[i] Using AP-catalyzed dioxetalle and components from commercial EIA kit, 20 min incubation time.[66]

is 4.17×10^{-16} mol/l. The number of SDs above the background is not reported. The concentration of AP that gives a CL signal at twice the background is reported as 7×10^{-15} mol/l.[53] This particular technique is quite promising as it combines the amplification effect of the enzyme with the low single-label detectability of CL. A TSH assay using AP-labeled antibody from a commercial kit and a 1,2-dioxetane phosphate derivative has an LOD of 0.0069 mIU/l (3.45×10^{-14} mol/l) for a 20-min substrate incubation period and 0.0045 mIU/l (2.25×10^{-14} mol/l) using 2 SDs above the blank signal.[66]

As a conclusion we try to establish which HIA configuration and detection method may lead to the highest assay sensitivity (lowest LOD). In spite of the variations in the quality of antibody preparations, incubations times, and washing efficiency, some general conclusions may be drawn from examination of literature data. Table 1 summarizes the LOD data for the various photophysical methods that are reviewed in this chapter. There is no unified criterion for the LOD calculation, but most of the studies used 2 SDs above the background. The fluorescein label PCFIA has the highest LOD of all assay methods. The use of high extinction coefficient label as phycobiliprotein does not significantly improve the LOD. Lower LOD is achieved using an enzyme-mediated immunofluorometric assay. The two examples in Table 1 use microparticles as a capture solid phase; however, the particles are not concentrated into a dot. The enzyme immunoassay method with absorption detection compares well with the fluorescence-based methods. However, the optical measurement is performed on the clear substrate solution without interference of the solid phase (coated 1/4" polystyrene bead). The final test solution volume is comparable to the original sample volume and the absorption measurement is performed in a fraction of this final test volume. The TRF DELFIA™ method has a lower LOD. It is an extractive assay in which the measurement is performed on the bulk of the enhancement solution away from the solid phase. The capture solid phase does not interfere with the optical measurement. As in the case of aborption, the TRF measurement samples a fraction of the total solution volume. The CLIA methods have the lowest LODs. The measurement is performed with the capture solid phase in the field of view of the detector. The volume of the final test solution is comparable to the original sample volume and the detector does not collect the total photon flux emitted from the triggered solution.

Concentrating the solid phase and the use of a fluorescence measurement to improve assay LOD may be countered by the scattered light detected as fluorescence and the variations in the solid phase contributions to the fluorescence background (SD of the zero dose response signal) in a front surface fluorescence measurement. The use of rate measurements in the enzyme-mediated fluoroimmunoassays minimizes the variable contribution of the solid phase to the signal. The TRF method eliminates this contribution to the assay background signal by rejecting the short-lived scattered excitation light. Similarly, CL measurement eliminates it because the signal generation process does not involve the use of a light source. It can be concluded that methods that eliminate the variable contributions of the solid phase to the photophysical measurement and the use of labels with high specific activity (signal per unit time per mole) generally produce lower limits of detection than other HIA methods. They seem to have a distinct advantage over methods that use particle or immunodot concentration and excitation and detection of short-lived fluorescent labels. These methods may not achieve the LODs required to perform high sensitivity immunoassays.

ACKNOWLEDGMENT

The author wishes to express his thanks to Dr. Stephen D. Stroupe for his comments on the manuscript and for measuring the LOD of fluorescein on Abbott TDx, to Professor John Butler for his critical editorial remarks, and to Dr. Larry Blecka for his continuous support and encouragement.

REFERENCES

1. ACS Committee on Environmental Improvement, Guidelines for data acquisition and data quality evaluation in environmental chemistry, *Anal. Chem.*, 52, 2242, 1980.
2. **Ekins, R. P.,** The precision profile: its use in assay design, assessment and quality control, in *Immunoassays for Clinical Chemistry,* Hunter, W. M. and Corrie, T. E. T., Eds., Churchill Livingstone, Edinburgh, 1983, 76.
3. **Jackson, T. M. and Ekins, R. P.,** Theoretical limitions on immunoassay sensitivity, *J. Immunol. Methods,* 87, 13, 1986.
4. **Yallow, R. S. and Berson, S. A.,** Radioimmunoassays, in *Statistics in Endocrinology,* McArthur, J. W. and Colton, T., Eds., MIT Press, Cambridge, MA, 1970, 327.
5. **Grum, F.,** *Optical Radiation Measurements,* Academic Press, New York, 1979, 173.
6. **Leuvering, J. H. W., Thal, P. J. H. M., van der Waart, M., and Schaus, A. H. W. M.,** Sol particle immunoassay (SPIA), *J. Immunoassays,* 1, 77, 1980.
7. **Grinbau, T. C. J., van Sommeren, A., and van Dinther, D. I.,** Disperse dye immunoassays, in *Affinity Chromatography and Biological Recognition,* Chaiken, I. M., Wilchek, M., and Parikh, I., Eds., Academic Press, New York, 1983, 375.
8. **Porstman, B. and Porstman, T.,** Chromogenic substrates for enzyme immunoassay, in *Nonisotopic Immunoassay,* Ngo, T. T., Ed., Plenum Press, New York, 1988, 57.
9. **Piketty, M. L., Talbot, J. N., Askienazy, S., and Michaud, G.,** Clinical significance of a low concentration of thyrotropin: five immunometric "kit" assays compared, *Clin. Chem.,* 33, 1237, 1987.
10. **Curry, R. E., Pardue, H. L., Mieling, G. E., and Santini, R. E.,** Design and evaluation of a filter fluorometer that incorporates photon counting, *Clin. Chem.,* 19, 1259, 1973.
11. **Curry, R. E., Heitzman, H., Reige, D. H., Sweet, R. V., and Simonsen, M. E.,** A system approach to fluorescent immunoassay, general principles and representative applications, *Clin. Chem.,* 25, 1591, 1979.
12. **Lim, C. S. and Miller, J. N.,** The use of polarizers to improve detection limits in fluorimetric analysis, *Anal. Chim. Acta,* 100, 235, 1978.

13. **Popelka, S. R., Miller, D. M., Holen, J. T., and Kelso, D. M.,** Fluorescence polarization immunoassay. II. Analyzer for rapid, precise measurement of fluorescence polarization with use of disposable cuvettes, *Clin. Chem.,* 27, 1198, 1981.
14. **Soini, E. and Hammila, I.,** Fluoroimmunoassay: present status and key problems, *Clin. Chem.,* 25, 353, 1979.
15. **O'Donnell, C. M., McBride, J., Suffin, S., and Broughton, A.,** A heterogeneous imrnunoassay for gentamicin using a second antibody separation, *J. Immunoassay,* 1, 375, 1980.
16. **Jolley, M. E., Wang, C.-H. J., Ekenberg, S. J., Zuelke, M. S., and Kelso, D. M.,** Particle concentration fluorescence immunoassay (PCFIA): a new rapid immunoassay technique with high sensitivity, *J. Immunol. Methods,* 67, 21, 1984.
17. **MaccCrindle, C., Schwenzer, K., and Jolley, M. E.,** Particle concentration fluorescence immunoassay: a new immunoassay technique for quantitation of human immunoglobulin in serum, *Clin. Chem.,* 31, 1487, 1985.
18. **Hirschfeld, T.,** Optical microscopic observation of single small molecules, *Appl. Opt.,* 15, 2965, 1976.
19. **Hassan, M., Landon, J., and Smith, D. S.,** Multiple fluorescein-substituted polymers as labels in fluoroimmunoassay, *FEBS Lett.,* 103, 339, 1979.
20. **Rowley, G. L., Henriksson, T., Louie, A., Nguyen, P. H., Kramer, M., Der-Balian, G., and Kameda, N.,** Sensitive fluoroimmunoassay for ferritin and IgE, *Clin. Chem.,* 33, 1563, 1987.
21. **Exley, D. and Ekeke, G. I.,** Fluoroimmunoassay of 5 alpha-dihydrotestosterone, *J. Steroid Biochemistry,* 14, 1297, 1981.
22. **Kronick, M. N.,** Phycobiliproteins as labels in immunoassay, in *Nonisotopic Immunoassay,* Ngo, T. T., Ed., Plenum Press, New York, 1988, 163.
23. **Oi, V. T., Glazer, A. N., and Stryer, L.,** Fluorescent phycobiliprotein conjugates for analysis of cells and molecules, *J. Cell Biol.,* 93, 981, 1982.
24. **Kronick, M. N. and Grossman, P.D.,** Immunoassay techniques with fluorescent phycobiliprotein conjugates, *Clin. Chem.,* 29, 1582, 1983.
25. **Glazer, A.N. and Stryer, L.,** Fluorescent tandem phycobiliprotein conjugates, *Biophy. J.,* 43, 383, 1983.
26. **Soini, E. and Kojola, H.,** Time-resolved fluorometer for lanthanide chelates — a new generation of nonisotopic immunoassays, *Clin. Chem.,* 29, 65, 1983.
27. **Hemmila, I., Dakubu, S., Mukkala, V.-M., Siitari, and Lovgren, T.,** Europium as a label in time-resolved immunofluorometric assays, *Anal. Biochem.,* 137, 335, 1984.
28. **Hemmila, I., Viljanen, M., and Lovgren, T.,** Time-resolved fluoroimmunoassays (TR-FIA) of tetanus antibodies, *Anal. Biochem.,* 137, 737, *1984.*
29. **Hemmila, I.,** Fluoroimmunoassays and immunofluorometric assays, *Clim. Chem.,* 31, 359, 1985.
30. **Soini, E. and Lovegren, T.,** Time-resolved fluoroimmunoassay, in *Nonisotopic Immunoassay,* Ngo, T.T., Ed., Plenum Press, New York, 1988, 231.
31. **Hammila, I.,** Lanthanides as probes for time-resolved fluorometric immunoassays, *Scand. J. Clin. Lab. Invest.,* 48, 389, 1988.
32. **Keelan, J. A., France, J. T., and Barling, P. M.,** An alternative fluorescence enhancement solution for use in lanthanide-based time resolved fluoroimmunoassays, *Clin. Chem., 33,* 2292, 1987.
33. **Bechaud, H., Bador, R., Claustrat, F., Desuzinges, C., and Mallein, R.,** New approach to competitive lanthanide immunoassay: time-resolved fluoroimmunoassay of progesterone with labeled analyte, *Clin. Chem.,* 34, 501, 1988.
34. **Diamandis, E. P.,** Immunoassays with time-resolved spectroscopy: principles and applications, *Clin. Biochem.,* 21, 139, 1988.
35. **Diamandis, E. P. and Morton, R. C.,** Time-resolved fluorescence using a europium chelate of 4,7-bis(chlorosulfophenyl)-1,10-phenanthroline-2,9-dicarboxylic acid (BCPDA), *J. Immuonol. Methods,* 112, 43, 1988.
36. **Evangelista, R. A., Pollak, A., Allore, B., Tempelton, E., Morton, R. C., and Diamandis, E. P.,** A new europium chelate for protein labelling and time-resolved fluorometric applications, *Clin. Biochem.,* 21, 173, 1988.
37. **Reichstein, E., Shami, Y., Ramjeesingh, M., and Diamandis, E.P.,** Laser-excited time-resolved solid-phase fluoroimmunoassays with the new europium chelate 4,7-bis(chlorosulfophenyl)-1,10-phenanthroline-2,9-dicarboxylic acid as label, *Anal. Chem.,* 60, 1069(1989).
38. **Diamandis, E. P., Morton, R. C., Reichstein, E., and Khosravi, M. J.,** Multiple fluorescence labeling with europium chelators. Application to time-resolved fluoroimmunoassays, *Anal. Chem.,* 61, 48, 1989.
39. **Carr, P. W.,** Kinetic and equilibrium assays based on the effect of extent of reaction and the uncertainity of rate parameters, *Anal. Chem.,* 50, 1602, 1978.
40. **Geigel, J. L., Brotherton, M. M., Cronin, P., et al.,** Radial partition immunoassay, *Clin. Chem.,* 28, 1894, 1982.
41. **Kang, J., Kaladas, P., Chen, S., et al.,** A highly sensitive immunoenzymatic assay involving "common-capture" particles and membrane filteration, *Clin. Chem.,* 32, 1682, 1986.

42. **Fiore, M., Mitchell, J., Doan, T., et al.,** The Abbott IMx benchtop immunoassay analyzer system, *Clin. Chem.,* 34, 1726, 1989.
43. **Harrick, N. J. and Loeb, G. I.,** Multiple internal reflection spectrometry, *Anal. Chem.,* 45, 687, 1973.
44. **Kronick, M. N. and Little, W. A.,** A new immunoassay based on fluorescence excitation by internal reflection spectroscopy, *J. Immunol. Methods,* 8, 235, 1975.
45. **Sutherland, R., Dahne, C., Place, J. F., and Ringrose, A. R.,** Immunoassay at quartz-liquid interface: theory, instrumentation and preliminary application to the fluorescent immunoassay of human immunoglobulin G, *J. Immunol. Methods,* 74, 253, 1984.
46. **Sutherland, R., Dahne, C., Slovacek, R., and Bluestein, B.,** Interface immunoassays using the evanescent wave, in *Nonisotopic Immunoassay,* Ngo, T. T., Ed., Plenum Press, New York, 1988, 331.
47. **Seitz, R. W.,** Chemiluminescence and bioluminescence analysis: fundamentals and biomedical applications, *Crit. Rev. Anal. Chem.,* 1981, 1.
48. **Schuster, G. B. and Schmidt, S. P.,** Chemiluminescence of organic compounds, *Adv. Phys. Org. Chem.,* 18, 187, 1982.
49. **McCapra, F. and Perring, K. D.,** General organic chemiluminescence, in *Chemi- and Bioluminescence,* Burr, J. B., Ed., Marcel Dekker, New York, 1986, 259.
50. **Woodhead, J. S. and Weeks, I.,** Chemiluminescence immunoassay, *Pure Appl. Chem.,* 57, 523, 1985.
51. **McCapra, F., Watmore, D., Sumun, F., Patel, A., Beheshti, I., Ramakrishnan, K., and Branson, J.,** Luminescent labels for immunoassay — from concept to practice, *J. Bioluminescence Chemiluminescence,* 4, 51, 1989.
52. **Thorpe, G. H. G., Kricka, L. J., Moseley, S. B., and Whitehead, T. P.,** Phenols as enhancers of the chemiluminescent horseradish peroxidase-luminol-hydrogen peroxide reaction: application in luminescence monitored enzyme immunoassays, *Clin. Chem.,* 31, 1335, 1985.
53. **Bronstein, I., Edwards, B., and Voyta, J. C.,** 1,2 Dioxetanes: novel chemiluminescent enzyme substrates. Applications to immunoassays, *J. Biolum. Chemilum.,* 4, 99, 1989.
54. **Wampler, J. E.,** Instrumentation: seeing the light and measuring it, in *Chemi- and Bioluminescence,* Burr, J. B., Ed., Marcel Dekker, New York, 1986, 1.
55. **Rich, E. S., Groover, C. H., and Wampler, J. E.,** The spatial distribution of light emission from liquid phase bio- and chemiluminescence, *Photochem. Photobiol.,* 33, 727, 1981.
56. **Freeman, T. M. and Seitz, W. R.,** Chemiluminescence fiber optic probe for hydrogen peroxide based on the luminol reaction, *Anal. Chem.,* 50, 1424, 1978.
57. **Thorpe, G. H. G., Whitehead, T. P., Penn, R., and Kricka, L. J.,** Photographic monitoring of enhanced luminescent immunoassays, *Clin. Chem.,* 30, 806, 1984.
58. **Bunce, R. A., Thorpe, G. H. G., Gibbons, J. E. C., Killeen, P. R., Ogden, G., Kricka, L. J., and Whitehead, T. P.,** Camera luminometer for use with luminescent assays, *Analyst,* 110, 657, 1985.
59. **Schroeder, H. R., Hines, C. M., Osborn, D. D., et al.,** Immunochemiluminometric assay for hepatitis B surface antigen, *Clin. Chem.,* 27, 1378, 1981.
60. **Krodel, E. and Unger, J.,** Future applications of Magic™ Lite technology, *J. Biolum. Chemilum.,* 4, 635, 1989.
61. **Law, S.-J., Miller, T., Piran, U., Klukas, C., Chang, S., and Unger, J.,** Novel poly-substituted aryl acridinium esters and their use in immunoassay, *J. Biolum. Chemilum.,* 4, 88, 1989.
62. **Parnham, A. J. and Tarbit, I. F.,** "Delfia" and "Amerlite": two sensitive non-isotopic immunoassay systems for assay of thyrotropin compared, *Clin. Chem.,* 33, 1421, 1987.
63. **Jago, P. H., Simpson, W. J., Denyer, S. P., et al.,** An evaluation of the performance of ten commercial luminometers, *J. Biolum. Chemilum.,* 3, 131, 1989.
64. **McCapra, F. and Behishti, I.,** Selected chemical reactions that produce light, in *Bioluminescence and Chemilurninescence: Instruments and Applications,* Vol. 2, van Dyke, K., Ed., CRC Press, Boca Raton, FL, 1985, chap. 2.
65. **Weeks, I., Beheshti, I., McCapra, F., Campbell, A. K., and Woodhead, S. J.,** Acridinium esters as high-specific-activity labels in immunoassay, *Clin. Chem.,* 29, 1474, 1983.
66. **Bronstein, I., Voyta, J. C., Thorpe, G. H. G., Kricka, L. J., and Armstrong, G.,** Chemiluminescent assay of AP applied in an ultrasensitive enzyme immunoassay of thyrotropin, *Clin. Chem.,* 35, 1441, 1989.
67. **Zucchelli, G. C., Pilo, A., and Masini, S., et al .,** Methodological evaluation of a new chemiluminescence immunoassay for thyrotropin using acridinium ester-labeled antibody, *J. Biolum. Chemilum.,* 4, 620, 1989.

Chapter 5

ENZYME SYSTEMS AND ENZYME CONJUGATES FOR SOLID-PHASE ELISA

That T. Ngo

TABLE OF CONTENTS

I. INTRODUCTION

A. ENZYMES AS SIGNAL AMPLIFIERS

Enzymes are proteins that catalyze most chemical transformations in biological systems; their catalysis is extraordinarily efficient and specific.[1,2] The rate of nonenzymatic-catalyzed reactions are enhanced 100 billion-fold by enzymes.[1] Because enzymes can catalyze the transformation of 1 million substrate molecules into products within 1 min, their enhancing properties have been used to advantage in biochemical analysis.[3-7] Let us use *Escherichia coli* β-galactosidase as an illustrative example. Under saturating substrate concentration, the rate of the enzyme-catalyzed reaction (v) can be used as a measure of the enzyme concentration ([Enzyme]) and is governed by the equation

$$v = k\ [\text{Enzyme}]$$

where k is the rate constant. When k = 500,000/min and the enzyme concentration is 0.1 n*M*, a reaction rate of 50 μ*M*/min is obtained. At this rate, the enzyme is capable of producing 50 μ*M* *o*-nitrophenol per minute by hydrolyzing the substrate *o*-nitrophenylgalactoside. Therefore, an absorbance of 0.6 optical density unit is achieved after 1 min of reaction (based on an extinction coefficient of 12,000 for *o*-nitrophenol). Operationally, an easily detectable quantity (50 μ*M* *o*-nitrophenol) of reaction product is generated by 0.1 n*M* of enzyme; this is tantamount to a chemical amplification of 500,000 fold in the signal detection.

B. THE DEVELOPMENT OF ENZYME-LINKED IMMUNOSORBENT ASSAY (ELISA)

The use of enzyme label in quantitative immunoassays dates to 1971. Engvall and Perlmann[8] developed a quantitative competitive assay for rabbit IgG which was also adapted for antibody measurement.[9,10] Simultaneously, Van Weemen and Schuurs independently developed ELISA for hapten and antigen.[12] Enzymes had been used by Avrameas as markers for antigens and antibodies in qualitative immunohistochemical techniques.[13] Since its discovery, ELISA has been applied to environmental monitoring,[15] the determination of plant constituents,[16] diagnostic tests in agriculture,[17-19] medical and veterinary diagnostics,[20,21] microbial detection and monitoring and quality control in the food industries.[22]

C. THE SELECTION OF AN ENZYME

Several enzymes, such as acetylcholinesterase, adenosine deaminase, alkaline phosphatase, amine oxidase, carbonic anhydrase, catalase, β-galactosidase, glucoamylase, glucose oxidase, glucose-6-phosphate dehydrogenase, lysozyme, malate dehydrogenase, penicillinase, peroxidase, and urease, have been used in ELISA and nonseparation (homogeneous) enzyme immunoassay techniques. However, only 3 enzymes, i.e., alkaline phosphatase, horseradish peroxidase, and *E. coli* β-galactosidase have have been extensively used in research applications and in commercially available diagnostic kits. General criteria for selecting a particular enzyme as a label in ELISA have been described[23-32] and can be summarized in three categories. The first category of criteria concerns the ultimate sensitivity of the ELISA. To achieve an assay with high sensitivity, it is necessary to use an enzyme with high turnover number that is capable of converting stable substrates into highly sensitive products. Acetylcholinesterase, alkaline phosphatase, catalase, β-galactosidase, horseradish peroxidase, and urease have high turnover numbers. Neither acetylcholinesterase nor urease have gained much popular uses because they do not have a conveniently assayable substrate system. On the other hand, the substrates for both alkaline phosphatase and horseradish peroxidase are not very stable. The nonenzymatic reaction of these substrates limits the ultimate sensitivity of the assay. The lack of a sensitive substrate for catalase has limited the widespread uses of this very active enzyme. β-Galactosidase substrates are stable and give rise to highly sensitive color or fluorescent products. This, together with the stability of the enzyme and its ease of conjugation with antigen or antibodies under controlled conditions, have made *E. coli* β-galactosidase the enzyme of choice as a label in many ultra-sensitive ELISA.[33] A second category concerns practical operations. Analytically, an enzyme label should be easy to assay, the enzyme substrate should have reasonable stability under the assay conditions, and similar enzyme activity should not be found in the sample. Horseradish peroxidase may not be suitable for sample derived from plant tissue with high peroxidative or catalatic activities. Another practical consideration is that the enzyme should possess nonessential reactive groups that can be used for conjugation with antigen or antibodies. Furthermore, conjugates should have a shelflife of greater than 12 months when kept at 4°C. The third category of importance is economics. Obviously, the enzyme label must be available at low cost in a soluble and preferably homogeneous and pure. The most important criterion for nonseparation (homogeneous) enzyme immunoassay is selection of an enzyme with the ability to be directly or indirectly modulated by the analyte antibody.[23,28,31,32,34]

D. THE LIMITATION OF ENZYME LABELS

Enzymes, being proteins, are denatured by known protein denaturing agents such as temperature extremes, acids, bases, organic solvents, detergents, salts, and chaotropic agents. Denatured enzymes generally lack catalytic activities. Enzyme activities can be inhibited by a number of inhibitors[35] or inactivated by protein-modifying reagents.[36] Destruction of enzymes by proteases or other microbial attacks can lead to an underestimation of the enzyme. Analytical methods that use enzymes must avoid or minimize exposure of enzymes to these agents. The rate of an enzymatic reaction is temperature dependent. Hence, proper temperature control during the assay is necessary for reproducible results.

E. THE CHEMISTRY OF ENZYME LABELING

The use of an enzyme in ELISA requires that it be stably linked to the ligand — a hapten, antigen, or antibody. This is generally accomplished through covalent bonding between the enzyme and the ligand. The chemistry of these covalent linkages is limited by the availability of the reactive groups on an enzyme and by the mild conditions required to maintain the proper function of the enzyme. Up to four possible types of reactive groups on a typical enzyme can be used for linkage to the ligand. Such groups include the epsilon amino group of a lysine, *N*-terminal amino groups, the imidazole of a histidine, 4-hydroxylphenyl of a tyrosine, and a thiol group of cysteine. Should a particularly desirable reactive group not be available on the enzyme, it can sometimes be introduced artificially by reacting the protein with a chemical reagent. For example, a thiol group can be introduced by reacting the amino group of the enzyme with iminothiolane[37,38] or alternatively by reacting the amino group with *N*-succinimidyl-3-(2-pyridyldithio)propionate and then with a reducing agent such as mercaptoethanol.[39] A hydroxyphenyl group can be introduced by reacting the amino group of the enzyme with 3-(4-hydroxyphenyl)propionic acid *N*-hydroxysuccinimide ester.[40]

1. Enzyme-Ligand Linkage via Amino Groups of the Enzyme

The amino group of an enzyme can be linked directly to ligand (L) carrying either aldehyde (Reaction 1), carboxylate or its activated ester (Reaction 2), isocyanate (Reaction 3), imidate (Reaction 4), epoxide (Reaction 5), aziridine (Reaction 6), activated halogen groups (Reaction 7), or a sulfonyl chloride (Reaction 8).[41] Alternatively, the amino group of an enzyme can also be linked indirectly to the amino or thiol group of a ligand via a bifunctional cross-linking reagent. Some of the amino groups reactive bifunctional cross-linking reagents are bis-imidates, bis-epoxides, diisocyanate, dialdehyde, *s*-triazine trichloride (Reaction 9), activated

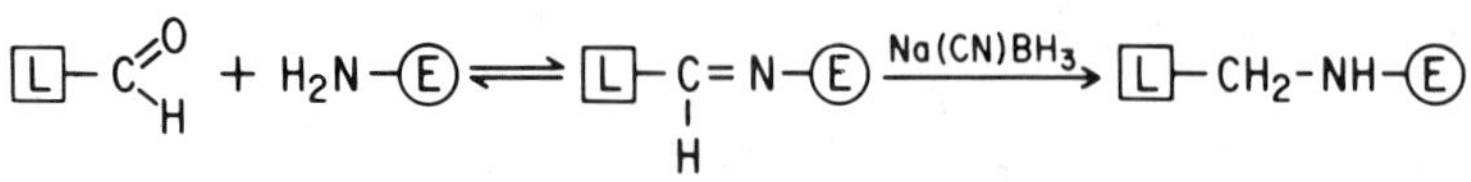
(1)

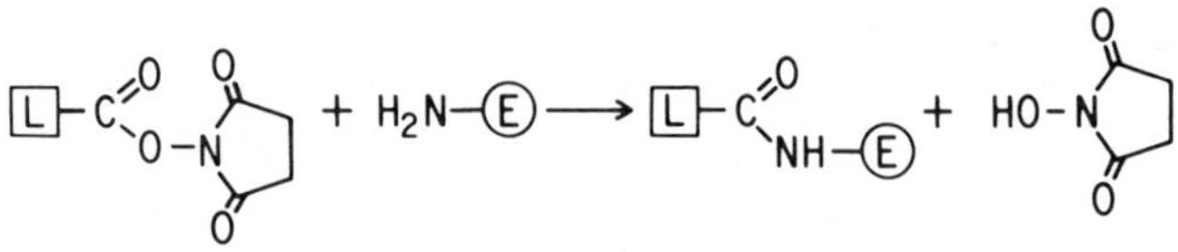
(2)

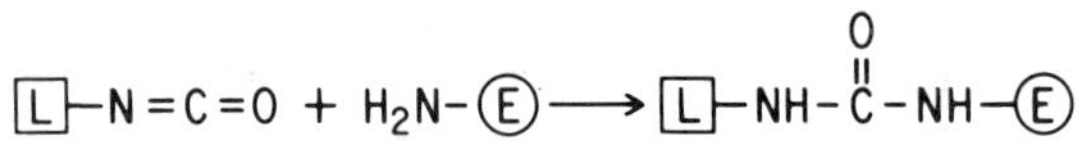
(3)

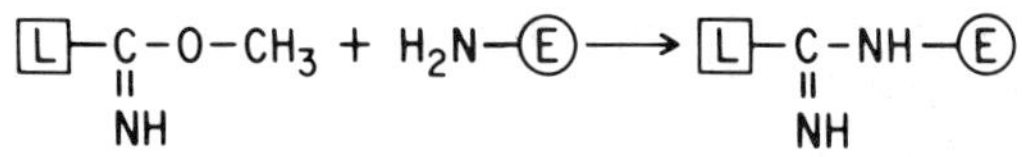
(4)

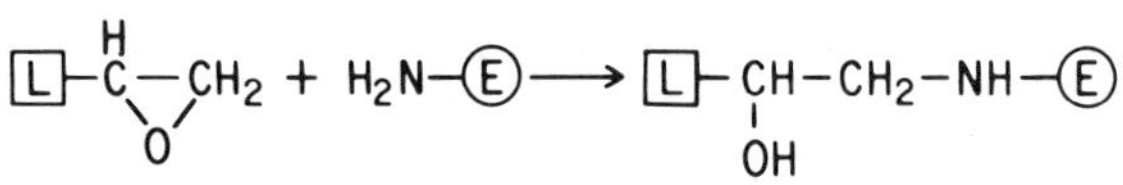
(5)

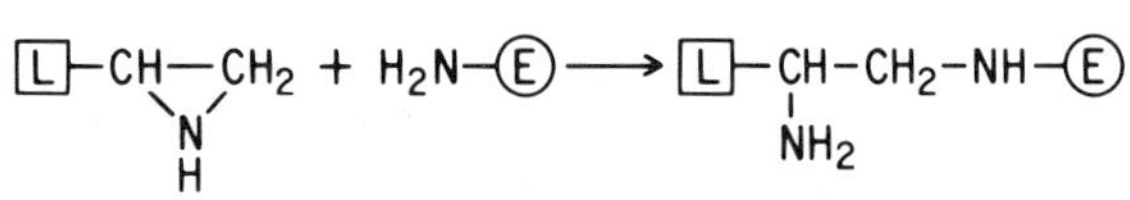
(6)

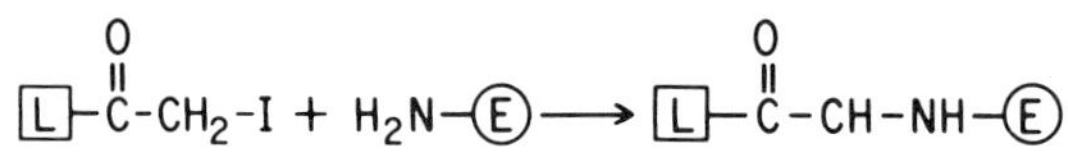
(7)

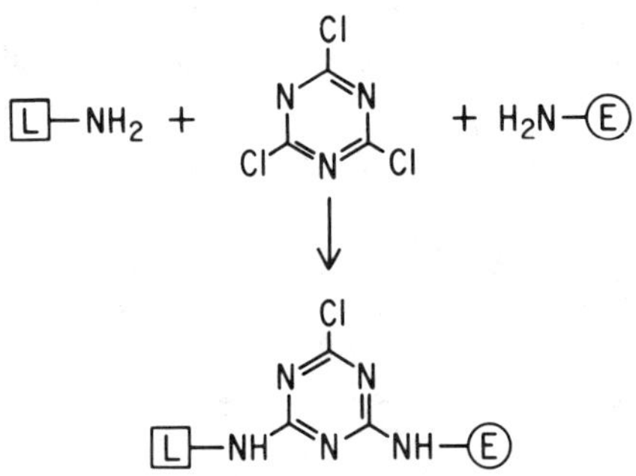
(8)

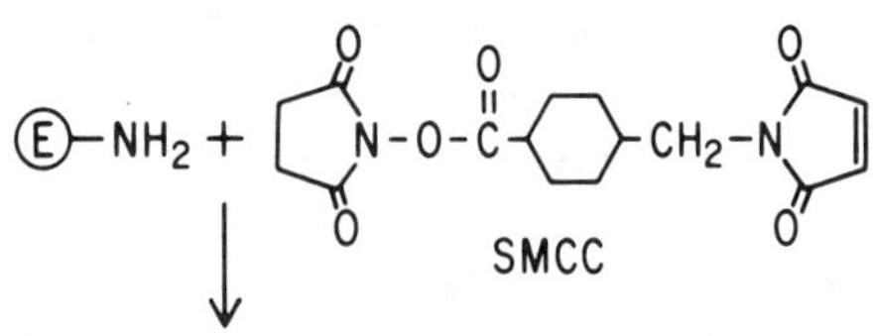
(9)

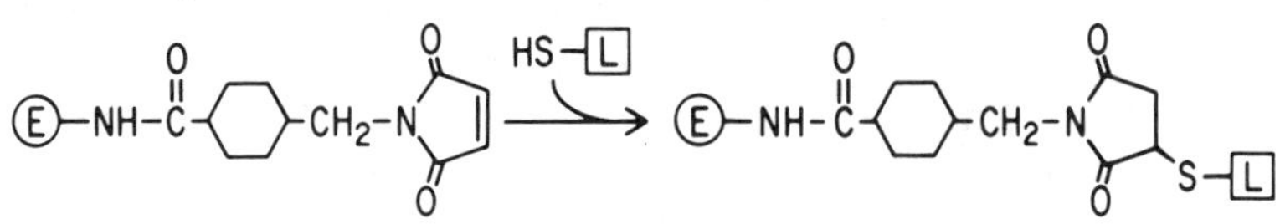
(10)

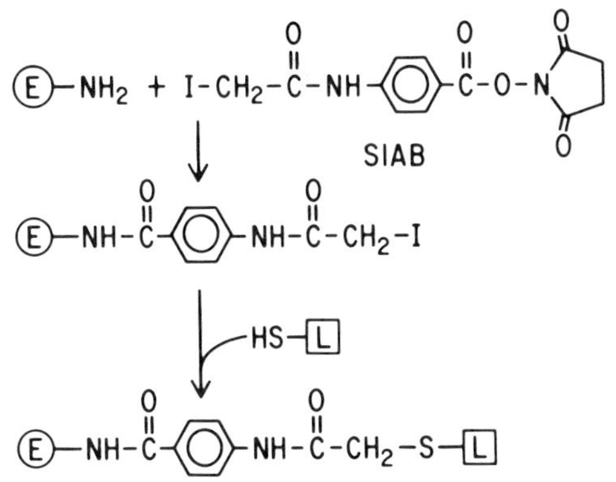

(11)

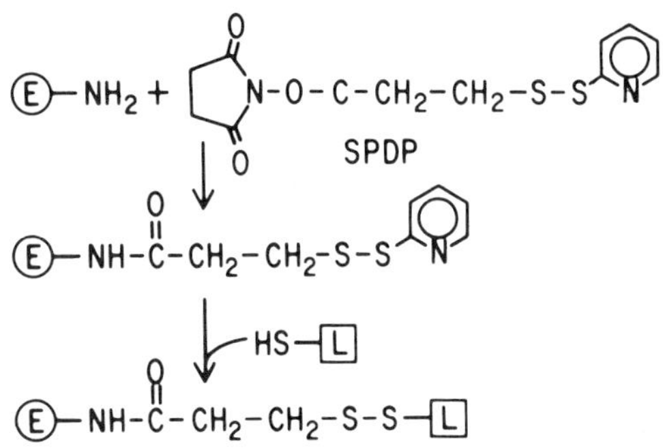

(12)

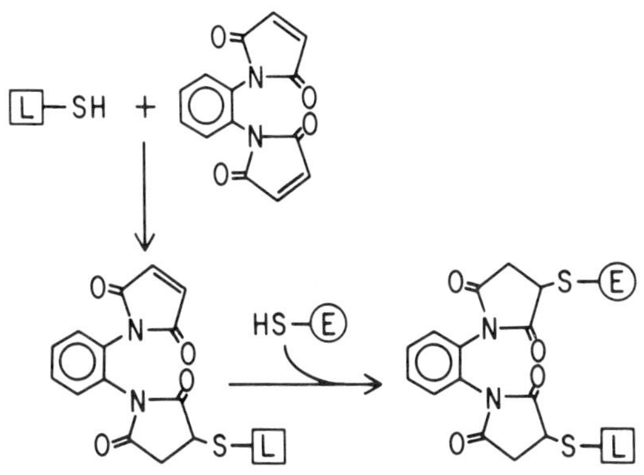

(13)

bis-carboxylate, and bis-activated halogen. The amino group of an enzyme can be linked to thiol of a ligand by using heterobifunctional reagents[42] such as succinimidyl-4-(*N*-maleimidomethyl)cyclohexane-1-carboxylate, SMCC (Reaction 10), *N*-succinimidyl(4-iodoacetyl)aminobenzoate, SIAB (Reaction 11), or *N*-succinimidyl-3-(2-pyridyldithio) propionate, SPDP.

2. Enzyme-Ligand Linkage via Thiol Groups of the Enzyme

The thiol group of an enzyme can be linked to a ligand carrying electrophilic groups such as activated halogens (see Reaction 11) or maleimido moiety (see Reaction 10). Alternatively, by using heterobifunctional cross-linking reagents, such as SMCC (see Reaction 10), the thiol can also be linked to an amino group or to a thiol group of the ligand via SPDP (see Reaction 12) or *N,N′-o*-phenylenedimaleimide (Reaction 13).[43] Thiol groups carrying enzymes can also be linked to the tyrosine or histidine residue of a polypeptide by using a novel heterobifunctional cross-linking agent, *N*-[B-(4-diazophenyl)ethyl]maleimide (DPEM) (Reaction 14).[44]

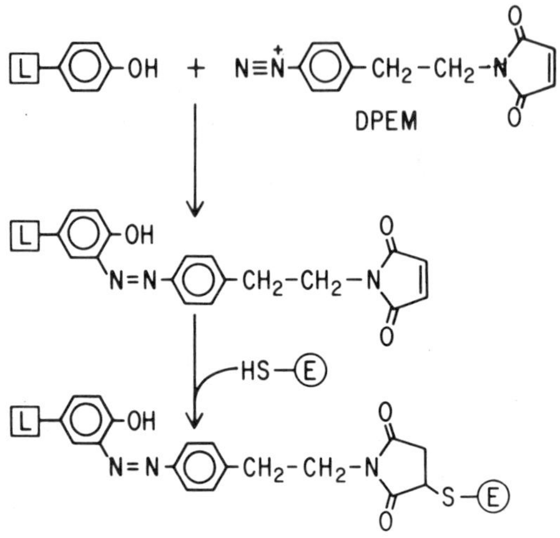

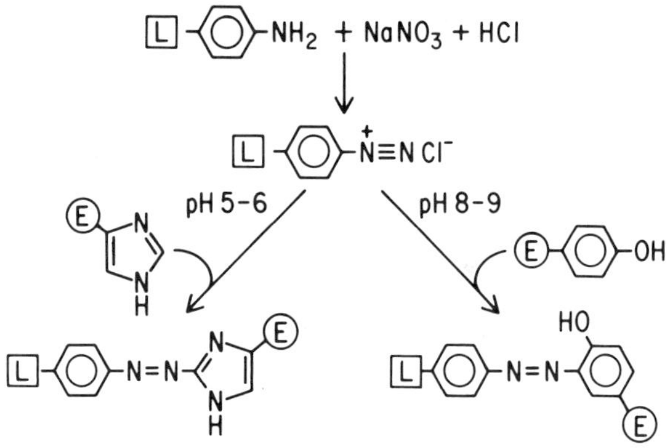

3. Enzyme-Ligand Linkage via Histidyl or Tyrosyl Residues of the Enzyme

The histidyl or tyrosyl residue of an enzyme can be linked to an aminophenyl group bearing ligand via an azo linkage. Alternatively, a primary amine-carrying ligand can be first reacted with a nitrophenyl and *N*-succinimidyl-containing heterobifunctional cross linker. The nitrophenyl group of the conjugate can then be reduced and, upon treatment with nitrous acid, the conjugate becomes reactive with histidine or tyrosine (Reaction 15).[44]

II. MOLECULAR PROPERTIES OF SOME ENZYMES USED AS LABELS IN ELISA

A. ALKALINE PHOSPHATASE (APASE)

1. Reaction and Mechanism

Alkaline phosphatase [E.C. 3.1.3.1, orthophosphoric-monoester phosphohydrolase (alkaline optimum)] catalyzes the hydrolysis of an orthophosphoric monoester to yield an alcohol and an orthophosphate. It has a optimal activity at alkaline pH and a wide substrate specificity. The enzyme also catalyzes transphosphorylation reactions. From the study on the substrate specificity of the enzyme, it was deduced that APase is able to hydrolyze P–F, P–O–C, P–O–P, P–S, and P–N bonds but not the phosphonate P–C bond. The enzyme has a strict specificity requirement for the phosphoryl part of the substrate but not for the nonphosphoryl part. Commonly used colorimetric substrates include *p*-nitrophenyl phosphate, phenolphthalein monophosphate, and thymophthalein monophosphate and the fluorogenic substrates include

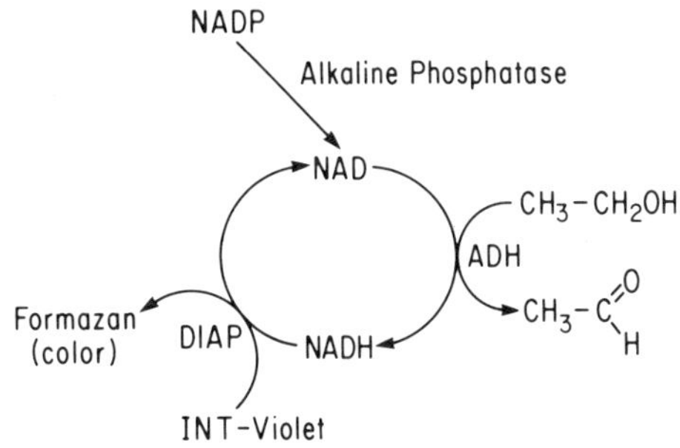

FIGURE 1. Sensitive assay for alkaline phosphatase using secondary enzyme redox amplifying system.

β-naphthyl phosphate, 4-methylumbelliferyl phosphate, and 3-*O*-methylfluorescein monophosphate.[45,46] As little as 1 amol of APase can be detected when using 4-methylumbelliferyl phosphate as the substrate in a 100-min fixed-time assay.[42] Recently, a highly sensitive, signal amplifying colorimetric detecting system for APase has been developed by Self and coworkers.[47,48] In this system the dephosphorylation of one substrate by APase resulted in one product capable of initiating cyclic reactions and therefore amplifying the formation of highly color products. A practical example of this system uses nicotinamide adenine dinucleotide phosphate (NADP) as the substrate for APase. The dephosphorylation of NADP by APase produces NAD which, together with ethanol, serves as the cosubstrate for alcohol dehydrogenase (ADH) that catalyzed oxidation of ethanol. A consequence of ethanol oxidation by ADH is the reduction of of NAD to form NADH. NADH in the presence of diaphorase reduces the colorless iodonitrotetrazolium violet INT-violet) to highly color formazan and concomitantly is itself oxidized to NAD. NAD can again be used in the ADH-catalyzed reaction and therefore repeating the cycle. The net result is an accumulation of formazan (amplification) by cycling of NAD and NADH in the presence of excess ADH, diaphorase, and their substrates. The amplification process is depicted in the above diagram (Figure 1). Dilute solution of APase is not stable enough to be useful for ELISA work. However, the activity of the diluted enzyme solution can be stabilized by adding inert proteins such as bovine serum albumin to the dilute solution.

A number of buffers such as trishydroxymethylaminomethane, diethyl barbiturate, glycine, borate, ethanolamine, and carbonate with pKa values range from 8.1 to 9.5 have been used in APase assay. Different buffers produce different pH-rate profiles. In general, both the Km and Vm of the enzyme increases with increasing pH and reach a limiting value at pH greater than 9. While the zinc ion is necessary for the enzyme activity, magnesium ion enhances the reaction rate. Phosphate ion, one of the products of APase reaction, is an effective competitive inhibitor. Arsenate is more potent an inhibitor than phosphate. Metal chelate, such as EDTA, inhibits the enzyme in a complex way. L-Phenylalanine, but not D-phenylalanine, is an uncompetitive inhibitor. The simplest kinetic mechanism of APase can be presented as

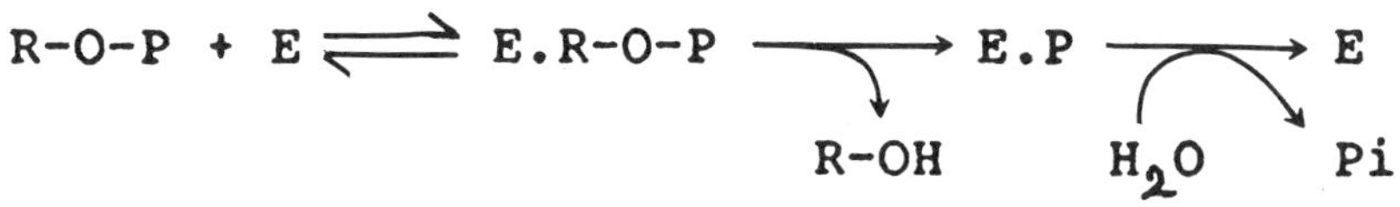

where R–O–P, E, E.R–O–P, E.P, R–OH, and Pi are, respectively, the substrate (orthophosphoric monoester), enzyme, Michaelis complex, phosphorylated intermediate, alcohol, and inorganic phosphate. The first step in the reaction is the formation of a Michaelis complex of enzyme and substrate, the phosphoryl transfer reaction with the subsequent formation of phosphorylated enzyme E.P intermediate. The rate of dephosphorylation appears to be similar to the reaction rate.[49] APase-catalyzed reaction shows a 10^9 to 10^{12} rate enhancement over the corresponding nonenzymatic reactions in the hydrolysis of both alkyl and aryl phosphate.

2. Physical and Chemical Properties

All mammalian APases are zinc metaloenzymes; the active site zinc ion is necessary for catalysis. The molecular weight of APase varies from 84,000 to 150,000 Da depending on the source of the enzyme.[46] They also contain variable amounts of carbohydrate. The extinction coefficient at 278 nm of 1% solution of APase was reported to be 7.2 to 7.8. An isoelectric point at pH 5.7 of APase indicated a high pecentage of acidic amino acids. There are approximately 50 lysine residues per molecule of the enzyme. Some of these lysine residues are "catalytically non-essential" and are available for reaction with cross-linkers without significantly affecting the activity of the enzyme. There seems to be no detectable thiol group. APase is stable at room temperature in solutions of slightly alkaline or neutral pH. It is unstable in acid pH and is irreversibly inactivated at pH less than 3.

3. Conjugation Methods

At least three general methods have been described for linking APase to either antibody or antigen (ligand). Cross-linking of APase to ligands by using either a one-step or two-step glutaraldehyde method was one of the early conjugation methods. This method inevitably leads to the formation of polymerized enzyme and ligand conjugates and results in severely diminished specific enzyme activity.[18,50,51] A more controlled and selective method of linking APase to an Fab fragment of an antibody was developed by Ishikawa et al.[42,43,52,53] using a heterobifunctional cross linker, SMCC. In a two-step reaction, at pH around neutrality, the succinimidyl part of SMCC reacts with the amino groups of APase leaving the maleimido group of SMCC free to react with thiol groups. At this pH the maleimido moiety is not reactive enough to react to the amino groups of the enzyme. APase is normally devoid of free thiol groups. The selective reactivity of the maleimido group can be reacted in a controlled manner with the thiol group of Fab fragment of an IgG in the second step. The following scheme (Reaction 10) summarizes the sequence of coupling reactions. For labeling APase to a protein without a free thiol group, a different linking strategy has been developed. Thiol groups were first introduced to the protein by reacting it with SPDP. The pyridyldithiopropionylated protein was then treated with dithiothreitol to generate the desired free thiol groups. The thiolated protein was then reacted with maleimido labeled APase (prepared by the procedure described above). The sequence of the reactions are summarized in Figure 2. This procedure proved very effective in preparing highly active APase-labeled anti-Salmonella conjugates.[54] A noncovalent method of labeling antibodies with APase has also been used in ELISA. Butler et al.[108] used soluble noncovalent enzyme immune complex (EIC) of APase and rabbit or guinea pig anti-APase in an amplified ELISA (a-ELISA) for the sensitive and reliable measurement of antigen-specific antibodies.

B. β-GALACTOSIDASE

1. Reaction and Mechanism

β-Galactosidase (E.C. β-D-galactoside galactohydrolase) catalyzes the hydrolysis of terminal nonreducing β-D-galactose residues in β-galactosides. The enzyme also catalyzes the transgalaclosidation, i.e., the transfer of galactosyl moiety of the substrate to an acceptor other than water. Monosaccharides, oliosaccharides, alkyl alcohol, and phenols have been used as acceptors.[55] β-Galactosidase has a strict requirement for D-galactopyranoside moiety and β-glycosidic linkage. Substitution at C-2, C-3, C-4, and C-6 by groups that are bulkier than hydroxyls prevents binding of the substrate to the enzyme. The replacement of oxygen by a sulfur atom in the glycosidic bond decreases the rate of enzymatic hydrolysis. The enzyme also catalyzes the hydrolysis of C–N and C–F bonds. A great number of variations in the

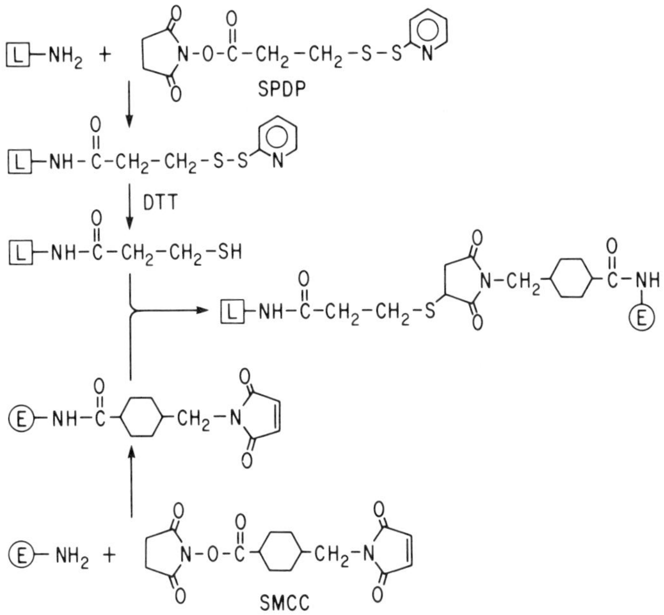

FIGURE 2. Thiolation of a protein ligand L and subsequent conjugation to SMCC-labeled enzyme E.

structure of the aglycon part of β-galactosides are tolerated by the enzyme. For example, *p*-nitrophenyl-β-D-galactoside, *O*-nitrophenyl-β-D-galactoside, 6-bromo-2-naphthyl-β-D-galactoside, indolyl-β-D-galactoside, methylumbelliferyl-β-D-galactoside, 2-naphthyl-β-D-galactoside, and fluorescein di-β-D-galactoside were shown to be substrate for β-galactosidase. The most commonly used colorimetric substrate is *o*-nitrophenyl β-D-galactopyranoside. Methylumbelliferyl-β-D-galactopyranoside is the most widely used fluorogenic substrate. The commonly used assay solution consists of 0.1 *M* Tris or sodium phosphate buffer pH 7 to 7.5 containing 1 m*M* magnesium chloride, 0.1 *M* reducing agent such as 2-mercaptoethanol and 5 m*M* of substrate such as *O*-nitrophenyl-β-D-galactoside. A three-step minimal kinetic mechanism for β-galactosidase reaction has been proposed by Wallenfels and Malhotra.[56] Here, the enzyme and substrate are assumed to be in rapid equilibrium with the Michaelis complex leading to the formation of a galactosyl-enzyme reaction intermediate with concomitant releases of the aglycon. The subsequent hydrolytic cleavage of the galactosyl-enzyme intermediate releases the galactose and the free enzyme. Nucleophiles such as alcohols or amines can replace water in the solvolysis of the galactosylenzyme intermediate and bring about transglycosidation reactions. The mechanism of β-galactosidase-catalyzed reactions resembles that of acid-catalyzed hydrolysis of β-D-galactosides in that the cleavage occurs at the galactosyl C-l oxygen bond. The proposed mechanism for the enzymatic reaction encompasses (1) a stereospecific solvolysis of a carbonium ion transition state in the active site that allows only one-sided attack by the solvent and (2) a double displacement mechanism that involves protonation of the glycosidic oxygen and attack by a nucleophilic group of the enzyme to form a galctosyl-enzyme intermediate and is then followed by a second displacement reaction by a nucleophile. The two consecutive displacement reactions result in the retention of configuration at the anomeric C atom. A number of substances such as thiogalactosides, heavy metal ions, metal chelators, and some alcohols and amines were shown to be potent inhibitors for β-galactosidases.

2. Physical and Chemical Properties

E. coli β-galactosidase is a multimeric enzyme with a molecular weight of approximately

500,000. Treatment of the enzyme with chaotropic agents, heavy metal, alkaline, or acid solution leads to dissociation of the enzyme into smaller subunits with molecular weight as low as 35,000. The enzyme has an isoelectric point of 4.6 and no carbohydrate or phosphate moiety. The amino acid composition shows greater than 20% acidic amino acids, 2.5% lysine, and 3 to 4% cysteine. The lysine and cysteine residues, about 100 and 60 mol per mole of the enzyme, respectively, can be conveniently used for cross linking with haptens, antibodies, or antigens. The *E. coli* enzyme is highly immunogenic capable of giving rise to precipitating antibodies. The antibodies to wild-type enzyme can activate an inactive mutant enzyme.

3. Conjugation Methods

Haptenic or protein ligands can be linked chemically to either the ϵ-amino group of lysine residues or to the thiol function of free cysteine residues of β-galactosidase. Carboxylic groups bearing ligands can be coupled to the amino functionalities of the enzyme by first activating the carboxyl group with, for example, *N*-hydroxysuccinimide and a carbodiimide (Reaction 2). For amino groups bearing ligands, either homo- or heterobifunctional cross-linking reagents can be used to cross link the enzyme with the ligands. For example, amino ligands can first be reacted with a bisimidate, e.g., dimethyl suberimidate, and then coupled to amino groups of the enzyme.[57,58] Alternatively, the amino groups of the ligand can first be reacted with a heterobifunctional cross linker such as SMCC and then the ligand-cross-linker conjugate can, in a controlled manner, be linked to the thiol group of the enzyme.[42,43] Thiol groups bearing ligands such as the Fab′ fragment of an immunoglobulin can first be reacted with a homobifunctional reagent, *N,N′-o*-phenylenedimaleimide; the conjugate can then be reacted with the thiol group of the enzyme.[33,59-62]

C. HORSERADISH PEROXIDASE (HRP)

1. Reaction and Mechanism

Peroxidase (E.C. 1.11.1.7, Donor: hydrogen peroxide oxidoreductase) catalyzes the oxidation of a variety of compounds by hydrogen peroxide or related compounds. A general reaction mechanism of peroxidase involves an oxidation of the enzyme by hydrogen peroxide to form intermediate compound I. Compound I is in turn reduced by a hydrogen donor via a one-electron transfer to form an intermediate compound II and a donor radical. Again, compound II is further reduced by a hydrogen donor via a one-electron process to regenerate a free enzyme and to produce a donor radical. Two donor radicals may recombine to a stable product. In the presence of excess substrate, additional intermediate compounds such as compounds III and IV, which are enzymatically less active, may be generated. HRP-I is a compound I forrned by reacting HRP with hydrogen peroxide in a diffusion-controlled reaction. The oxidation state of Fe in HRP-I is IV which indicates that an additional oxidizing equivalent of HRP-I is stored somewhere in HRP but not in the Fe atom. Such a deduction was made on the basis that adding an oxygen atom to Fe (III) would convert it formally to Fe (V) rather than Fe (IV) as it was observed in the case of HRP-I. With the exception of iodide, sulfite, and luminol, which reduce HRP-I directly to the native enzyme, most hydrogen donors reduce HRP-I to HRP-II (equivalent to compound II of the general mechanism) via a one-electron transfer process. The reduction of HRP-II to the native enzyme by a hydrogen donor also proceed via a one-electron transfer process. An overall kinetic mechanism of a HRP-catalyzed reaction is schematically shown in Figure 3. The pH-rate profile of HRP-catalyzed reaction is dependent on the nature of the hydrogen donor used. In general, the rate constant changes very little with variation of pH from 4 to 7. The Km, however, appears to have an optimal value at about pH 5. A more detailed review on the kinetics and mechanism of HRP-catalyzed reactions can bc found in References 64 and 65. A number of aromatic phenols and amines serve as hydrogen donor and are being used as chromogenic, fluorogenic, and chemiluminogenic substrates for HRP. Some examples of HRP chromogenic substrates are *O*-

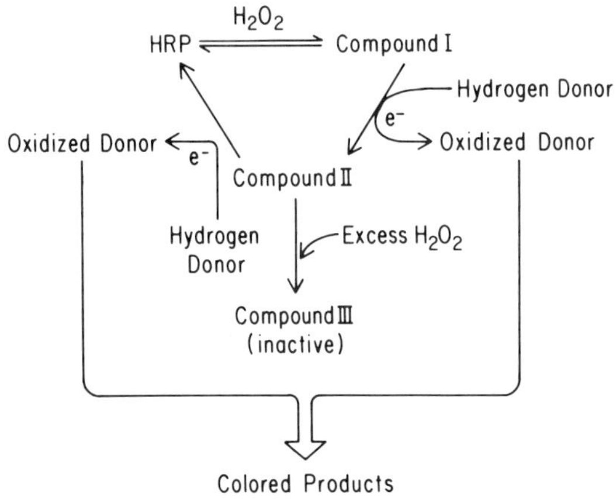

FIGURE 3. Mechanism of HRP catalysis.

phenylenediamine (OPD), 2,2′-azino-di (3-ethyl-benzthiazoline) sulfonic acid (ABTS), *O*-dianisidine (ODIA), 5-aminosalicylic acid (5AS), dicarboxidine, 3,3′,5,5′-tetramethylbenzidine (TMB), phenol and aminoantipyrine (Trinder-Emerson reagent), and 3-methyl-2-benzothiazolinone hydrazone (MBTH) and 3-dimethylaminobenzoic acid (DMAB) (Ngo-Lenhoff reagent.[66-68] A detailed description of assay conditions for the above substrate systems is provided by Porstmann and Porstmann.[69] Luminol, isoluminol, and acridinium ester are well-known luminogenic HRP substrates.[70-72] Examples for fluorogenic substrates are 3-(4-hydroxyphenyl)propionic acid, 4-hydroxyphenylacetic, homovanillic acid, tyramine, and 3′,6′-diacetyl-2′,7′-dichlorofluorescein.[73,74] Cyanide, sulfide, fluoride, azide, hydrazine, hydroxamic acid, etc. are known inhibitors of HRP. A number of widely used HRP substrates, such as ABTS, OPD, *O*-tolidine, and benzidine, were shown to be mutagenic and are likely to be carcinogenic.[78,79] Caution in handling these substrates should be exercised. An electrochemical method for measuring HRP activity based on a fluoride-selective electrode for sensing fluoride ions liberated from HRP-catalyzed oxidation of fluoroaromatic compounds has been described.[80-84]

2. Physical and Chemical Properties

HRP is a holoenzyme with hemin (ferriprotoporphyrin IX) as the prosthetic group. The apoenzyme is a glycoprotein consisting of some 300 amino acid residues with a molecular weight of approximately 40,000. Seven isoenzymes of HRP have been isolated by ion-exchange chromatography and are denoted as A1, A2, and A3 for the anionic form, and B, C, D, and E for the cationic form. Isoenzyme C is the predominant cationic isoenzyme; it comprises the bulk of commercially available highly purified HRP and more than 50% of the total HRP in horseradish. The covalent structure of HRP C has been determined. It consists of 308 amino acid residues and 8 neutral carbohydrate side chains attached through asparagine residues. The molecular weight of the polypeptide chain alone was 33,890. The enzyme has, per molecule, 4 disufide linkages, 1 tryptophan, 3 histidine, 6 lysines, and 21 arginines and an isoelectric point of 8.8. A commonly used criterion for purity of HRP is the R.Z. (Rheinheitzahl) or purity number, the ratio of maximum absorbance in the Soret region to the absorbance of aromatic amino acid at 280 nm. For HRP C, R.Z. values of 3 or greater indicate high purity. In a lyophilized, dry state HRP is stable for several years when it is kept at 4°C. In a solution form, HRP is stable for months when kept at 4°C at a concentration of about 0.1 to 1 mg/ml. A dilute solution of HRP (less than 25 n*M*), however, is rapidly and irreversibly inactivated by the presence of 100 m*M* phosphate buffer. The inactivation can be minimized by using citrate buffer or phosphate buffer at concentrations of less than 20 m*M*.[85] Recent study showed that HRP contained 2 mol of calcium ions per mole of the enzyme and that the

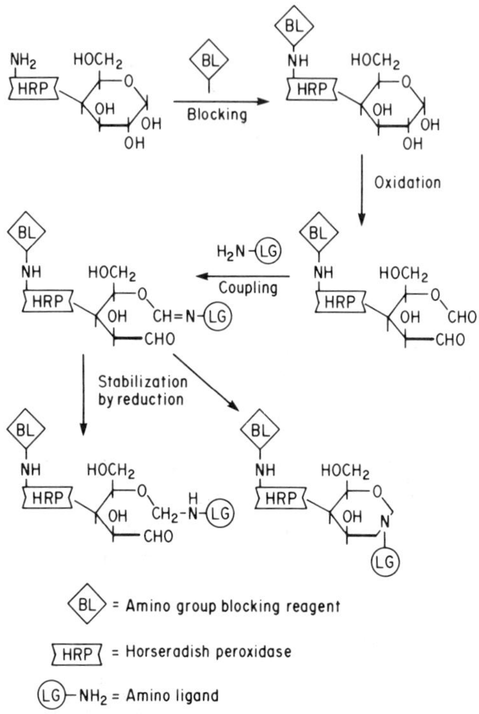

FIGURE 4. Coupling of amino ligand to periodate-oxidized horseradish peroxidase.

removal of these calcium ions from the enzyme caused a 2-fold decrease in its catalytic activity.[86-89] It was shown that calcium ions functioned by maintaining the protein structure in the heme environments as well as the spin state of the heme iron which promoted the enzyme activity of HRP.[86-89] The reduced enzymatic activity of HRP was mainly brought about by a decrease in the compound II reduction process.[86,87]

3. Conjugation Methods

HRP has been conjugated to a variety of antibodies, antigens, and haptens. Conjugation of ligands to the carbohydrate portion of HRP can be carried by first blocking the amino groups of HRP with fluorodinitrobenzene to minimized self-conjugation and then the cis 1,2-diol groups of the carbohydrate are oxidized by using sodium periodate. The aldehyde groups generated by oxidation of HRP are reacted with amino groups of the ligands to form Schiff's base linkages which are subsequently stabilized by a reduction step.[90-93] The reduction of the HRP-ligand Schiff's base leads to two possible stabilized conjugates as shown in Figure 4. Alternatively HRP can be oxidized under acidic conditions (pH 4 to 5) without prior blocking of its amino groups. In acid conditions the formation of Schiff's base is not favored and therefore the extent of self-aggregation is minimal.[94] Since the number of available amino groups in HRP is low (less than 6), it was possible to treat HRP with glutaraldehyde without causing excessive self-cross-linking. Glutaraldehyde-treated HRP can be coupled to proteins with (two-step method) or without (one-step method) prior removal of unreacted glutaraldehyde.[50,96] However, the recovery of both immunological and enzyme activities by either the periodate oxidation or glutaraldehyde method was low. The labeling of antibody by periodate-oxidized HRP often leads to the formation of polymeric species. Tsang et al.[98] studied the molecular properties of two kinds of HRP-IgG conjugates prepared by the glutaraldehyde and the periodate methods. They concluded that conjugates prepared by the periodate method were capable of providing very high sensitivity detection but with rather narrow assay linearity range, which was more useful for a "plus-minus" type of assay. Conjugates prepared by the glutaraldehyde method, on the other hand, gave wider range of assay linearity which was useful for quantitative measurement purposes. HRP-labeled protein conjugates with a more defined ratio of HRP to the protein have been prepared by disulfide interchange methods, where the two proteins were held together by disulfide bonds.[99,100] Here, both HRP and the protein were reacted separately with SPDP, then the SPDP on one labeled protein was reduced to thiol group, and finally the thiolated protein was mixed with the SPDP-labeled protein to form the conjugate. Arends[101] described a method for purifying HRP-labeled antibody for enzyme immunoassay by affinity chromatography on Conconavalin A. Methods for preparing HRP-labeled antibodies or antigens that give a high recovery of both immunological and enzymatic activities and a defined ratio of enzyme and ligand in the conjugates have been developed by Ishikawa and co-workers.[102-104] The reaction of HRP with 40 molar excess of SMCC results in the incorporation of 1 to 2 mol of maleimido groups per mole of HRP. The maleimido-labeled HRP can be facilely conjugated to Fab′ fragment of an antibody or to a thiolated IgG or any thiol group containing ligand. Thiolation of an amino ligand can be carried out in several ways, e.g., reaction of an amino ligand with *S*-acetyl succinic anhydride and then with hydroxylamine or reaction of an amino group with *N*-succinimidyl-3-(2-pyridyldithio)propionate and then with dithiothreitol. Berkowitz and Webert[105] reported that HRP and its antibody conjugates were inactivated by a polystyrene surface and could be prevented by addition of Tween (polyoxyethylenesorbitan). Porstmann et al.[106] found that the inclusion of 0.1% nonionic detergent enhanced the activity of HRP by delaying the inactivation that occurred during the course of oxidative substrate reaction.

III. SUMMARY

The three enzymes most widely used as labels in ELISA are calf intestinal alkaline phosphatase, *E. coli* β-galactosidase, and HRP. The properties of these enzymes that are important to ELISA have been described.

Alkaline phosphatase has been used extensively. For example, Halbert and Lin[107] summarized its use in ELISA for detecting specific antibodies in autoimmune diseases, e.g., rheumatoid factor, DNA, nucleoprotein, and immune complexes and antibodies to infectious agents such as the Rubella virus, *Toxoplasma* gondii, Cytomegalo virus, Herpes simplex, and *Entamoeba histolytica*. Butler et al.[108,121] used APase to develop a-ELISA with improved sensitivity and flexibility. The a-ELISA was used in measuring the isotypic distribution of specific immune responses in many species, the distribution and quantification of antigens, and to study the stoichiometry of solid-phase assays for antibodies. APase was the enzyme label in the rapid and sensitive immunoassay system called "radial partition enzyme immunoassay". APase was also used in commercially available enzyme immunoassay kits together with monoclonal antibodies in immunoenzymetric formats.[109] Kohen et al.[110] used APase-labeled avidin for developing highly sensitive monoclonal antibody-based ELISAs for hGH. A number of commercially available in-office and in-home self-testing ELISA kits for pregnancy and ovulation have been developed using APase.[111,112] Highly sensitive dot-ELISAs for detecting infectious diseases have also been developed using APase.[113]

E. coli β-galactosidase has been used as an enzyme label in both nonseparation (homogeneous) and separation-required (heterogeneous) enzyme immunoassays. The enzyme and its substrate, *o*-nitrophenyl-β-D-galactoside, have a long shelf life. Using this enzyme as the label, a number of ELISAs with a sensitivity comparable to radioimmunoassay (RIA) have been developed for antigens and low molecular weight compounds.[58,114-117] Ornithine-aminotransferase at a concentration as low as 0.03 fmol per test tube was successfully measured by ELISA using this enzyme and a fluorogenic substrate.[118] Improvement in the ELISA sensitivity has resulted in a detection limit for ornithine aminotransferase of 1 amol per test tube.[33] Castro and Monji[119] developed a steric hindrance enzyme immunoassay (SHEIA) for thyroxine by using β-galactosidase. Freytag[120] used β-galactosidase to develop highly sensitive and rapid affinity column-mediated immunoenzymometric assays (ACMIA) for digoxin and hCG. It appears that *E. coli* β-galactosidase is the enzyme to be used as a label for developing highly sensitive ELISA and the work of Ishikawa and co-workers should be used as the starting point.[33,42,43,52,53,59,60,62,63,102,104,115,118]

HRP is the smallest of the 3 most popular enzyme labels with a molecular weight of 40,000. Each molecule of HRP is glycosylated by eight neutral carbohydrate residues which renders the enzyme less sticky and therefore gives fewer nonspecific signals than other enzymes need for ELISA. HRP was one of the first enzymes used in developing ELISAs for low molecular weight compounds, antigens, and infectious agents.[11-13,26,122-125] HRP-labeled antibody remains the reagent of choice for visualizing antigens blotted on a membrane. Tsang et al.[126] used HRP-labeled antibody in enzyme-linked immunoelectrotransfer blot technique. HRP-labeled antibody is also widely used in immunohistochemical staining.[127] HRP-labeled antibody conjugates appear to be superior to APase- or β-galactosidase-labeled conjugates in terms of assay sensitivity and time.[128]

REFERENCES

1. **Laidler, K. J. and Bunting, P. S.,** *The Chemical Kinetics of Enzyme Action,* Clarendon Press, Oxford, 1973, chap. 9.
2. **Fersht, A.,** *Enzyme Structure and Mechanism,* W.H. Freeman, Reading, England, 1977, chap. 10.
3. **Bergmeyer, H. U., Ed.,** *Methods of Enzymatic Analysis,* Vol. 1-10, 3rd ed., VCH Publishers, Weinheim, West Germany, 1985.
4. **Carr, P. W. and Bowers, L. D.,** *Immobilized Enzymes in Analytical and Clinical Chemistry,* Wiley-Interscience, New York, 1980.
5. **Ishikawa, E., Kawai, T., and Miyai, K.,** *Enzyme Immunoassay,* Igaku-Shoin, Tokyo, 1981.
6. **Guilbault, G. G.,** *Handbook of Enzymatic Methods of Analysis,* Marcel Dekker, New York, 1976.
7. **Ngo, T. T. and Lenhoff, H. M.,** *Enzyme-Mediated Immunoassay,* Plenum Press, New York, 1985.
8. **Engvall, E. and Perlmann, P.,** Enzyme-linked immunosorbent assay (ELISA), Quantitative assay of immunoglobulin G, *Immunochemistry,* 8, 871, 1971.
9. **Engvall, E., Jonsson, K., and Perlmann, P.,** Enzyme-linked immunosorbent assay. II. Quantitative assay of protein antigen, immunoglobulin G by means of enzyme-labeled antigen and antibody-coated tubes, *Biochim. Biophys. Acta,* 251, 427, 1971.
10. **Engvall, E. and Perlmann, P.,** Enzyme-linked immunosorbent assay, ELISA. III. Quantitation of specific antibody by enzyme labeled anti-immunoglobulin in antigen coated tubes, *J. Immunol.,* 109, 129, 1972.
11. **Van Weemen, B. K. and Schuurs, A. H. W. M.,** Immunoassay using antigen enzyme conjugates, *FEBS Lett.,* 15, 232, 1971.
12. **Van Weemen, B. K. and Schuurs, A. H. W. M.,** Immunoassay using hapten enzyme conjugates, *FEBS Lett.,* 24, 77, 1972.
13. **Avrameas, S.,** Immunoenzyme techniques: enzymes as markers for the localization of antigens and antibodies, in *International Review of Cytology,* Vol. 27, Bourne, G. H., Danielli, J. F., and Jeon, K. W., Eds., Academic Press, New York, 1970, 349.
14. **Hunter, K. W. and Lenz, D. E.,** Detection and quantification of the organophosphate insecticide paraoxon by competitive inhibition enzyme immunoassay, *Life Sci.,* 30, 1982, 355.
15. **Van Emon, J., Hammock, B., and Seiber, J. N.,** Enzyme-linked immunoassay for Paraquat and its application to exposure analysis, *Anal. Chem.,* 58, 1866, 1986.
16. **Mansell, R. L. and McIntosh, C. A.,** Enzyme immunoassay of plant constituents, in nonisotopic immunoassay, Ngo, T. T., Ed., Plenum Press, New York, 1988, 147.

17. **Halpern, G. M., Moss, R. B., Hsu, Y. P., and Blessing-Moore, J.,** Enzyme immunoassay (ELISA) for honey bee venom specific IgG4, in *Immunoenzymatic Techniques,* Avrameas, S., Druet, P., Masseyeff, R., and Feldman, G., Eds. Elsevier Science, Amsterdam, 1983, 395.
18. **Gugerli, P.,** Use of enzyme immunoassay in phytopathology, in *Immunoenzymatic Techniques,* Avrameas, S., Druet, P., Masseyeff, R., and Feldman, G., Eds., Elsevier Science, Amsterdam, 1983, 369.
19. **Klausner, A.,** Immunoassays flourish in new markets, *Biotechnology,* 5, 551, 1987.
20. **Wright, P. F. and Nielsen, K. H.,** Application of enzyme immunoassay in veterinary medicine: serodiagnosis of bovine Brucellosis, in *Nonisotopic Immunoassay,* Ngo, T. T., Ed., Plenum Press, New York, 1988, 129.
21. **Bommeli, W. R.,** Use of enzyme immunoassay in veterinary medicine, in *Immunoenzymatic Techniques,* Avrameas, S., Druet, P., Masseyeff, R., and Feldman, G., Eds., Elsevier Science, Amsterdam, 1983, 395.
22. **Aleixo, J. A. G., Swaminathan, B., and Minnich, S. A.,** Salmonella detection in foods and feeds in 72 hours by an enzyme immunoassay, *J. Immunol. Methods,* 2, 135, 1984.
23. **Wisdom, G. B.,** Enzyme-immunoassay, *Clin. Chem.,* 22, 1234, 1976.
24. **Scharpe, S. L., Cooreman, W. M., Blomme, W. J ., and Laekeman, G. M.,** Quantitative enzyme immunoassay: current status, *Clin. Chem.,* 22, 733, 1976.
25. **O'Sullivan, M. J., Bridges, J. W., and Maeks, V.,** Enzymeimmunoassay: a review, *Ann. Clin. Biochem.,* 16, 221, 1979.
26. **Schuurs, A. H. W. M. and Van Weemen, B. K.,** Enzyme-immunoassay, *Clin. Chim. Acta,* 81, 1, 1977.
27. **Maggio, E. T.,** Enzymes as immunochemical labels, in *Enzyme-Immunoassay,* Maggio, E. T., Ed., CRC Press, Boca Raton, FL, 1980, 53.
28. **Ngo, T. T. and Lenhoff, H. M.,** Enzymes as versatile labels and signal amplifiers for monitoring immunochemical reactions, *Mol. Cell. Biochem.,* 44, 3, 1982.
29. **Blake, C. and Gould, B. J.,** Use of enzymes in immunoassay techniques, a review, *Analyst,* 109, 533, 1984.
30. **Oellerich, M.,** Enzyme-immunoassay, *J. Clin. Chem. Clin. Biochem.,* 22, 895, 1984.
31. **Ngo, T. T.,** Enzyme mediated immunoassay: an overview, in *Enzyme-Mediated Immunoassay,* Ngo, T. T. and Lenhoff, H. M., Eds., Plenum Press, New York, 1985, 3.
32. **Gould, B. J. and Marks, V.,** Recent developments in enzyme immunoassay, in *Nonisotopic Immunoassay,* Ngo, T. T., Ed., Plenum Press, New York, 1988, 3.
33. **Ishikawa, E. and Kato, K.,** Ultrasensitive enzyme immunoassay, *Scand. J. Immunol.,* 8(Suppl. 7), 43, 1978.
34. **Ngo, T. T.,** Enzyme modulator mediated immunoassay (EMMIA), *Int. J. Biochem.,* 15, 583, 1983.
35. **Webb, J. L.,** *Enzyme and Metabolic Inhibitors,* Vol. 1 to 3, Academic Press, New York, 1963.
36. **Means, G. E. and Feeney, R. E.,** Chemical modification of proteins, Holden-Day, San Francisco, 1971.
37. **Traut, R. R., Bollen, A., Sun, T. T., Hershey, J. W. B., Sundberg, J., and Pierce, L. R.,** Methyl 4-mercaptobutyrimidate as a cleavable cross-linking reagent and its application to the *Escherichia coli* 30S ribosome, *Biochemistry,* 12, 3266, 1973.
38. **Jue, R., Lambert, J. M., Pierce, L. R., and Traut, R. R.,** Addition of sulfhydryl groups to *Escherichia coli* ribosomes by protein modification with 2-iminothiolane (methyl 4-mercaptobutyrimidate), *Biochemistry,* 17, 5399, 1978.
39. **Carlsson, J., Drevin, H., and Axen, R.,** Protein thiolation and reversible protein-protein conjugation. *N*-Succinimidyl 3-(2-pyridyldithio)propionate — a new heterobifunctional reagent, *Biochem. J.,* 173, 723, 1978.
40. **Bolton, A. E. and Hunter, W. M.,** The labeling of proteins to high specific radioactivities by conjugation of a ^{125}-I-containing acylating agent, *Biochem. J.,* 133, 529, 1973.
41. **Kennedy, J. H., Kricka, L. J., and Wilding, P.,** Protein-protein coupling reactions and the applications of protein conjugates, *Clin. Chim. Acta,* 70, 1, 1976.
42. **Ishikawa, E., Imagawa, M., Hashida, S., Yoshitake, S., Hamagichi, Y., and Ueno, T.,** Enzyme-labeling of antibodies and their fragments for enzyme immunoassay and immunohistochemical staining, *J. Immunoassay,* 4, 209, 1983.
43. **Ishikawa, E., Hashida, S., Kohno, T., and Tanaka, K.,** Methods for enzyme-labeling of antigens, antibodies and their fragments, in *Nonisotopic Immunoassay,* Ngo, T.T., Ed., Plenum Press, New York, 1988, 27.
44. **Fujiwara, K. and Saita, T.,** The Use of *N*-[B-(4-diazophenyl)ethyl]maleimide as a heterobifunctional agent in developing enzyme immunoassay for neurotensin, *Anal. Biochem.,* 161, 157, 1987.
45. **Reid, T. W. and Wilson, I. B.,** *E. coli* alkaline phosphatase, in *The Enzymes,* Vol. 4, 3rd ed., Boyer, P.D. Ed., Academic Press, New York, 1971, 373.
46. **Fernley, H. N.,** Mammalian alkaline phosphatase, in *The Enzymes,* Vol. 4, 3rd ed., Boyer, P.D. Ed., Academic Press, New York, 1971, 417.
47. **Self, C. H.,** Enzyme amplification — a general method applied to provide an immunoassisted assay for placental alkaline phosphatase, *J. Immunol. Methods,* 76, 389, 1985.
48. **Stanley, C. J., Paris, F., Plumb, A., Webb, A., and Johannsson, A.,** Enzyme amplification: a new technique for enhancing the speed and sensitivity of enzyme Immunoassays, *Am. Clin. Prod. Rev.,* 4, 34, 1985.
49. **Fernley, H. N. and Bisaz, S.,** Studies on alkaline phosphatase. Phosphorylation of calf-intestinal alkaline phosphatase by P932)-labeled pyrophosphate, *Biochem. J.,* 107, 279, 1968.

50. **Avrameas, S.,** Coupling of enzymes to proteins with glutaraldehyde. Use of the conjugates for the detection of Antigens and antibodies, *Immunochemistry,* 6, 43, 1969.
51. **Engvall, E.,** Preparation of enzyme-labeled staphylococcal protein A and its use for detection of antibodies, *Scand. J. Immunol.,* 8(Suppl. 7), 25, 1978.
52. **Ishikawa, E., Yamada, Y., Hamaguchi, Y., Yoshitake, S., Shiomi, K., Ota, T., Yamamoyo, Y., and Tanaka, K.,** Enzyme-labeling with maleimides and its application to immunoassay of peptide hormones, in *Enzyme Labeled Immunoassay of Hormones and Drugs,* Pal, S. B., Ed., Walter de Gruyter, Berlin, 1978, 43.
53. **Yoshitake, S., Yamada, Y., Ishikawa, E., and Masseyeff, R.,** Conjugation of glucose oxidase from *Aspergillus niger* and rabbit antibodies using *N*- hydroxysuccinimide ester of *N*-(4-carboxycyclohexylmethyl)-malaimide, *Eur. J. Biochem.,* 101, 395, 1979.
54. **Ngo, T. T.,** unpublished data.
55. **Wallenfels, K. and Weil, R.,** B-Galactosidase, in *The* Enzymes, Vol. 7, 3rd. ed., Boyer, P. D., Ed., Academic Press, New York, 1972, 617.
56. **Wallenfels, K. and Malhotra, O. P.,** *Adv. Carbohydrate Chem.,* 16, 239, 1961.
57. **O'Sullivan, M. J., Gnemmi, E., Morris, D., Al Bassam, M. N., Simmons, M., Bridges, J. W., and Marks, V.,** An enzyme immunoassay for triiodothyronine, in *Enzyme Labeled Immunoassay of Hormones and Drugs,* Pal, S. B., Ed., Walter de Gruyter, Berlin, 1978, 301.
58. **Al-Bassam, M. N., O'Sullivan, M. J., Gnemmi, E., Bridges, J. W., and Marks, V.,** Nortriptyline enzyme immunoassay, in *Enzyme Labelled Immunoassay of Hormones and Drugs,* Pal, S. B., Ed., Walter de Gruyter, Berlin, 1978, 375.
59. **Kato, K., Hamaguchi, Y., Fukui, H., and Ishikawa, E.,** Enzyme-linked immunoassay: conjugation of rabbit anti-(human immunoglobulin G) antibody with B-D-galactosidase from *Escherichia coli* and its use for human immunoglobulin G assay, *Eur. J. Biochem.,* 62, 285, 1976.
60. **Hamaguchi, Y., Kato, K., Ishikawa, E., Kobayashi, K., and Katunuma, N.,** Enzyme-linked sandwich immunoassay of macromolecular antigens using the rabbit antibody-loaded silicone pieces as a solid phase, *FEBS Lett.,* 69, 11, 1976.
61. **Weston, P. D., Devries, J. A., and Wrigglesworth, R.,** Conjugation of enzymes to immunoglobulins using dimaleimides, *Biochim. Biophys. Acta,* 612, 40, 1980.
62. **Ishikawa, E., Imagawa, M., Yoshitake, S., Niitsu, Y., Urushizaki, I., Inada, M., Imura, H., Kanazawa, R., Tachibana, S., Nakazawa, N., and Ogawa, H.,** Major factors limiting sensitivity of sandwich enzyme immunoassay for ferritin, immunoglobulin E, and thyroid-stimulating hormone, *Ann. Clin. Biochem.,* 19, 379, 1982.
63. **Inoue, S., Hashida, S., Kohno, T., Yanaka, K., and Ishikawa, E.,** A micro-scale method for the conjugation of affinity-purified Fab′ to B-D-galactosidase from *Escherichia coli, J. Biochem.,* 98, 1387, 1985.
64. **Saunders, B. C., Holmes-Siedel, A. G., and Stark, B. P.,** *Peroxidase,* Butterworth, London, 1964.
65. **Dunford, H. B. and Stillman, J. S.,** *Coord. Chem. Rev.,* 19, 187, 1976.
66. **Ngo, T. T. and Lenhoff, H. M.,** A sensitive and versatile chromogenic assay for peroxidase and peroxidase-coupled reactions, *Anal. Biochem.,* 105, 389, 1980.
67. **Geoghegan, W. D., Struve, M. F., and Jordon, R. E.,** Adaptation of the NGO-Lenhoff peroxidase assay for ELISA, *J. Immunol. Methods,* 60, 61, 1983.
68. **Geoghegan, W. D.,** The Ngo Lenhoff (MBTH-DMAB) peroxidase assay, in *Enzyme-Mediated Immunoassay,* Ngo, T. T. and Lenhoff, H. M., Ed., Plenum Press, New York, 1985, 451.
69. **Porstmann, B. and Porstmann, T.,** Chromogenic substrates for enzyme immunoassay, in *Nonisotopic Immunoassay,* Ngo, T. T., Ed., Plenum Press, New York, 1988, 57.
70. **Wood, W. G., Fricke, H., and Strasburger, C. J.,** Solid-phase luminescence immunoassays using kinase and aryl hydrazide labels, in *Nonisotopic Immunoassay,* Ngo, T. T., Ed., Plenum Press, New York, 1988, 257.
71. **Whitehead, T. P., Thorpe, G. H. G., Carter, T. J. N., Groucutt, C., and Kricka, L. J.,** Enhanced luminescence procedure for sensitive determination of peroxidase- labelled conjugates in immunoassay, *Nature,* 305, 158, 1983.
72. **Thorpe, G. H. G., Haggart, R., Kricka, L. J., and Whitehead, T. P.,** Enhanced luminescent enzyme immmunoassays for Rubella antibody, immunoglobulin E and digoxin, *Biochem. Biophys. Res. Commun.,* 119, 481, 1984.
73. **Guilbault, G. G., Brignac, P. J., and Juneau, M.,** New substrates for the fluorometric determination of oxidative enzymes, *Anal. Chem.,* 40, 1256, 1968.
74. **Guilbault, G. G., Brignac, P. J., and Zimmer, M.,** Homovanilic acid as a fluorometric substrate for oxidative enzymes. Analytical applications of the peroxidase, glucose oxidase, and xanthine oxidase systems, *Anal. Chem.,* 40, 190, 1968.
75. **Kato, N., Naruse, H., Irie, M., and Tsuji, A.,** Fluorophotometric enzyme immunoassay of thyroid-stimulating hormone, *Anal. Biochem.,* 96, 419, 1979.
76. **Guilbault, G. G., Sadar, M. H., and Peres, K.,** Fluorometric determination of carbohydrates, *Anal. Biochem.,* 31, 91, 1969.

77. **Kelley, T. A.,** Separation-free enzyme fluorescence immunoassay by flow injection analysis, in *Enzyme-Mediated Immunoassay,* Ngo, T. T., Ed., Plenum Press, New York, 1985, 191.
78. **Voogd, C. E., Van der Stel, J. J., and Jacobs, J. J. J. A. A.,** On the mutagenic action of some enzyme immunoassay substrates, *J. Immunol. Methods,* 36, 55, 1980.
79. **Al-Kaissi, E. and Mostratos, A.,** Assessment of substrate for horseradish peroxidase in enzyme immunoassay, *J. Immunol. Methods,* 58, 127, 1983.
80. **Siddiqi, I. W.,** An electrochemical assay system for peroxidase and peroxidase-coupled reactions based on a fluoride ion selective electrode, *Clin. Chem.,* 28, 1962, 1982.
81. **Alexander, P. W. and Maitra, C.,** Enzyme-linked immunoassay of human immunoglobulin G with the fluoride ion selective electrode, *Anal. Chem.,* 54, 68, 1982.
82. **Alexander, P. W. and Maitra, C.,** Continuous-flow potentiometric determination of horseradish peroxidase with a fluoride-selective electrode, *Anal. Chim. Acta,* 208, 173, 1988.
83. **Alexander, P. W.,** Enzyme-linked immunoassay with a fluoride ion selective electrode, in *Electrochemical Sensors in Immunological Analysis,* Ngo, T. T., Ed., Plenum Press, New York, 1987, 203.
84. **Boitieux, J. L., Desmet, G., and Thomas, D.,** Heterogenous potentiometric enzyme immunoassay for antigens and haptens with iodide selective electrode, in *Electrochemical Sensors in Immunological Analysis,* Ngo, T. T., Ed., Plenum Press, New York, 1987, 211.
85. **Bovaird, J. H., Ngo, T. T., and Lenhoff, H. M.,** Optimizing the *o*-phenylenediamine assay for horseradish peroxidase: effects of phosphate and pH, substrate and enzyme concentrations, and stopping reagents, *Clin. Chem.,* 28, 2423, 1982.
86. **Haschke, R. H. and Friedhoff, J. M.,** Calcium-related properties of horseradish peroxidase, *Biochem. Biophys. Res. Commun.,* 80, 1039, 1978.
87. **Ogawa, S., Shiro, Y., and Morishima, I.,** Calcium binding by horseradish peroxidase C and the heme environmental structure, *Biochem. Biophys. Res. Commun.,* 80, 674, 1979.
88. **Shiro, Y., Kurono, M., and Morishima, I.,** Presence of endogeneous calcium ion and its functional and structural regulation in horseradish peroxidase, *J. Biol. Chem.,* 261, 9382, 1986.
89. **Morishima, I., Kurono, M., and Shiro, Y.,** Presence of endogeneous calcium ion in horseradish peroxidase. Elucidation of metal-binding site by substitutions of divalent and lanthanide ions for calcium and use of metal-induced NMR resonace, *J. Biol. Chem.,* 261, 9391, 1986.
90. **Nakane, P. K. and Kawaoi, A.,** Peroxidase-labeled antibody. A new method of conjugation, *J. Histochem. Cytochem.,* 22, 1084, 1974.
91. **Boorsma, D. M. and Streſkerk, J. G.,** Peroxidase-conjugate chromatography isolation of conjugate prepared with glutaraldehyde or periodate using polyacrylamide-agarose gel, *J. Histochem. Cytochem.,* 24, 481, 1976.
92. **Boorsma, D. M., Streſkerk, J. G., and Kors, N.,** Peroxidase and fluorescein isothiocyanate as antibody markers. A quatitaive comparison of two peroxidase conjugates prepared with Glutaraldehyde or periodate and a fluorescein conjugate, *J. Hist. Chem. Cytochem.,* 24, 1017, 1976.
93. **Farr, A. G. and Nakane, P. K.,** Immunohistochemistry with enzyme labeled antibodies: a brief review, *J. Immunol. Methods,* 47, 129, 981.
94. **Wilson, M. B. and Nakane, P. K.,** Preparation and standardization of enzyme-labeled conjugates, in *Immunoassays in Clinical Laboratory,* Nakamura, R. M., Ditto, W. R., and Tucker, E. S., III, Alan R. Liss, New York, 1979, 81.
95. **Avrameas, S. and Ternynck, T.,** Peroxidase labeled antibody and fab conjugate with enhanced intracellular penetration, *Immunochemistry,* 8, 1175, 1971.
96. **Van Weeman, B. K. and Schuurs, A. H. W. M.,** Immunoassay using antibody-enzyme conjugates, *FEBS Lett.,* 43, 215, 1974.
97. **Sarkar, M. and Mahato, S. B.,** Homogeneous enzyme immunoassay of diosgenin and its glycosides, *Anal. Biochem.,* 136, 168, 1984.
98. **Tsang, V. C. W., Hancock, K., and Maddison, S. E.,** Quantitative capacities of glutaraldehyde and sodium *m*-periodate coupled peroxidase-anti-human IgG conjugates in enzyme-linked immunoassays, *J. Immunol. Methods,* 70, 91, 1984
99. **Pain, D. and Surolia, A.,** Preparation of protein A-peroxidase monoconjugate using a heterobifunctional reagent, and its use in enzyme immunoassays, *J. Immunol. Methods,* 40, 219, 1981.
100. **Nilsson, P., Bergquist, N. R., and Grundy, M. S.,** A technique for preparing defined conjugates of horseradish peroxidase and immunoglobulin, *J. Immunol. Methods,* 41, 81, 1981.
101. **Arends, J.,** Purification of peroxidase-conjugated antibody for enzyme immunoassay by affinity chromatography on conconavalin A, *J. Immunol. Methods,* 25, 171, 1979.
102. **Imagawa, M., Hashida, S., Ishikawa, E., and Sumiyoshi, A.,** Evaluation of Fab′-horseradish peroxidase conjugates prepared using pyridyl disulfide compounds, *J. Appl. Biochem.,* 4, 400, 1982.
103. **Yoshitake, S., Imagawa, M., and Ishikawa, E.,** Efficient preparation of rabbit Fab′-horseradish peroxidase conjugates using maleimide compounds and its use for enzyme immunoassay, *Anal. Lett.,* 15, 147, 1982.
104. **Ruan, K.-H., Hashida, S., and Ishikawa, E.,** A micro-scale preparation of affinity-purified Fab′ enzyme conjugates with high purity for enzyme immunoassay, *Anal. Lett.,* 17, 2075, 1984.

105. **Berkovitz, D. B. and Webert, D. W.,** The inactivation of horseradish peroxidase by a polystyrene surface, *J. Immunol. Methods,* 47, 121, 1981.
106. **Porstmann, B., Porstmann, T., Gaede, D., Nugel, E., and Egger, E.,** Temperature dependent rise in activity of horseradish peroxidase caused by non-ionic detergents and its use in enzyme immunoassay, *Clin. Chim. Acta,* 109, 175, 1981.
107. **Halbert, S. P. and Lin, T.-M.,** Enzyme immunoassay of antibody, in *Enzyme-Mediated Immunoassay,* Ngo, T. T. and Lenhoff, H. M. Eds., Plenum Press, New York, 1985, 223.
108. **Butler, J. E., Peterman, J. H., and Kroertge, T. E.,** The amplified enzyme-linked immunosorbent assay (a-ELISA), in *Enzyme-Mediated Immunoassay,* Ngo, T. T. and Lenhoff, H. M., Eds., Plenum Press, New York, 1985, 241.
109. **Shimizu, S. Y., Kabakoff, D. S., and Sevier, E. D.,** Monoclonal antibodies in immunoenzymetric assays, in *Enzyme-Mediated Immunoassay,* Ngo, T. T. and Lenhoff, H. M., Eds., Plenum Press, New York, 1985, 433.
110. **Kohen, F., Amir-Zaltsman, Y., Strasburger, C. J., Bayer, E. A., and Wilchek, M.,** The avidin-biotin reaction in immunoassay, in *Complementary Immunoassays,* Collin, W. P., Ed., John Wiley & Sons, New York, 1988, 57.
111. **Auditore-Hargraves, K. and Hargreaves, W. R.,** In-office and self-testing, in *Complementary Immunoassays,* Collin, W. P., Ed., John Wiley & Sons, New York, 1988, 71.
112. **May, K.,** In-home testing, in *Complementary Immunoassays,* Collin, W. P., Ed., John Wiley & Sons, New York, 1988, 83.
113. **Papas, M. G.,** Dot enzyme-linked immunosorbent assays, in *Complementary Immunoassays,* Collin, W. P., Ed., John Wiley & Sons, New York, 1988, 113.
114. **Kitagawa, T. and Aikawa, T.,** Enzyme coupled immunoassay of insulin using a novel coupling reagent, *J. Biochem.,* 79, 233, 1976.
115. **Kato, K., Hamaguchi, Y., Fukui, H., and Ishikawa, E.,** Enzyme-linked immunoassay. I. Novel method for synthesis of the insulin-B-D-galactosidase conjugate and its applicability for insulin assay, *J. Biochem.,* 8, 235, 1975.
116. **Exley, D. and Abuknesha, R.,** The preparation and purification of a B-D-galactosidase-oestradiol-17B conjugate for enzyme-immunoassay, *FEBS Lett.,* 79, 301, 1977.
117. **Exley, D. and Abuknesha, R.,** A highly sensitive and specific enzyme-immunoassay method for oestradiol-17B, *FEBS Lett.,* 91, 162, 1978.
118. **Hamaguchi, Y., Kato, K., Fukui, H., Shirakawa, I., Okawa, S., Ishikawa, E., and Katunuma, N.,** Enzyme-linked sandwich immunoassay of macromolecular antigens using rabbit antibody-coupled glass rod as a solid phase, *Eur. J. Biochem.,* 71, 459, 1976.
119. **Castro, A. and Monji, N.,** Steric hindrance enzyme immunoassay (SHEIA), in *Enzyme-Mediated Immunoassay,* Ngo, T. T. and Lenhoff, H. M., Eds., Plenum Press, New York, 1985, 291.
120. **Freytag, J. W.,** Affinity column mediated immunoenzymometric assays, in *Enzyme-Mediated Immunoassay,* Ngo, T. T. and Lenhoff, H. M., Eds., Plenum Press, New York, 1985, 277.
121. **Butler, J. E.,** The immunochemistry of ELISAs: principles and applications for the quantitative determination of immunoglobulins, in ELISA and other solid phase immunoassays, Kemeny, D. M. and Challacombe, S. J., Eds., John Wiley & Sons, Chichester, England, 1988, 107.
122. **Van Weemen, B. K. and Schuurs, A. H. W. M.,** Immunoassay using antibiody enzyme conjugates, *FEBS Lett.,* 43, 215, 1974.
123. **Van Hell, H., Brqamds, J. A. M., and Schuurs, A. H. W. M.,** Enzyme-immunoassay of human placental lactogen, *Clin. Chim. Acta,* 91, 309, 1979.
124. **Schall, R. F., Jr., Fraser, A. S., Hansen, H. W., Kern, C. W., and Tenoso, H. J.,** A sensitive manual enzyme immunoassay for thyroxine, *Clin. Chem.,* 24, 1801, 1978.
125. **de Savigny, D. and Speiser, F.,** Enzyme immunoassays for tropical parasitic diseases, in *Enzyme-Mediated Immunoassay,* Ngo, T. T. and Lenhoff, H. M., Eds., Plenum Press, New York, 1985, 415.
126. **Tsang, V. C. W., Bers, G. E., and Hancock, K.,** Enzyme-linked immunoelectrotransfer blot, in *Enzyme-Mediated Immunoassay,* Ngo, T. T. and Lenhoff, H. M., Eds., Plenum Press, New York, 1985, 389.
127. **Hsu, S.-M.,** Immunoperoxidase techniques using the avidin-biotin system, in *Enzyme-Mediated Immunoassay,* Ngo, T. T. and Lenhoff, H. M., Eds., Plenum Press, New York, 1985, 467.
128. **Porstmann, B., Porstmann, T., Nugel, E., and Evers, U.,** Which of the commonly used marker enzymes gives the best results in colorimetric and fluorimetric enzyme immunoassays: horseradish peroxidase, alkaline phosphatase or B-galactosidase, *J. Immunol. Methods,* 79, 27, 1985.

Practical Applications

Chapter 6

COMPETITIVE, NONCOMPETITIVE, AND MULTI-ANALYTE MICROSPOT IMMUNOASSAYS

Roger P. Ekins

TABLE OF CONTENTS

I. INTRODUCTION

A. RADIO-ISOTOPIC "BINDING ASSAYS": A HISTORICAL OVERVIEW

All immunoassay methods rely on observation of the products of the binding reaction between an antigen (the analyte) and an antibody, from which the concentration of the analyte is inferred. (Note: the term "immunoassay" is also applied to methods designed to measure the amount of antibody in the system, in which circumstance the antibody itself constitutes the analyte. However, for the sake of clarity it will be assumed in this presentation that antibody is employed as a specific binding reagent to measure an analyte that is not itself an antibody.) Immunoassays fall within the wider class of techniques — frequently described as "binding assays" — which rely on observation of the reaction between the analyte and a specific binding substance, the latter being generally of biological origin and typically a specific binding protein. For example, hormone receptors have been used in this manner in techniques generally referred to as "receptor assays"; likewise, specific hormone binding proteins present in blood and other biological fluids have been similarly exploited. Antibodies have, in practice, comprised the group of binding proteins most extensively used in this context, primarily because they can be generated conveniently and in large amount (either by conventional *in vivo* methods or, more recently, by the *in vitro* techniques pioneered by Köhler and Milstein[1]) against a wide range of substances of biological interest, including both high molecular weight substances (e.g., proteins polypeptides, viral particles, oligonucleotides) and substances of relatively small molecular size ("haptens") such as drugs, steroid and thyroid hormones, vitamins, etc. Nevertheless, naturally occurring specific binding proteins continue to be used in "protein binding" assays relying on the same analytical principles as those underlying immunoassay, for example, in assays for vitamin D, vitamin B_{12}, folic acid, and (albeit occasionally) thyroxine. More recently, binding assays also essentially identical in principle to current immunoassay methods, but relying on DNA "probes" as the specific binding reagent, have been introduced, though they are restricted in application to the detection and/or measurement of complementary nucleotide sequences, and have not as yet achieved the widespread usage or commercial importance of antibody-based assay methodologies.

Immunoassays characteristically display very high "structural specificity", reflecting the extreme ability of appropriately selected antibodies to distinguish and to bind to antigenic sites (epitopes) on particular analyte molecules. Meanwhile, the use of high specific activity radioisotopic markers such as ^{125}I or ^{3}H (or, in the case of DNA probes, ^{32}P) to label either the antibody or (tracer) analyte employed in the assay system has permitted binding reactions between very small numbers of molecules to be readily monitored, providing the basis of these methods' high sensitivity. The combination of high specificity, sensitivity, and simplicity has underlain the ubiquitous use of radioisotopically based immunoassays such as radioimmunoassay (RIA) and immunoradiometric assay (IRMA, see below) in research and diagnosis in medicine and in many other fields (e.g., the food industry, agriculture, forensic science, and environmental monitoring), and methods of this genre have now been described for thousands of substances of biological importance.

Radioisotopic labels are relatively easily measured using sample-counting equipment which is now widely available, familiar, and inexpensive; moreover, radioactive decay is unaffected by the chemical or physical environment in which the radioisotope is situated. These attributes underlie the continuing popularity of isotopically based immunoassay methods. Nevertheless, a number of factors have led to the development of analogous nonisotopic techniques, this representing the principal focus of methodological interest in the immunoassay field in recent years. The reasons for this interest are addressed in Section I.D. However, before discussing this topic, it is useful to briefly examine the strategies available for observation of the basic binding reaction on which all binding assays are based.

B. LABELED ANTIGEN IMMUNOASSAY METHODS

The original isotopically based immunoassay methods relied on the use of radiolabeled antigen (i.e., radiolabeled analyte) which, when added to the reaction medium, revealed the extent of binding of the unlabeled analyte to a fixed amount of antibody present in the system. The first immunoassay conforming to this concept was described in 1960 by Berson and Yalow;[2] however, binding assays relying on identical analytical principles (albeit using naturally occurring binding proteins as the specific binding reagent) were independently developed at about the same time for thyroxine and vitamin B_{12} by the present author and colleagues.[3,4] Antibody-based methods relying on radiolabeled antigen are generally referred to as "radioimmunoassays" (RIA), analogous methods using other labels (e.g., fluorescent or enzyme labels) being described by corresponding terms, e.g., fluoroimmunoassay (FIA) and enzymoimmunoassay (EIA).

C. LABELED ANTIBODY IMMUNOASSAY METHODS

Radiolabeled antibody assays emerged a few years later than the RIA methods, largely because of the technical difficulties involved in the preparation, in reasonably pure form, of radiolabeled antibodies possessing the specificity and affinity characteristics required for high sensitivity analyte measurements. Wide in Sweden,[5] and Miles and Hales in the UK,[6] originally pioneered this particular approach in the later 1960s, albeit labeled antibody methodology per se was not itself novel, antibodies coupled to other markers (e.g., red cells) having previously been used for assay purposes. Radiolabeled antibody methods are generally known as "immunoradiometric assays" (IRMA) to distinguish them from those relying on labeled antigen (RIA), and corresponding terms (e.g., immunofluorometric, IFMA, and immunochemiluminometric assay, ICMA) are used to describe analogous methods using, for example, fluorescent and chemiluminescent markers to label the antibody on which the system relies.

For some time following their development, considerable discussion centered on the claim that labeled antibody methods (such as IRMA) are intrinsically more sensitive than the corresponding labeled analyte (e.g., RIA) methods.[7] In retrospect, it is evident that this debate — by focussing on the distinction between labeled antibody and labeled antigen methods — diverted attention from the more fundamental reasons why certain assay designs are likely to yield higher sensitivity than others. Inevitably, some of the arguments made in support of the above claim were fallacious; indeed, it is not the case that "immunometric" methods per se are more sensitive. This issue is discussed in greater detail in Section II.

D. REASONS FOR THE DEVELOPMENT OF NONISOTOPIC "BINDING ASSAYS"

1. "Nonscientific" Reasons (e.g., Environmental, Economic, Legal, Logistic)

Several "nonscientific" advantages of nonisotopic methods may be grouped under this heading, including, for example, avoidance of the problems of radioisotope disposal, the legal constraints imposed on the use of radioisotopes in many countries, and the limited shelf life of radioisotopically labeled immunoassay reagents, etc. Nonisotopic labels also permit the development of "nonseparation" or "homogenous" methods (i.e., methods avoiding the inconvenience of physical separation of the immunological reaction products prior to label measurement), and also of methods yielding colored reaction products, thereby obviating the need for complex electronic signal-measurement equipment and facilitating the development of simple diagnostic kits for home use (e.g., for pregnancy testing).

2. The Demand for "Ultra-Sensitive" Immunoassay Methods

Radioisotopically based techniques may be shown to be restricted to the measurement of substances which occur in test samples at concentrations in the order of 10^7 molecules per milliliter and above (other than by the use of large sample volumes and complex analyte-extraction and concentration techniques).[8] This limitation arises in part from physicochemical and statistical considerations unrelated to the nature of the label per se; however, in certain assay designs, the inability to measure radioisotopic labels at very low concentrations emerges as a key factor constraining assay sensitivity. The attainment of greater sensitivity — necessary, for example, in certain medical contexts — requires the use of labels of even higher specific activity than radioisotopes, and which therefore permit labeled reagents to be measured at significantly lower concentrations.

3. The Development of "Immunosensors"

Immunosensors comprise solid "probes" or "sensors" coated with specific antibody whose reaction with analyte in the test medium is transduced into a measurable electronic or optical signal. Transduction methods and devices currently under study rely on the measurement of surface plasmon resonance, evanescent wave detection, field-effect transistors, piezo-electric and surface acoustic wave devices, etc. Ideally, immunosensors should respond to changing analyte concentrations in the surrounding medium; however, the slow dissociation kinetics of high affinity antigen-antibody reactions constitute a major obstacle to the attainment of this particular objective. Other problems associated with immunosensor development are a lack of sensitivity and an inevitable reliance on single analyte-binding-site recognition by antibodies situated on the immunosensor surface, implying a lack of assay specificity. At best, therefore, such devices appear likely to prove of value only for the measurement of high-concentration analytes in special situations e.g., as bedside tests in hospital wards.

4. The Development of Multianalyte Immunoassay Methods

Developments under this heading reflect an increasing need to measure many different substances in the same sample, both in medicine and in many other biologically related fields, e.g., the food industry. The conventional approach to multianalyte immunoassay has relied on the use of different labels, each label being used to monitor a different immunological reaction. However, the number of independent immunological reactions that can be concurrently monitored in this way is restricted, due to limitations on the number of different labels that can be measured in the same sample. A more promising approach, permitting, in principle, the simultaneous measurement of tens or hundreds of different analytes in the same sample, relies on spatial separation of the products of the immunological reactions. This possibility arises as a consequence of the availability of very high specific activity, nonisotopic, reagent labels (e.g., fluorescent labels) which permit the development of "microspot" immunoassays in which a "sensor" or "capture" antibody is located within a very small area (in the order of 100 μm^2 or less) on the surface of a solid support.[9] This in turn permits the construction of multi-microspot arrays, each antibody microspot being directed against a different analyte, thus allowing a virtually unlimited number of analytes in samples of small volume to be measured.

II. IMMUNOASSAY DESIGN

A. PRINCIPLES OF IMMUNOASSAY DESIGN

Considerable controversy has arisen in the past regarding the optimal design of immunoassay systems (i.e., the choice of optimal concentrations and physicochemical characteristics of the principal reagents employed). This has stemmed primarily from disagreement regarding the meaning of terms commonly used to describe assay performance such as "sensitivity", "precision", "accuracy", etc. Much of the confusion regarding these terms has now been resolved (albeit they continue to be formally defined by different official bodies in contradictory ways); nevertheless, misconceptions persist in regard to immunoassay design, representing a legacy of past confusion regarding the concepts of assay sensitivity and precision, and hence the central objectives of assay optimization. These misconceptions are of considerable practical importance, since they bear upon such issues as the amount of antibody that should ideally be attached to the surface of solid supports used in an immunoassay context, and hence to the nature and properties of the solid supports themselves. To dispel some of the misunderstandings which persist in this area, it is necessary first to clarify the concepts underlying the proper assessment of immunoassay performance.

The objective of any immunoassay system is clearly to measure the amount or concentration of the substance of interest (the analyte) in a test sample. The objective of assay *design* is ensure that the analyte concentration is measured as "accurately" as possible (i.e., with minimal error, though see below) within the practical constraints imposed by time, cost, availability of reagents, equipment, etc. If the analyte concentration is relatively "low", the assay system must clearly be constructed (if possible) in such a way that it is capable of measuring "low" analyte concentrations accurately, i.e., it must possess higher "sensitivity" than a system designed to measure relatively high concentrations.

Although these propositions are unlikely to provoke dissent, they nevertheless involve fundamental concepts which, as indicated above, have been the subject of past confusion and controversy, and which therefore merit closer examination.

1. Concepts of Assay Sensitivity, Precision, Accuracy, Bias

In common English usage, an *accurate* measurement is one characterized by small error. The term will be used in this sense throughout this chapter, though certain authorities (e.g., the International Federation of Clinical Chemistry) formally define an "accurate" measurement as one which is unbiased (see below). Errors of measurement may usefully be categorized as either "systematic" or "random". A systematic error causes all measurements to be deflected from the true value, resulting in assay *bias*. Conversely, random errors cause measurements to scatter about a mean value which, assuming the assay system to be unbiased, approximates the true value. The extent to which results scatter about the mean — usually, but not necessarily, represented by the standard deviation (s.d.) or coefficient of variation (c.v.) of replicate measurements — reflects the *precision* of the analyte measurement. An assay system may clearly be both biased and imprecise, i.e., individual analyte concentration measurements may scatter randomly about a mean value which is itself deflected from the true or correct value. Meanwhile, a single determination of an (otherwise known) analyte concentration may be demonstrably incorrect (i.e., it may be "inaccurate"), but such a single, isolated, observation cannot by itself reveal whether the error incurred reflects imprecision or bias (or both) of the assay system. Only by repeating the measurement many times can estimates be made of each of these parameters.

In general, both the bias and (im)precision of analyte measurements vary in a consistent manner as a function of analyte concentration. (Note: both bias and precision may also vary between individual samples, because of variation in their individual chemical compositions and/or physical characteristics. These variations can, however, be disregarded in a discussion of basic concepts.) In principle, it is possible to represent the relationship between both bias and precision and the analyte concentration by means of bias and precision "profiles". In practice, however, it is usually impossible to estimate assay bias across the analyte concentration range of interest (since the true analyte concentrations in test samples are likely to be unknown); indeed, if the true "bias profile" were known, all measurements could be appropriately corrected, implying that errors due to assay bias could be eliminated. In the case of precision, however, the scatter of replicate measurements about a mean can invariably be estimated (assuming sufficiency of the test sample), implying that a "precision profile" can be constructed by estimating the standard deviation (or c.v.) of replicate measurements spanning the entire analyte concentration range of interest, albeit without exact knowledge of the true analyte concentrations the samples contain.[10]

Knowledge of the precision of measurement of any specified analyte concentration is of importance because, *inter alia,* it permits assessment of the probability that a measured con-

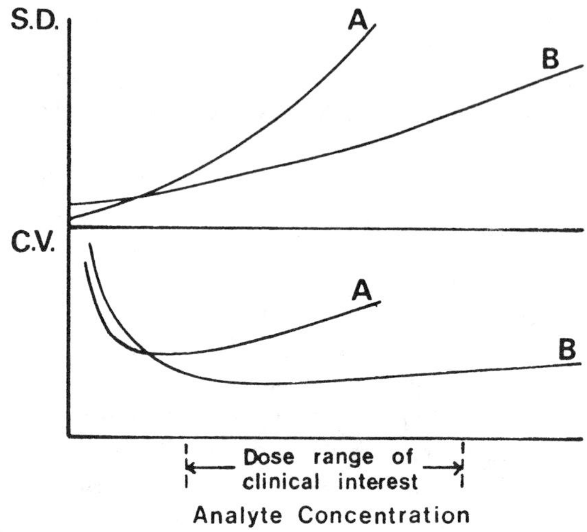

FIGURE 1. The "precision profile" of an assay portrays the error in the analyte concentration ("dose") estimate as a function of dose. This may be represented, *inter alia,* by the absolute error (e.g., the s.d. of the dose measurement (SD_D) or the relative error (e.g., the coefficient of variation (CV_D). The precision of measurement of zero dose represents the lower limit of detection, or sensitivity, of the assay. The working range may be defined as the range of dose estimates within which CV_D is less than a value acceptable to the investigator. Curves A and B represent precision profiles likely to be observed using, for example, low and high antibody concentrations, respectively in a typical RIA. Note that the assay system yielding profile A is the more sensitive, but that an assay system yielding profile B is more precise over the dose range of clinical interest. *Any* change in immunoassay procedure is likely to alter the shape of the precision profile in an analogous manner, thus only by observing the change in the shape and position of the profile can the effect on assay performance of a change in assay design be assessed.

centration *apparently* differing from the specified concentration *truly* differs from it. It defines, in short, the "minimal detectable difference" from the specific "basal" concentration. Of particular interest is the *lower limit of detection* of the assay system. This quantity essentially reflects the precision of the zero concentration (or "zero dose") measurement, which in turn determines the probability that a measurement genuinely differs from zero. It thus represents a special case of the general concept represented by the "minimal detectable difference" between any two analyte concentrations. The lower limit of detection or "detection limit" is now widely accepted as representing assay sensitivity. Thus, the more "sensitive" of any two assay systems is that yielding the smaller detection limit, i.e., the system which displays the greater precision of measurement of zero dose. In other words, the sensitivity of any assay system is essentially given by the zero dose intercept of the (s.d.) precision profile (see Figure 1).

Although this concept of sensitivity is embedded in much current scientific literature (as evidenced by the common representation of assay sensitivity in units of analyte amount or concentration), the term is also frequently used in a fundamentally different sense, i.e., as the slope of the dose response curve. Indeed, sensitivity has, in the past, been formally defined in this manner by many national and international bodies, some of whom currently continue to do so. Objections to this definition have been expressed by workers in diverse scientific fields (ranging from gravimetric analysis to radiation measurement), primarily on the grounds that a system characterized by a large response to a given stimulus or "dose" is not, *ipso facto,* one capable of detecting a small stimulus. Nevertheless, it represents a concept that is superficially persuasive, and has proved difficult to eradicate, particularly in relation to immunoassay design.

Paradoxically, the fallacy of equating sensitivity with the slope of the dose response curve is particularly well illustrated in the immunoassay field. Immunoassay data can be portrayed equally legitimately in terms of a wide variety of dose and response metameters; moreover,

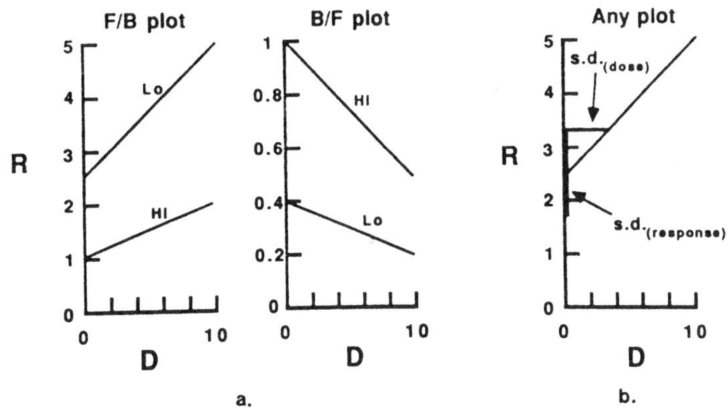

FIGURE 2. a. Diagrammatic representation of response curves yielded in a typical RIA system using "hi" and "lo" concentrations of antibody. Note that increasing the antibody concentration *increases* the slope of the bound/free (B/F) labeled antigen response curve (right), but *decreases* the slope when data are plotted in terms of F/B (left). b. Irrespective of the manner in which the dose response curve is plotted, the precision of measurement of zero dose (given by error in the measurement of the zero response divided by the slope of the response curve) remains unchanged.

the effects on the slope of the response curve of altering the design of the system will generally differ, depending on the particular dose and response metameters employed. For example, reduction of the antibody concentration in a conventional RIA *decreases* the slope of the curve when this is plotted in terms of the bound/free (B/F) labeled antigen ratio, but *increases* the slope when the curve is plotted in terms of F/B (see Figure 2a). The notion that the "sensitivity" of the assay decreases or increases depending on the manner in which the response curve is graphically represented is clearly nonsensical, and highlights the absurdity of this definition.

Despite its obvious invalidity, the notion that immunoassay sensitivity is represented by the response curve slope has exerted a profound effect on immunoassay development during the past 30 years, and continues to influence the way in which immunoassays (including most commercial immunoassay kits) are currently constructed. For example, this concept led Berson and Yalow to claim that maximal (RIA) sensitivity is attained using an antibody concentration given either by 1/K or by 0.5/K, where K is the antibody affinity constant.[11,12] (Not surprisingly, Berson and Yalow's conclusions varied during the 1960's and early 1970's essentially in consequence of their replacement of the {B/F antigen vs. dose} response curve by the {percent antigen bound v. dose} curve as the basis of their theoretical analyses.) The use of these antibody concentrations in an RIA procedure implies that, at zero dose, fractional binding of labeled antigen approximates to 50 and 33% respectively, assuming the labeled antigen concentration to be "vanishingly small".[11,12] Likewise, Berson and Yalow's RIA design theory was based on the concept that the precision of an immunoassay is given by the slope of the {percent antigen bound vs. log dose} curve.[11,12] Though frequently contested by the present author,[13,14,15] these precepts have been accepted by many workers in the field and most currently available commercial RIA kits continue to be designed in broad accordance with them.

2. Statistical/Physicochemical Concepts Underlying the Optimal Choice of Immunoassay Reagents

Clearly, the fundamental distinction between definitions of sensitivity and precision relying exclusively on the slope of the dose response curve, and those now more generally accepted in the field, is that the former totally disregards random errors arising in the measurement of the response. For example, as shown in Figure 2b, the s.d. of the zero dose measurement depends both upon the slope of the response curve and the s.d. of the zero-dose response measurement. In short, it is neither possible to estimate the detection limit of an assay, nor to take logical steps to maximize an assay's sensitivity, without appropriate statistical analysis of the errors arising in the measurement of the assay response.

An obvious prerequisite of an "accurate" immunoassay method is that the antibody should react only with the analyte among the substances present in test and standard samples, i.e., the antibody on which the assay is based should be "specific". This clearly constitutes an important aspect of assay design. However, it is inappropriate to discuss here the techniques of antibody selection and other measures conventionally employed to ensure that, as far as possible, immunoassay results are unaffected by the presence of "competing" substances and/or the influence of other extraneous factors. Of greater conceptual importance — since they underlie major differences in immunoassay methodology — are the strategies adopted to maximize immunoassay precision (an objective which embraces, as indicated above, the particular aim of maximizing immunoassay sensitivity). These include, for example, the choice of optimal concentrations of antibody and, where appropriate, labeled analyte, and the order (and timing) of addition of these key reagents to the assay system.

From the foregoing discussion, it is evident that an optimal assay design can only be selected on the basis of observation of the precision yielded by alternative designs of the analyte concentration, or range of concentrations, which the analyst wishes to measure. This *necessarily* requires the determination of the precision profiles yielded by the various assay designs under consideration. For example, if the analyst wishes to determine whether use of a greater or lesser amount of antibody in the assay system is advantageous, observation of the effects on the precision profile of varying the antibody concentration is essential (as illustrated in Figure 1).

Unfortunately, determination of the precision profile of an assay system necessitates the collection of relatively large amounts of data and appropriate statistical analysis;[15] this difficulty — coupled with past misunderstanding of the concepts of precision and sensitivity — has meant that, in practice, the choice of (supposedly) optimal immunoassay designs has almost invariably been based on observation of the shape and slope of the assay response curve. This has inevitably been the source of considerable mythology in the immunoassay field. A typical example illustrates the misconceptions that can derive from this approach. Figure 3a portrays the (theoretically calculated) dose response curves relating the percentage of radiolabeled antibody bound to analyte plotted as a function of analyte concentration.[16] The authors of this publication — pioneers in the development of immunoradiometric assays — concluded from this figure that "as the antibody concentration is increased, so the sensitivity of the assay is reduced," albeit they also commented that "the antibody concentration can be increased tenfold, i.e., from 0.1 to 1.0, without altering significantly the sensitivity of the assay."

These conclusions are immediately overturned simply by plotting identical data in terms of the *total amount* of radiolabeled antibody bound to the analyte, rather than the *percentage* of antibody so bound (see Figure 3B). This leads, *inter alia,* to the contrary finding that assay "sensitivity" is *increased* as the antibody concentration is increased. In reality, neither of these contradictory conclusions is likely to be valid, since only by taking errors in the determination of the response variable into account can the true effect of varying the antibody concentration on assay sensitivity be assessed.

This example illustrates a general rule, i.e., that virtually *any* conclusion relating to

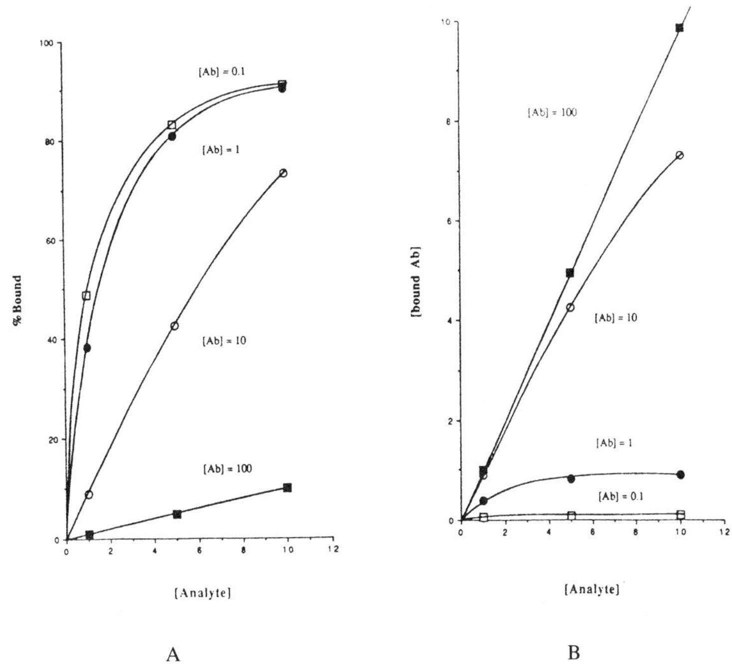

FIGURE 3. Effect of change in labeled antibody concentration on response curves relating % antibody bound to analyte concentration.[16] The authors identified the slope of the response curve with sensitivity, and thus judged a labeled antibody concentration of 0.1 as yielding highest sensitivity. b. Identical data plotted as concentration of antibody bound against analyte concentration. Note that a labeled antibody concentration of 0.1 now yields the response curve of least slope. (All concentrations expressed in units of 1/K, where K = the antibody affinity constant.)

immunoassay design based on observations of the shape and slope of the dose response curve alone can be reversed simply by plotting the curve in a different coordinate frame. Nevertheless, the development of the immunoassay field during the past 30 years has been profoundly influenced by fallacious reasoning of the kind exemplified above, its basic failing being a disregard of the statistical considerations which underlie indicators of assay performance such as sensitivity, precision, etc. It should be borne in mind in this context that, if all steps in an immunoassay were error free, no optimal assay design would exist. Thus, only by analysis of the statistical errors arising in an immunoassay procedure can a genuinely optimal design be identified.

3. "Competitive" and "Noncompetitive" Immunoassay Designs

The introduction of many new immunoassay methodologies during the past 2 decades has resulted in a corresponding proliferation in the nomenclature by which the various methodologies are commonly known. Though certain of the acronyms (e.g., RIA, IRMA, ELISA, IFMA) which now exist in this field provide a useful, albeit purely empirical, indication of the nature of the assay system (for example, the type of label on which the system relies and the particular component which is labeled), many reflect little more than trivial technical features characterizing assay kits marketed by individual manufacturers. In particular, they generally disregard one of the more fundamental distinctions relating to immunoassay methodology, i.e., that which exists between assay systems of so-called "competitive" and "noncompetitive" design. This unsatisfactory terminology historically derives from the manner in which the principles governing the original RIA and related methods (relying on other binding proteins) have often been portrayed. Such portrayals embody the notion that RIA depends on the "competition" between unlabeled analyte molecules and a fixed number of labeled analyte molecules for a limited number (also fixed) of antibody binding sites. This concept — extended to assay systems in which binding proteins other than antibodies comprise the specific binding reagent — led to the more general description "competitive protein-binding assay" (CPBA). Though various objections have been raised to this terminology (e.g., the criticism that the labeled analyte is often chemically identical to the unlabeled material, thus serving as a "tracer" of the latter's reaction with the binding reagent, and not as a true chemical competitor) the description "competitive assay" has nevertheless become firmly established in the immunoassay field.

In contrast to assays of this genre, those in which the antibody itself is labeled (typified by IRMA) are often described as "noncompetitive". Such assays are not generally portrayed as involving competition of labeled and unlabeled molecular species, and do not (necessarily) depend on the use of a limited amount of antibody. For this reason, the descriptions "labeled antibody assay" (or "immunometric assay") and "noncompetitive immunoassay" are commonly regarded as essentially synonymous. This simple convention is contradicted, however, by the description of certain labeled antibody methods as "competitive" (e.g., "competitive enzyme-linked immunoassay", CELIA[17]), apparently on the grounds that, like RIAs, they rely on the use of a restricted amount of labeled antibody in the system.

In short, the term "competitive" as commonly applied to immunoassay methodology encompasses two different concepts. The first is that labeled and unlabeled analyte molecules "compete" for a limited number of antibody binding sites; the second, that the amount of (labeled) antibody employed in the assay system is "small". However, the fundamental reasons underlying the existence of two alternative immunoassay strategies, and the essential differences between them, are obscured by these concepts, leading to confusion regarding the fundamental distinctions between what, for want of better terms, may be described as "competitive" and "noncompetitive" systems. For example, the optimal amount of antibody employed in any immunoassay is dependent on several factors, including the analyte concentration range on which the system is targeted, the specific activity of the label used, the random experimental errors incurred in the assay procedure, etc. This effectively precludes a clear distinction between "competitive" and "noncompetitive" assays simply on the basis of the amount of antibody used in the system (albeit certain assay designs often permit, in practice, the use of considerably higher concentrations of antibody than others).

To clarify the fundamental reasons underlying the categorization of immunoassays in this way, it is useful to portray the principles of immunoassay somewhat differently from the manner in which they are customarily represented.

4. The Antibody Occupancy Principle of Immunoassay

All immunoassays essentially depend upon measurement of the "fractional occupancy" by analyte of antibody "receptor" binding sites following reaction of analyte with a "sensor" or "capture" antibody CAb; see Figure 6. Thus, if the fractional occupancy of (monoclonal) antibody binding sites by analyte at thermodynamic equilibrium is represented by F, it may readily be shown that:

$$F^2 - F(1/[Ab] + [An]/[Ab] + 1) + [An]/[Ab] = 0 \tag{1}$$

where [An] = analyte concentration, [Ab] = antibody concentration (both expressed in units of 1/K).*

Fractional occupancy of antibody may be determined in one of two basically different ways: (1) the occupied sites may be determined *indirectly*, i.e., by measurement of *unoccupied*

* It is frequently useful, in discussing basic immunoassay theory, to represent concentrations in terms of units of 1/K, where K is the affinity constant of the antibody. This has the effect of generalizing the conclusions reached.

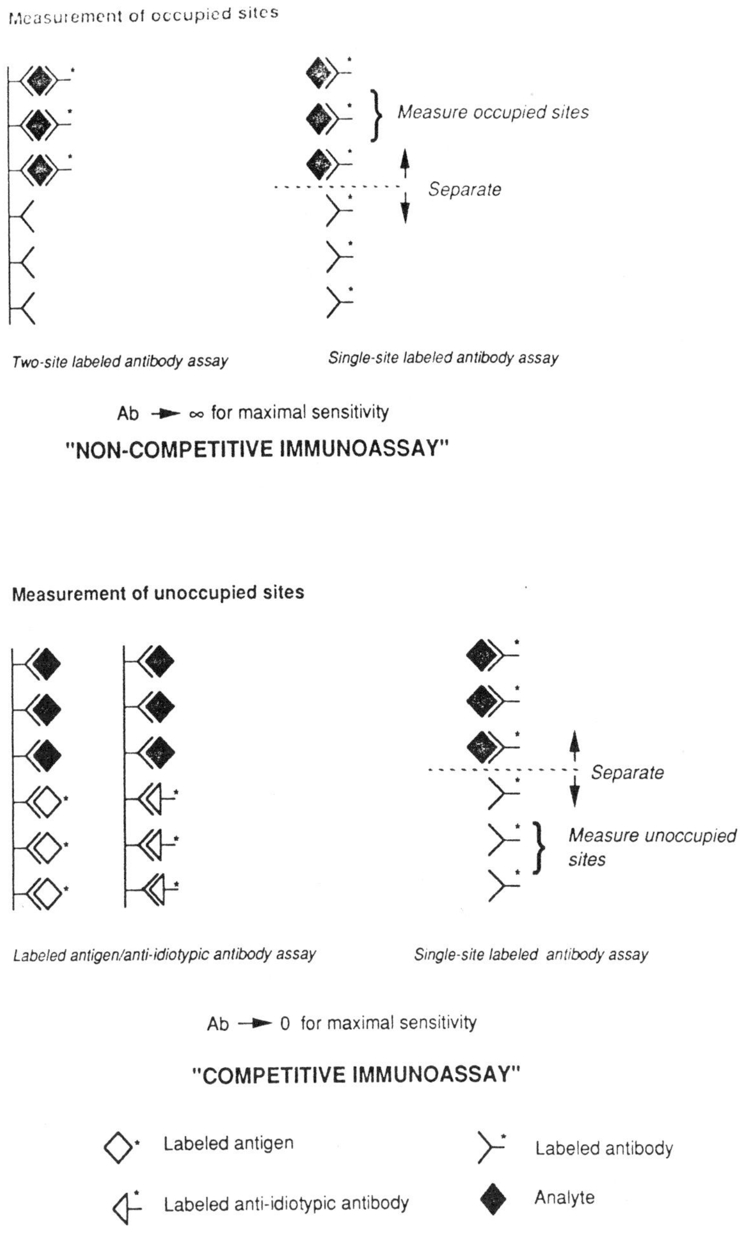

FIGURE 4. "Noncompetitive" assays rely on direct measurement of antibody binding sites occupied by analyte; "competitive" assays rely on measurement of unoccupied sites, from which the occupied sites are estimated essentially by subtraction. Labeled antigen (below left) or labeled anti-idiotypic antibody methods (below center) rely on measurement of sites unoccupied by analyte, and are therefore of "competitive" design. Single site labeled antibody methods (right) may be either of noncompetitive or competitive design, depending on whether labeled antibody bound or unbound to analyte is measured following separation of the two fractions, e.g., by immunosorbtion.

sites (from which the fraction of occupied sites is inferred, essentially by subtraction) or (2) the occupied sites are measured *directly* (see Figure 4). Viewed in this light, it is evident that, because of experimental errors, methods relying on measurement of residual, unoccupied, antibody binding sites optimally necessitate capture-antibody concentrations tending towards zero as the analyte concentration itself approaches zero (under which circumstance fractional occupancy is maximal, and its value is measured with maximal precision). Such techniques may be termed "competitive". Conversely, those techniques in which occupied sites are *directly* measured often permit (but do not necessarily require) the use of relatively high CAb concentrations and may thus be termed "noncompetitive".

Figure 4 also illustrates the basic strategies currently employed for the measurement of unoccupied or occupied antibody binding sites, In practice, minor variations in immunoassay design arise in consequence, for example, of differences in the order in which the key reagents are introduced into the assay system. Thus, in the case of conventional labeled antigen methods (Figure 4, bottom left), labeled antigen (or labeled anti-idiotypic antibody) may be introduced into the system either together with, or some time following, the introduction of the analyte itself. This difference in protocol affects the detailed design of the assay (permitting, in particular, the use of a higher concentration of labeled analyte), and may result in slightly enhanced assay sensitivity. Nevertheless, whatever order of addition of the reagents is adopted, the basic function of the labeled antigen is the determination of binding sites *unoccupied* by unlabeled analyte; such methods are therefore always of "competitive" design. Simple labeled antibody methods (Figure 4, right) may, in contrast, be described as "competitive" or "noncompetitive", depending on whether, following separation of unoccupied from occupied sites (using, for example, a solid phase immunosorbant), occupied sites are either measured "directly" (i.e., by measurement of the labeled antibody remaining in solution) or "indirectly" (i.e., by measurement of labeled antibody attached to the immunosorbant). The differing statistical implications of these alternative approaches to antibody-occupancy measurement determine, *inter alia,* the antibody concentrations which are optimal in each case, and underlie the description of immunoassays as "noncompetitive" or "competitive".

"Two site" or "sandwich" assays of the kind portrayed in Figure 4 (top left) can be considered from two points of view, since they rely on the use of two antibodies, each being (usually) directed against a different antigenic site on the analyte molecule. For the purpose of the present discussion, it is convenient to designate the unlabeled antibody situated on the solid support as the "CAb", whose occupancy by analyte is determined by reaction with a labeled antibody (i.e., a "developing antibody") which is used to quantify occupied capture antibody binding sites. However, it is equally legitimate to regard the labeled antibody itself as the "CAb", the function of the solid-phase antibody being to sequester occupied labeled-antibody binding sites. In either case, if the label attached to the solid support (i.e., that present in the antibody-analyte-antibody "sandwich") is measured, the assay can be described as "noncompetitive". (Note, however, that if the labeled antibody *remaining in solution* were measured, the design of the assay, i.e., the optimum concentration of labeled antibody used, would radically differ. In this case, the assay could be described as "noncompetitive" if the solid phase antibody were designated as the CAb, but as "competitive" if the labeled antibody were so regarded.)

Such considerations demonstrate that the fundamental distinction between so-called "competitive" and "noncompetitive" immunoassay strategies (and the key factor which underlies the differing sensitivities and other performance characteristics they yield) is unrelated to which component, if any, of the reaction system is labeled. Indeed immunosensors (which do not involve the use of a reagent label in the conventional sense) may likewise be of "noncompetitive" or "competitive" design depending on whether a transduced signal is generated when antibody binding sites are occupied or unoccupied (see Figure 5).

A further corollary of these concepts is that, were measurement of antibody occupancy to be entirely error free, no antibody concentration would be optimal, nor would any sensitivity

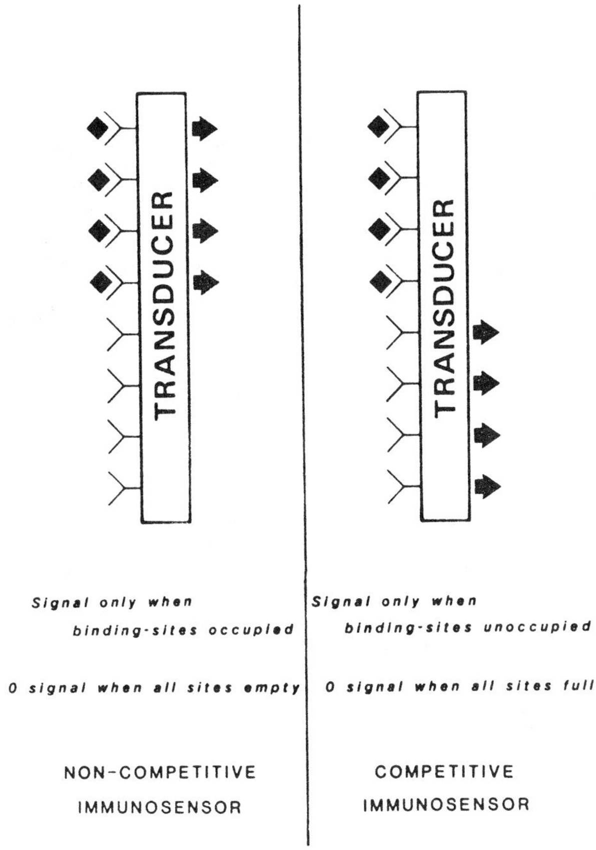

FIGURE 5. Noncompetitive and competitive immunosensors. A transduction system can be regarded as a form of label, and an immunosensor as a labeled antibody system. Note the correspondence between this figure and single-site labeled antibody systems illustrated in Figure 4 (right).

differences stem from the adoption of "competitive" or "noncompetitive" approaches to the measurement. In short, optimal immunoassay designs and measurement strategies arise primarily in response to the need to minimize imprecision in assay results deriving from the random errors arising in the determination of antibody occupancy, a particular aim being maximization of precision of the zero analyte measurement (i.e., maximization of assay sensitivity). Consequently, when an immunoassay system relies on the determination of *unoccupied* binding sites, the amount of CAb used must be small (tending, in principle, to zero) to minimize error in the resulting estimate of occupied sites by analyte concentrations themselves tending to zero. Conversely, when occupied sites are measured "directly", considerable advantage *may* derive from using relatively large amounts of CAb in the system (but see below). Indeed, in an "ideal" system (i.e., one in which, following reaction between analyte and antibody, *only* occupied antibody sites emit a signal, and are measured), the optimal capture antibody concentration yielding maximal sensitivity tends to infinity.

Theoretical models (incorporating estimates of such factors as experimental errors in pipetting, nonspecific binding of labeled antibody or antigen, etc.) reveal that competitive and noncompetitive immunoassays are likely to differ significantly in many of their performance

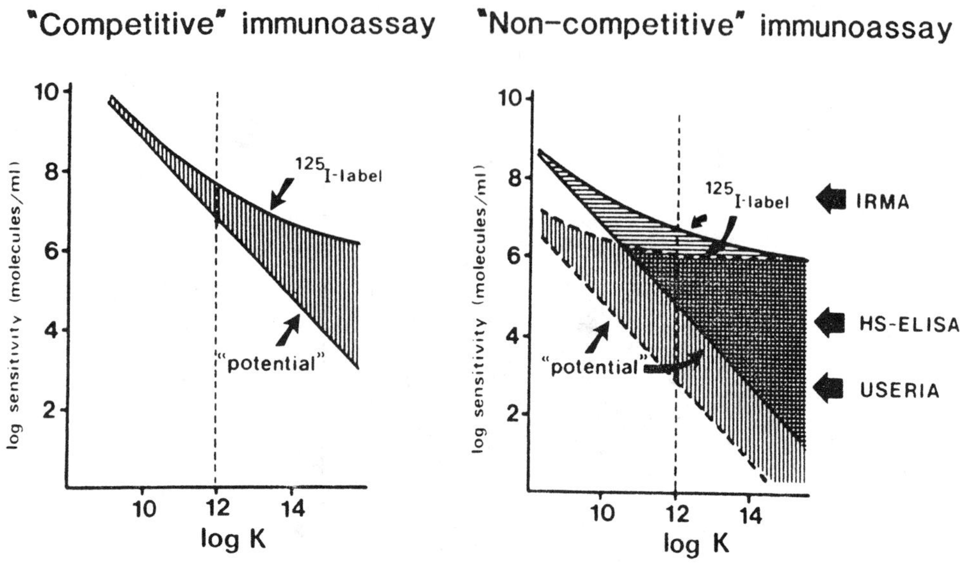

FIGURE 6. Theoretically predicted sensitivities of competitive and non-competitive immunoassay methods plotted as the logarithm of the detection limit (molecules/ml) against log K. (Note: in the case of noncompetitive sandwich assays, K refers to the affinity of the labeled antibody.) Calculations relating to competitive assays assume that pipetting and other manipulation errors result in a 1% error in the measurement of the response variable. The "potential" sensitivity curve assumes the use of a label of infinite specific activity, implying that measurement of the label per se is without error. The "^{125}I label" curve indicates sensitivities attainable taking into account the statistical error incurred in counting ^{125}I disintegrations for a "reasonable" sample counting time. The shaded area indicates the loss in sensitivity arising in consequence of statistical counting errors. Thus, little increase in sensitivity can be gained by using labels of higher specific activity than ^{125}I using antibodies with affinities $\leq 10^{12}$ l/M. In the case of noncompetitive assays, the "potential" sensitivity curves relate to nonspecific binding of labeled antibody of 1% (upper curves) and 0.01% (lower curves), and illustrate the increase in sensitivity derived by minimizing nonspecific binding to (for example) solid CAb supports. The "^{125}I label" curves emphasize the much greater loss in sensitivity (relative to that potentially attainable using a noncompetitive approach) resulting from the use of a radioisotopic label of relatively low specific activity, and the greater advantages of using high specific activity labels in this context. Arrows indicate sensitivities reported for noncompetitive immunoassays based on ^{125}I (IRMA), and enzyme relying on fluorogenic (HS-ELISA[23]) and radioactive (USERIA[22]) substrates. (Note: the measuring instrument background has been disregarded in the calculations on which this figure is based.)

characteristics, including their potential sensitivities (see Figure 6). They also differ in the times required to attain (approximate) thermodynamic equilibrium, noncompetitive methods generally requiring shorter incubation times in consequence of the permitted use of higher CAb concentrations. However, increasing the antibody concentration in an immunoassay system also results in a loss of specificity, since other antigens cross-reacting with antibody (albeit with lower affinity) exhibit an increasing potency relative to that of the analyte itself as the antibody concentration is raised.[13,18]

As indicated in Figure 6, an important factor governing the sensitivity of noncompetitive

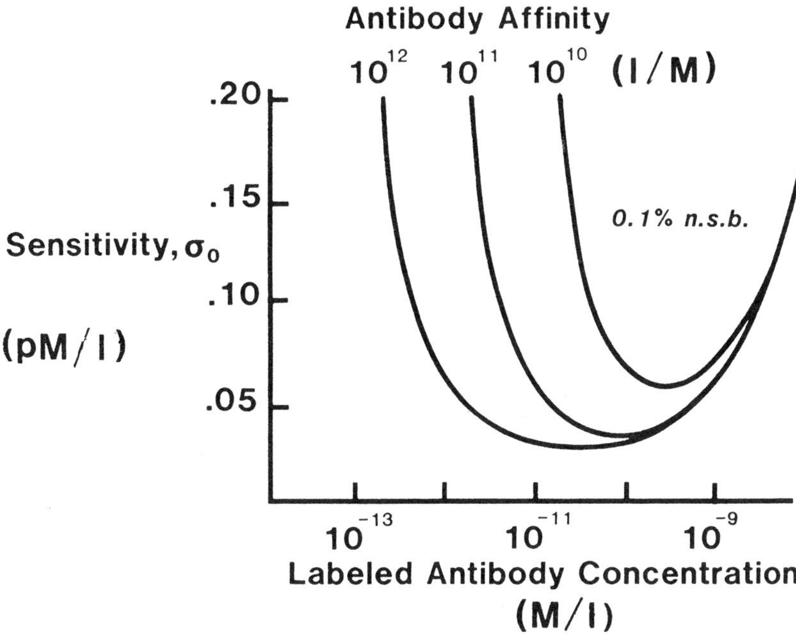

FIGURE 7. Theoretical sensitivity [represented by the standard deviation of the zero dose measurement (σ_0)] of a noncompetitive "sandwich" assay as a function of labeled antibody concentration, assuming nonspecific binding of the labeled antibody to the solid support to be 0.1%. Note that the detection limit falls to a minimum and thereafter increases (i.e., sensitivity is reduced) as the labeled antibody concentration is increased. (from Jackson, T. M., Marshall, N. J., and Ekins, R. P., *Immunoassays for Clinical Chemistry,* Hunter, W. M. and Corrie, J. E. T., Eds., Churchill Livingstone, Edinburgh, 1983, 557. With permission.)

methods is the "blank" or "background" signal emitted in the absence of analyte, since error in the measurement of this signal is clearly the determinant of the error in measurement of zero analyte. Among contributors to the background signal are the "noise" of the measuring instrument itself, signal generators (such as, in "sandwich" immunoassays, solid CAb supports or, in the case of radioisotopic methods, cosmic ray and other extraneous radiation sources) and "nonspecifically bound" labeled antibody (i.e., labeled antibody appearing in the analyte-bound fraction in the absence of analyte). Minimization of each of these components is clearly essential to the attainment of maximal sensitivity, mere arithmetic subtraction of the blank signal being of no benefit. Nonspecific binding of labeled antibody in sandwich assay designs is of particular importance in this context, since, assuming a constant fraction of the total labeled antibody in the system is nonspecifically bound, an increase in the labeled antibody concentration results in a proportional increase in the blank signal arising from this source, albeit the "specific" signal (deriving from labeled antibody bound to occupied CAb binding sites) increases to a lesser extent. Thus, an increase in the labeled antibody concentration above a certain value generally causes a fall in sensitivity (see Figure 7), in consequence of the resulting progressive reduction in the signal/noise ratio.[19]

The optimal CAb concentration used in a noncompetitive assay (and, in sandwich assay designs, the optimal labeled antibody concentration) are thus crucially dependent on the effect on the background noise (deriving from the measuring instrument itself, solid CAb supports, nonspecific binding of labeled antibody, etc.) as antibody concentrations are varied. The nature of the solid support used in sandwich assay systems, and the mode of attachment of antibodies to its surface, are of crucial importance in this context, since such factors as the surface density of CAb, and the background signal deriving both from the solid support itself, and labeled antibody nonspecifically adsorbed to it, are key determinants of optimal CAb and labeled antibody concentrations, and the sensitivity of the system. However, before discussing this point in greater detail, it is appropriate to consider another analytical concept of importance in immunoassay design.

5. "Ambient Analyte" (Sample Volume Independent) Immunoassay

The recognition that all immunoassays essentially rely on measurement of antibody occupancy leads to a second important analytical principle, termed "ambient analyte immunoassay".[20] This term represents a form of immunoassay system which, unlike conventional methods, quantifies the analyte *concentration* in the medium to which the antibody is exposed, being essentially independent both of sample volume, and of the amount of antibody present. The basis of this concept is illustrated in Figure 8 in which the fractional occupancy of ("monospecific" or "monoclonal") antibody binding sites in the presence of varying analyte concentrations, plotted against antibody concentration, is portrayed. This figure reveals that, when an antibody concentration less than (say) 0.01/K (preferably, but not essentially, coupled to a solid support) is exposed to an analyte-containing medium, the resulting (fractional) occupancy of antibody binding sites is independent of the antibody concentration per se, and solely reflects the "ambient" analyte concentration in the medium to which the antibody is exposed. (Note: for the sake of generality, all concentrations in Figure 8 are expressed in terms of 1/K, where K = the affinity constant of the antibody. For example, if $K = 10^{11}$ l/M, a concentration of $0.1 \times 1/K$ represents 0.1×10^{-11} M/l, or $0.1 \times 10^{-11} \times 10^{-3} \times 6.02 \times 10^{23} = 6.02 \times 10^{8}$ molecules/ml.) Analyte binding by antibody inevitably causes analyte depletion in the surrounding medium but, provided the proportion bound to antibody is small, reduction in the ambient analyte concentration can be ignored. For example, if the amount of capture antibody used is less than 0.01/K, then analyte depletion in the medium is invariably less than 1% (regardless of the initial analyte concentration), and the system is thus effectively sample-volume independent. (Note: although an upper antibody-concentration limit of 0.01/K has been specified to illustrate these concepts, higher antibody concentrations are clearly permissible, albeit they cause greater analyte depletion and, since such depletion may significantly differ in different samples, a greater potential source of error in assay measurements. Nevertheless, greater errors may be acceptable in many situations, and the antibody concentration limit is therefore somewhat flexible.)

The physicochemical concepts underlying "ambient analyte" immunoassay are of considerable potential value in a variety of contexts. For example, they provide the basis for the development of "immunosensors" capable of monitoring ambient analyte concentrations in fluids in which they are immersed, assuming the ultimate emergence of transduction systems of sufficient sensitivity to reveal the fractional occupancy of antibody molecules situated on the sensor surface. Identical concepts also underlie the development of the dual-label, "multianalyte" immunoassay systems discussed below. However, the question arises: are the capture-antibody concentrations appropriate to ambient analyte immunoassay (which differ considerably from those conventionally employed in both "competitive" and "noncompetitive" assay designs) reconcilable with the achievement of high assay sensitivity? In other words, is a concentration of antibody of 0.01/K (or less) far removed from the optimal concentration required to maximize immunoassay sensitivity?

As indicated earlier, the answer to this question is complex, being dependent on a number of factors (such as the specific activity of the label, the properties of solid supports for capture antibodies, etc.) whose implications cannot be examined in detail here. Nevertheless, examination of this issue is particularly appropriate to the subject of this volume, since it illustrates the possibilities opened up by the use of solid-phase antibody systems in the light of a clear understanding of immunoassay design. In particular, the proposition that, contrary to generally accepted beliefs, the use of exceedingly low concentrations of capture antibody is not incom-

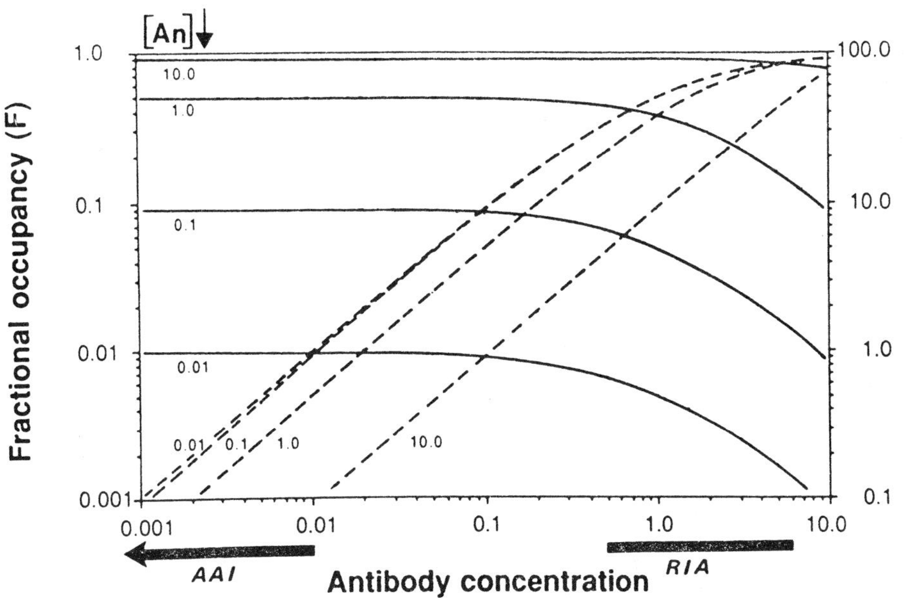

FIGURE 8. Fractional antibody binding-site occupancy (F) plotted as a function of antibody binding-site concentration for different values of analyte (antigen) concentration [An]. The percentage binding of analyte to antibody is also shown. All concentrations are expressed in units of 1/K. Note that for antibody concentrations of less than 0.01/K (approximately), percentage binding of analyte is <1%; also that fractional binding site occupancy is essentially unaffected by variations in antibody concentration extending over several orders of magnitude, being goverened solely by [An]. Note that radioimmunoassays and other "competitive" immunoassays are commonly designed using antibody concentrations approximating 0.5/K to 1/K or above (implying >30% binding of analyte concentrations ≈0), in accordance with the precepts of Berson and Yalow.[12]

patible with the achievement of very high immunoassay sensitivity can be illustrated by consideration of a model solid-phase capture antibody assay system.

For the sake of simplicity, let us postulate that CAb molecules in the system are attached to a solid support in such manner that their binding sites remain exposed to the analyte, and that their affinity vis-a-vis the analyte is unchanged. The antibody "concentration" in the system (given by the number of molecules on the support divided by the incubation volume) is unaffected by such attachment, and antibody occupancy by analyte at equilibrium will be identical to that occurring were the antibody to be distributed uniformly throughout the incubation mixture.*

Let us also suppose that the antibody molecules exist as a monolayer of maximal density on the solid surface. Thus, a change in the antibody concentration in the assay system merely implies a corresponding change in the total surface area over which antibody is distributed. If, for example, the antibody affinity constant is 10^{11} l/M, then a concentration of 0.1/K represents approximately 6×10^8 molecules per milliliter as described above; moreover, if the incubation volume is 1ml, and the maximal antibody surface density on the solid support is 6000 molecules per square micrometer, then a surface area of 10^5 μm^2 (i.e., 0.1 mm^2) would entirely accommodate antibody molecules corresponding to a concentration in the system of 0.1/K. Likewise, an area of 0.01 mm^2 would accommodate antibody molecules corresponding to a concentration of 0.01/K, etc.

Let us further postulate that, following exposure of the antibody to a solution of analyte at a concentration of 0.01/K (i.e., 6×10^7 molecules per milliliter) we measure "noncompetitively" the occupancy of the solid-phase CAb (e.g., by exposing the antibody to a second, labeled, antibody directed against the analyte, forming a typical antibody sandwich). Let us also suppose that essentially all occupied sites react with the labeled antibody, while labeled antibody binds "nonspecifically" to the solid support itself, at a surface density of 1 molecule per square micrometer.

We may now consider the effects of progressive reduction of the antibody-coated surface area from, for example, 1 mm^2 (effective antibody concentration 1/K) through 0.1mm^2 (0.1/K) to 0.01 mm^2 (0.01/K) and below. From Equation 1, the value of F for the 1mm^2 area is 4.98 $\times 10^{-3}$. Thus, the total number of labeled antibody molecules specifically bound to the area = 2.99×10^7, while the number nonspecifically bound = 10^6. Thus, assuming the "field of view" of the labeled-antibody detecting instrument is restricted to the area on which the capture antibody is deposited (see Figure 9a), and assuming the background (or "noise") of the instrument itself is zero (i.e., the only source of background is the nonspecifically bound antibody within the instrument's field of view), the signal/noise ratio observed for the 1 mm^2 area approximates 30. Similarly, the value of F for a 0.1 mm^2 area is 9.02×10^{-3}, the number of labeled antibody molecules specifically bound to the area = 5.41×10^6, the number nonspecifically bound = 10^5, and the signal/noise ratio approximates 54. Likewise the signal/noise ratio for a 0.01 mm^2 area can be shown to approximate 59. Thus the signal/noise ratio *increases* as the antibody-coated surface area is decreased, approaching a maximal (plateau) value of 60 as the area of CAb falls below 0.01mm^2 and approaches zero. Thus, if, albeit simplistically, we accept the signal/noise ratio as indicative of the precision of the antibody-occupancy measurement, these considerations suggest that it is advantageous to reduce the antibody-coated surface area (and hence the CAb concentration) to zero, albeit little advantage accrues from reducing the concentration below 0.01/K.

An obvious paradox underlies the first of these conclusions, since if the antibody-coated

* It is readily demonstrable that antibody binding site occupancy by analyte at equilibrium is unaffected by the location of antibody molecules within an incubation mixture. Thus, if the molecules are situated on the surface of an inert solid support within the mixture, their binding to analyte at equilibrium is unaffected, although the rate at which equilibrium is reached is reduced.

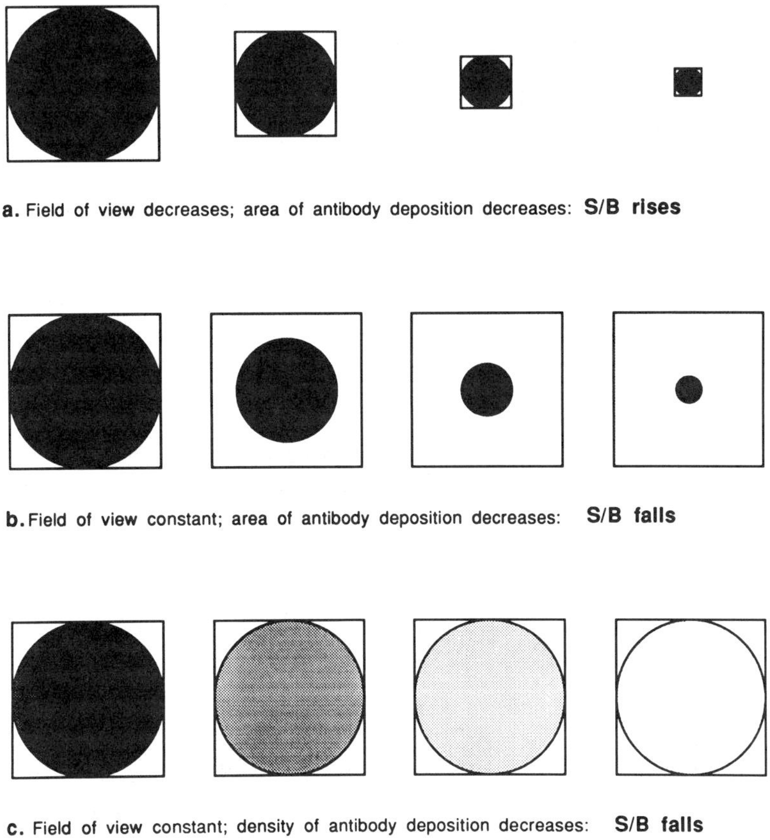

FIGURE 9. CAb is assumed coated on circular (shaded) areas; the field of view of the signal-measuring instrument being represented by square (shaded) areas. In situation (a), reduction of both the area of deposition of CAb and the field of view results in an increase in the signal/noise (S/B) ratio. If the CAb is reduced either by reducing the antibody-coated area (b), or the density of antibody coating (c), while the field of view remains unchanged, S/B falls.

area were reduced to zero, both signal and noise would likewise reduce to zero, implying that nothing would be measured. In reality, other statistical factors come into play if the antibody coated area is so reduced that the number of individual events observed by the detecting instrument becomes very small, prohibiting reduction of the antibody concentration to zero in the manner that the above considerations appear to suggest. The point at which further reduction in the antibody-coated area causes such loss of measurable signal that precision deteriorates depends on the specific activity of the labeled antibody used to measure occupied CAb binding sites, the higher the specific activity, the smaller the permissible area. It is evident, in short, that circumstances exist in which, even in a noncompetitive immunoassay system, the optimal CAb concentration in the system may be exceedingly low, lying well within the concentration range defining an "ambient analyte" immunoassay system. This concept underlies the "microspot" immunoassay methodology currently under development in the author's laboratory, discussed in greater detail below.

It should be noted, however, that if, in the model system considered above, reduction in the antibody-coated surface area were *not* accompanied by a corresponding reduction in the field of view of the detecting instrument, reduction in "signal" would not be accompanied by a corresponding reduction in the background caused by nonspecific binding (see Figure 9b,c).

Therefore, although reduction in the coated area would result in an increase in the fractional occupancy of the CAb, this would not necessarily result in an increase in the signal/noise ratio. In such circumstances it might prove desirable to *increase* the amount of antibody in the system. Likewise, if the surface *density* of CAb were decreased (the deposition area being held constant) similar considerations would apply. They would also apply if the inherent noise of the instrument itself were both relatively high and remained unchanged irrespective of the instrument's field of view.

These considerations exemplify the complex factors underlying the choice of an optimal concentration of CAb. In particular, it should be noted that each of the three approaches to antibody reduction shown in Figure 9 results in identical antibody concentrations in the incubation mixture, but that the *optimal* concentration is likely to differ markedly in the case of approach *a*. Furthermore, given appropriate instrumentation and a very high specific activity label, it is evident that this approach is not only likely to yield highest assay sensitivity, but will do so as the amount of CAb in the system approaches zero. This contradicts traditional beliefs which have grown up in this area, which have been largely based on practical experience with radioisotopic labels and counting equipment. Indeed, it is evident that certain accepted ideas relating to high sensitivity immunoassay are of dubious validity, and are likely to require substantial revision.

B. IMMUNOASSAY DESIGN: SOME PRACTICAL ASPECTS

1. Alternative Labels and Signal Detection Systems

a. General

A variety of approaches have been used to measure antibody binding site occupancy, some relying on a labeled reagent (e.g., the analyte itself, a closely related analogue, or an anti-idiotypic antibody) to identify unoccupied sites; other relying on a second labeled antibody directed against the analyte itself to identify occupied sites; and yet others in which the capture antibody is itself labeled, immunosorbents being used to separate unoccupied from occupied sites. However, the fundamental principles governing immunoassay do not depend on the use of a labeled reagent, labels being employed in this context solely to render the reactions between small numbers of analyte and antibody molecules more readily "visible". Clearly, for high sensitivity, the specific activity of labeled reagents should be high, implying that the binding reactions between small numbers of molecules may be monitored with high precision. However, increase in specific activity beyond a certain point is likely to yield diminishing returns in regard to assay sensitivity, since other factors also affect the precision of binding measurements, and ultimately become dominant as the specific activity of the label is progressively increased.

Such factors generally differ in relative importance depending on the basic design of the assay. In competitive designs, variations in the total amount of antibody used in the system, and other experimental (or "manipulation") errors incurred in the determination of unoccupied sites, imply that, even using a label of infinite specific activity, it is impossible to achieve a detection limit lower than ε/K, where ε = the c.v. in the measurement of the response variable (e.g., percentage of labeled analyte bound to antibody), and K = the antibody affinity constant. For example, if the overall "manipulation" error incurred in measurement of the assay response is 1% ($\varepsilon = 0.01$), and $K = 10^{11}$ l/M, the limiting sensitivity attainable in a competitive assay (using an infinite specific activity label) is 10^{-13} M/l.[15] In practice, sensitivities closely approaching this limit are attainable using radioisotopes as labels, implying that little or no sensitivity advantage is gained, in competitive assays, by using nonisotopic labels of higher specific activity. However, the principal sensitivity constraints applying to noncompetitive methods are of a different nature, relating primarily to the detectability of the specific signal generated by occupied antibody binding sites against the "blank" or background signal observed in the complete absence of analyte. Thus, as illustrated in Figure 6, by adopting a "noncompetitive" strategy, and minimizing the background signal from all possible sources (including, for example, that from nonspecifically bound labeled antibody), labels of specific activity higher than those of commonly used isotopes can, in practice, yield assays with sensitivities considerably greater than those achievable using a competitive approach.

b. Types of Label

A wide variety of labels have been used in the present context, some of which are discussed in detail in other chapters of this volume. Of those in common use, radioisotopic, chemiluminescent, enzyme, and fluorescent (see Chapter 4) labels all display specific activities which are sufficiently high not to limit, in practice, the sensitivity of competitive assays, albeit the potential sensitivities of noncompetitive methods can only be fully exploited using the nonisotopic labels.

The radioisotope most commonly employed in this context is ^{125}I, although other radioisotopes (e.g., ^{3}H, ^{57}Co) are used in certain circumstances. The continuing popularity of radioimmunoassay (RIA) and immunoradiometric assay (IRMA) methods derives in part from the ease and comparative reliability with which such isotopes can be measured, in part from the flexibility inherent in the fact that, given access to suitable isotope counting facilities, the individual researcher or clinical chemist can either synthesize his/her own radiolabeled reagents, or purchase such reagents or immunoassay kits from a variety of alternative manufacturers. Although the high specific activities of radioisotopes constituted one of the key factors underlying the sensitivities of RIA and IRMA, and hence to the importance these methods have acquired in analytical science, they impose major constraints on the sensitivities attainable using the noncompetitive approach. It should be borne in mind that antibody molecules labeled with ^{125}I yield detectable signals (generated by radioisotopic decay) at a rate of approximately one signal per second per 10^7 molecules; this signal generation rate implies, in practice that, irrespective of assay design, analyte concentrations of below approximately 10^7 molecules/ml lie beyond the sensitivity range of isotopically based methods (see Figure 6).

The conversion of many substrate molecules by a single enzyme molecule implies amplification of the activity of an enzyme-labeled reagent (assuming high sensitivity of detection of the reaction product). Enzymes are consequently being increasingly used in the present context,[21] particularly as the basis of ultrasensitive noncompetitive techniques. Early examples include the USERIA technique developed by Harris et al[22] and the method of Shalev et al.,[23] relying on radioactive and fluorogenic substrates respectively (see Figure 6). More complex methods of amplification of the enzyme-generated signal rely on observation of the latter's catalytic effect on a secondary enzyme system which remains essentially dormant until thus activated.[24,25]

Chemiluminescent labels, in spite of quantum efficiencies generally considerably lower than 100%, are also, in principle, capable of yielding much higher specific activities than radioisotopes, and hence higher (noncompetitive) immunoassay sensitivities (Chapter 4). For example, the acridinium esters[26] produce chemiluminescence under mild conditions, and are, in practice, more luminescent than substances such as luminol.[27,28] Employed in noncompetitive assay designs, these esters yield sensitivities considerably higher than those of isotopically based techniques, enabling, for example, TSH to be measured at concentrations of 0.01 μU/ml or lower.[29,30] Another approach combines the attributes of luminometric measurement with the signal amplification properties of enzyme labels. Based on the studies of Whitehead et al,[31] assays utilizing antibodies labeled with horseradish peroxidase and luminescent substrates have been developed by Amersham International. Luminescence enhancement may be achieved by the use of phenols[32] resulting in further signal amplification and increase in assay sensitivity. Commercially available kits based on this approach are of comparable sensitivity to other forms of nonisotopic, noncompetitive immunoassay developed by other manufacturers.

Fluorescent labels also yield very high specific activities since each fluorophor molecule may be induced to yield many photons in response to exposure to a high intensity light input. Fluorescent markers also possess other attractive features, including the possibility of repeated measurements on the same sample, and of observation of their spatial distribution on a solid surface (thus permitting development of multiple immunoassays, see below). Major problems associated with conventional fluorescence measurements are the background fluorescence generated by many biological substances, plastics, solvents, etc. and the bleaching effects deriving from continuous fluorphore exposure to light. These effects conspire to limit the assay sensitivities normally attainable by such methods. However, these disadvantages may be largely overcome by time-resolution techniques.

These exploit the fundamental nature of fluorescence. When a fluorophore is excited by pulsed radiation, fluorescence is emitted following each incident pulse with an intensity which decreases exponentially with time in a characteristic manner.[8] An electronically "gated" detection system may be used to accumulate photons emitted over selected time intervals following extinction of the incident light source, enabling fluorescent signals emitted by fluorophores possessing different decay times to be distinguished. Thus, when a sample contains two major fluorophores, one of which displays a fluorescence decay time much longer than the other, the signal emitted by the slowly decaying fluorophore may be identified and measured by permitting the more rapidly decaying fluorescence to die away before commencing photon measurement. This simple principle can be exploited to reduce or effectively eliminate background fluorescence if this is characterized by a much shorter lifetime than that of the fluorophore of interest.

The fluorescent lanthanide chelates[33] form a unique group of fluorophores of particular importance in this context because of their long fluorescence decay times, which are some 3 to 4 orders of magnitude greater than those characterizing most background fluorescence. Another useful property of the lanthanide chelates is their large Stokes shift, i.e., the difference in the wave lengths of the fluorescent and (optimal) exciting radiations.[34] These attributes provide the basis for the construction of relatively simple and inexpensive pulse-light, time-resolving fluorometers capable of yielding the high signal/noise ratios essential for high sensitivity fluorometrically based immunoassay.

The most widely used methodology of this type is that known as DELFIA, developed by the author and colleagues in collaboration with the instrument manufacturers LKB/Wallac, and now marketed by Pharmacia. This is based on the use of europium chelates which, although not themselves fluorescent, may be attached to antibodies without significant impairment of their immunochemical properties. Antibodies so labeled may be used in conventional noncompetitive "sandwich" assays or in "competitive" assays of the type shown in Figure 4 (bottom right), relying on adsorption of labeled antibody to antigen linked to a solid support. In both cases, following reaction with antigen on the solid support, the adsorbed antibody is exposed to an appropriately formulated β-diketone "cocktail", resulting in transference of the metal ion into the highly fluorescent chelate solution, this procedure, developed in the author's laboratory by S. Dakubu,[35] being described as "Dissociation Enhancement".[36]

Though this approach yields high sensitivity, it fails to fully exploit the potential of the time-resolved fluorescence approach (Chapter 4). By dispersing the fluorophore in a relatively large volume of "enhancement" solution the signal/noise ratio is reduced in comparison with that which would be obtained were the fluorophore confined to a small area on a solid surface. Though the europium (and occasionally terbium) chelates presently used in the DELFIA kits are unsuitable for direct *in situ* surface fluorescence measurements, other compounds (such as the rare earth cryptates developed by Lehn and co-workers[37]) permit direct fluorescence measurement without the necessity for prior dissociation from antibody. Not only do such compounds offer the possibility of far higher immunoassay sensitivity, but permit the development of multianalyte immunoassay methods of the kind discussed below.

2. The Immunoreagent Separation System

a. "Conventional" Systems (Electrophoresis, Precipitation, Adsorption, etc.)

All immunoassay systems require separation and measurement of the products of the immunological reaction between antibody and analyte, enabling identification of the occupancy by analyte of CAb binding sites. For example, in conventional RIA methods, labeled analyte bound to unoccupied sites is separated from residual, "free", labeled analyte, following which either or both labeled analyte fractions are counted, thus revealing antibody binding-site occupancy. Many separation methods have been described, relying typically on adsorption of the free-labeled analyte fraction by solid adsorbants such as charcoal or silica, precipitation of the antibody-bound fraction using a second antibody or a chemical precipitating agent, chromatographic or electrophoretic separation techniques, etc. All have been extensively reviewed in RIA textbooks, and do not require further discussion here. Nonisotopic assay systems relying on labeled analytes possess a potential advantage in that antibody-bound labeled analyte may be distinguishable from residual free analyte without physical separation of the two fractions. For example, the activity of an enzyme label attached to the analyte molecule (which, in practice, must be "small") may be diminished or enhanced when the latter is bound to antibody, causing the measured signal to vary depending on the analyte concentration in the test sample. However, for a variety of reasons, the sensitivity of such assay systems is relatively low; moreover, they are susceptible to nonspecific interference by constituents of test samples of the activity of the signal-generating system. For these reasons, their use has been confined to the measurement of nonprotein analytes of small molecular size, such as therapeutic drugs, present in biological fluids at relatively high concentration.

b. Solid-Phase Antibody and Antigen Systems

Attachment of one of the key components of an immunoassay system to a solid support offers certain practical advantages in the present context. Aside from permitting easy and rapid separation of the reaction products, the solid-phase component may be repeatedly washed following reaction, thus ensuring minimal contamination by labeled reaction products remaining in the liquid phase. CAb, analyte (or anti-idiotypic antibody), or "precipitating" antibody (i.e., a second antibody reactive with the CAb, used to sequester antibody-bound labeled analyte from free analyte in a competitive labeled analyte system) may be attached to the solid support. The solid support may itself be particulate in nature (e.g., microcrystalline cellulose, Sephadex, latex, glass, etc.) or may comprise the surface of a plastic bead, sample tube, or well of a microtiter plate. The immunological reagent may be attached to the solid support covalently or may be simply adsorbed to the surface and held in place by noncovalent attractive forces. It may be attached directly to the solid support or via an intermediary such as protein A (which binds antibody) or avidin (which in turn binds biotinylated CAb or antigen), and which itself may be linked to a protein adsorbed to the surface of the solid material. (See Chapters 12 and 17.)

Such is the variety of functions that the solid-phase reagent must fulfill in immunoassay systems of different design that it is impossible to generalize when specifying the required properties and physicochemical characteristics of the solid support. For example, if "precipitating antibody" (see above) is attached to the support, it serves merely to bind (and hence sequester) the CAb (together with any attached labeled antigen) from the incubation mixture following the immunological reaction. Provided, therefore, the solid-phase precipitating antibody is present in such excess that it binds essentially all the CAb in the system, its affinity, the manner in which it is linked to the solid support, and variations in its amount are of relatively little consequence. Conversely, in a conventional "competitive" labeled antigen RIA relying, for example, on antibody-coated incubation tubes, the amount of CAb coated to the surface of each tube must be carefully standardized in order to ensure high assay precision and sensitivity. It is also desirable in these circumstances that alterations in molecular structure of the antibody when situated within the surface microenvironment should not significantly diminish the affinity of the antibody vis-a-vis analyte, thereby also reducing assay sensitivity and precision. The possibility of such effects is reduced, however, by attachment of the antibody to the solid support by means of an intermediary, such as protein A, avidin, or other linking molecule. (See Chapters 11, 12, and 17.)

In general, the principal perceived advantage deriving from the solid phasing of any of the primary reagents involved in an immunoassay procedure has been the added convenience this confers on the separation of the reaction products following the immunological reaction, coupled with an associated reduction in the "misclassification" of these products, i.e., the inclusion of free labeled reagent in the bound fraction and *vice versa.* Reagent-product misclassification and "misclassification errors" (caused by intersample variation in the extent of misclassification) leads to a loss of assay sensitivity, precision, and working range. The crucial importance of minimizing "nonspecific binding" of labeled antibody (i.e., the misclassification of free labeled antibody as bound) in the particular context of ultrasensitive noncompetitive immunoassay has been emphasized above, but elimination of misclassified reaction products in the particular labeled reagent fraction (free or bound) that is measured at the termination of the immunological reaction is an important requirement in all immunoassay designs (albeit it is one that has frequently been neglected in the past). The use of appropriate solid-phase reagents facilitates very "clean" separation of reaction products; for example, the nonspecific binding of labeled antibody by certain solid supports such as ultrafine Sephadex and microcrystalline cellulose is, following careful washing, on the order of 0.01%, a characteristic which underlies the very high sensitivity and wide working range of immunoassays based on these materials.

The principal perceived *disadvantage* associated with solid phasing any of the primary reagents is the reduction in reaction velocities consequent upon the inevitable increase in the diffusion distances by which the reactants are separated, causing the time to reach equilibrium to be prolonged (see Chapter 18). This in turn implies that, after any specified time interval, the reaction between antigen and antibody will have proceeded further towards equilibrium in a homogenous liquid phase system that in which one of the reactants is solid phased. The "apparent" equilibrium constant governing the reaction as measured at this time will therefore be lower in the heterogenous than in the homogenous system. Nevertheless, when equilibrium is ultimately reached, there is no theoretical reason for supposing that the antigen and antibody fractions that are respectively free and bound (the latter forming the antigen/antibody complex) will differ in the two systems (except as a result of any partial deactivation of the solid-phased reactant) so that the equilibrium constant as measured at equilibrium will be identical. However, it is, of course, a corollary of the reduction in the kinetics of the *forward* binding reaction that a corresponding reduction in the kinetics of the *backward* (dissociation) reaction will occur; for example, antigen bound to solid-phase antibody will inevitably appear to dissociate more slowly than when the antibody is distributed uniformly throughout the liquid phase. A variety of reasons have been advanced for this entirely expected phenomenon, and it has also led (*inter alia*) to the notion that binding reactions on solid surfaces are "irreversible".[38] (See Chapter 19.)

Were this to be true, the affinity constant of antibody/antigen reactions occurring on solid surfaces would be increased (implying a potential increase in immunoassay sensitivity), and though this has sometimes been suggested to be the case,[39] past experience in the author's laboratory has not been in accordance with this claim. In short, the sensitivity increases deriving from the use of solid phase reagents are essentially due to associated improvements in the efficiency of separation of the reaction products, not to any (physicochemically unexpected) increase in the affinity of the binding reaction.

Because of the reduction in reaction velocity between antigen and antibody when either reactant is attached to a solid support (see Chapter 18), many experienced immunoassayists prefer to allow the primary reaction to proceed in a "homogenous" liquid-phase incubation mixture (i.e., one in which all reactants are mixed in the same liquid phase), following which a solid phase reagent (e.g., precipitating antibody) is introduced into the system to separate the reaction products. This strategem combines the reaction speed advantages deriving from homogenous reactions with the efficiency and "cleanliness" of a solid-phase separation system, albeit with some loss of convenience.

Notwithstanding the added convenience of immunoassay systems involving the use of solid supported reagents, and the added sensitivity that, for the reasons given above, may derive from the resulting increased efficiency of reagent separation, one of the major potential advantages offered by solid phase systems, that of spatial separation of a number of immunoassays in the same incubation mixture, has so far not been fully exploited. The possibility of constructing assay methods capable of simultaneous measurement of many different analytes in the same sample stems from some of the concepts discussed earlier in this chapter.

III. MICROSPOT IMMUNOASSAY

A. BASIC PRINCIPLE

The recognition that an ambient analyte immunoassay system may be constructed relying on the use of a vanishingly small CAb concentration (e.g., $<0.01/K$, but see above), and measurement of its fractional occupancy (using a high specific activity labeled reagent by one of the alternative approaches shown in Figure 4), leads to two further important concepts. First, the antibody may be located on a small area or "microspot" on the surface of a solid support, the total number of (functionally active) antibody molecules within the microspot being less than $v/K \times 10^{-5} \times N$ (where v = the sample volume to which the microspot is exposed (milliliters) and N = Avogadro's number (6×10^{23})). For example, if $v = 1$ and $K = 10^{12}$ l/M, then the maximum number of antibody binding sites located in the microspot causing insignificant disturbance (<1%) to the analyte concentration in the sample is 6×10^{6} (this maximum clearly being greater the lower the effective affinity constant of the antibody).

One important advantage of restricting the number of CAb molecules on the microspot so that ambient analyte immunoassay conditions are fulfilled is that, under these conditions, the ratio of occupied (or unoccupied) to total binding sites is solely a function of the analyte concentration to which the antibody has been exposed, and is independent of the number of antibody molecules located within the microspot.

Meanwhile, measurement of fractional occupancy of the CAb is readily effected by measurement of the ratio of signals emitted by two different labels respectively attached to the CAb and to a "developing" antibody recognizing either occupied or unoccupied capture antibody binding sites. This concept can be described as dual label, "ratiometric" microspot immunoassay.

The principles underlying this form of immunoassay may be best illustrated by supposing that the procedure is carried out in two steps. The first involves exposure to an analyte-containing fluid of labeled CAb in a microspot resulting in occupancy of antibody binding sites reflecting the ambient analyte concentration (see Figure 10a). The antibody-bearing probe may then be removed and exposed to a solution containing a high concentration of the "developing" antibody, labeled with a second label, directed against either a second epitope on the analyte molecule if this is large (i.e., the occupied site), or against unoccupied antibody binding sites in the case of small analyte molecules (see Figure 10b). (Note: an antibody which simulates antigen, and reacts with unoccupied sensor-antibody binding sites, is described as a "mirror-image anti-idiotypic antibody"; the use of such an antibody instead of labeled antigen is not essential, but is suggested here merely to illustrate the basic concepts involved.)

Occupancy of the CAb may thus be determined by measurement of the ratio of the signals emitted by the two labeled antibodies forming the dual-antibody "couplets". Fluorescent

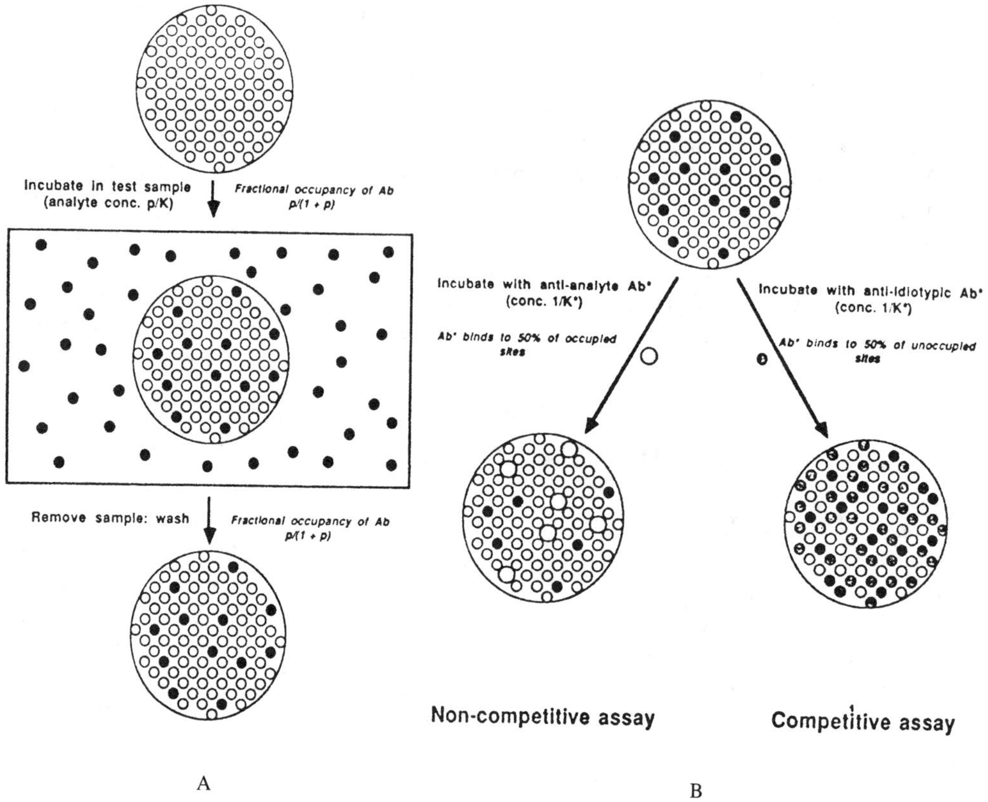

FIGURE 10. (A) Microspot immunoassay: first incubation. The fractional occupancy of CAb binding sites (p/1 + p) reflects the analyte concentration (p) 1/k to which the microspot has been exposed (where p is the analyte concentration expressed in molar units of 1/K). (B) Second incubation. The microspot is exposed to a second "developing" antibody (concentration 1/K*) reactive with either occupied sites (noncompetitive assay) or unoccupied sites (competitive assay), where K* is the affinity constant of the developing antibody with respect to occupied and unoccupied sites, respectively.

labels are clearly particularly useful in this context because, by the use of optical scanning techniques, they readily permit arrays of different antibody "microspots" distributed over a surface, each directed against a different analyte, to be scanned, thus enabling multiple analyte assays to be simultaneously performed on the same small sample. Several advantages stem from a dual fluorescence measurement approach of this kind. For example, neither the amount nor the distribution of the CAb within the measuring instrument's field of view are critical, since the ratio of the signals emitted is unaffected. Likewise, fluctuations in the intensity of the incident (exciting) light beam are of little importance.

B. MICROSPOT IMMUNOASSAY SENSITIVITY: SOME THEORETICAL CONSIDERATIONS

Since "microspot immunoassay" of the kind described above challenges some of the basic concepts which have dominated immunoassay design and development in the past 2 to 3 decades, the issue of the sensitivity attainable using this approach is of primary importance. Earlier in this chapter, it has been shown that, under certain conditions, reduction of the amount of CAb in an immunoassay system to values of 0.01/K and below may increase the signal/noise ratio (and hence assay sensitivity) even in assays of "noncompetitive" design. A number of factors, such as the characteristics of the signal measuring equipment, the density

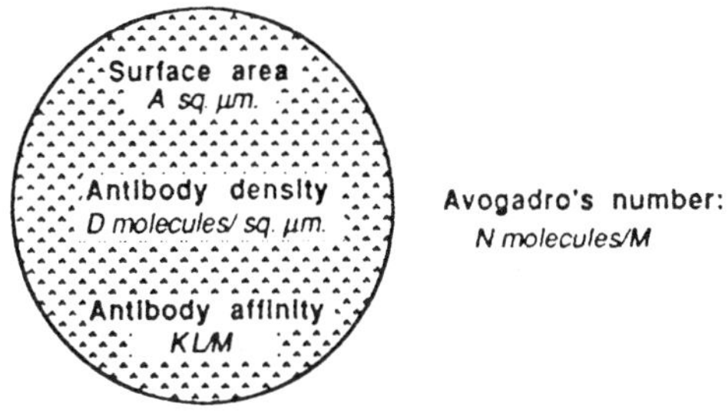

$$\text{Minimum test sample volume (mls): } \frac{A \times D \times K \times 10^5}{N}$$

FIGURE 11. Microspot ambient analyte immunoassay. The microspot shown is assumed to be uniformly coated with antibody, though, if the dual-labeled antibody "ratiometric" approach is adopted, uniform coating is not essential. The minimum test sample volume for ambient analyte assay conditions to prevail (permitting adoption of the ratiometric approach) is shown.

of CAb molecules on the surface of the solid support, the specific activity of the labeled "developing antibody", the properties of the capture antibody support, etc. clearly influence microspot immunoassay sensitivity. To simplify discussion of these issues, it may be assumed that the CAb is coated as a monolayer on a solid support at a uniform density, implying that only the "developing" antibody need be labeled and measured in order to ascertain fractional occupancy of sensor antibody binding sites.

Figure 11 illustrates the surface of an antibody microspot of surface area A (square micrometer) that is uniformly coated with (active) CAb of affinity K (l/M) in a monolayer of density D (molecules per square micrometer). We assume that the spot is exposed to an analyte-containing medium of volume vs. mls, containing an analyte concentration C molecules/ml. The concentration of antibody in the system is thus given by AD/v.

From Equation 1, for antibody concentrations $\rightarrow 0$ (i.e., $[Ab] < 0.01$), $F \approx [An]/1 + [An]$); (see Figure 8).

Meanwhile, the fraction of analyte bound to antibody (f) at equilibrium is given by

$$f^2 - f(1/[An] + [Ab]/[An] + 1) + [Ab]/[An] = 0 \tag{2}$$

Thus, for analyte concentrations $\rightarrow 0$ (i.e., $[An] < 0.01$), $f \approx [Ab]/(1 + [Ab])$; (see Figure 8). Furthermore, when $[Ab] < 0.01$, and when $[An] \geq 0$, $f < 0.01$.

Expressed in units of 1/K, the concentration ([Ab]) of CAb located on the microspot is given by $DAK/(v \times 6 \times 10^{20})$, (since Avogadro's constant, expressed as number of molecules/mM, is 6×10^{20}, approximately). The fraction of an analyte concentration $\rightarrow 0$ bound to the spot is thus $DAK/(v \times 6 \times 10^{20} + DAK)$, so that the number of analyte molecules bound to the spot is given by $vCDAK/(v \times 6 \times 10^{20} + DAK)$.

Case 1: Sandwich (Two-Site) Assay (Noncompetitive)

Following incubation of sample with antibody, we assume the sample is removed and the microspot then exposed to a volume V mls of a solution of a second, labeled "developing" antibody of affinity K* (l/M) at a concentration given by Q (expressed in units of 1/K*).

The fraction of analyte bound by labeled antibody (F*) at equilibrium is given by

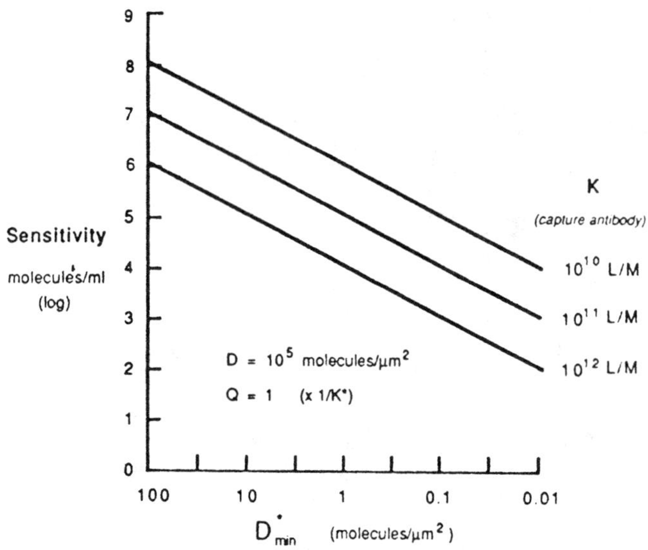

FIGURE 12. Theoretically predicted competitive microspot immunoassay sensitivity plotted against the limit of detection of developing antibodies on the surface of the solid support, assuming different values of CAb affinity. The surface density of CAb on the solid support is assumed to be 10^5 molecules per square micrometer.

$$F^{*2} - F^{*}(1/[An^{*}] + [Ab^{*}]/[An^{*}] + 1) + [Ab^{*}]/[An^{*}] = 0 \quad (3)$$

where [An*] represents the analyte concentration in the developing-antibody solution, expressed in units of 1/K*, i.e., $vCDAKK^*/[(v \times 6 \times 10^{20} + DAK)V \times 6 \times 10^{20}]$.

Assuming $[An^*] < 0.01$, $F^* \approx [Ab^*]/(1 + [Ab^*])$. (For example, if [Ab*] = 1, the fraction of analyte molecules bound by labeled antibody = 0.5 approx.) Thus, since the number of analyte molecules bound to the spot is given by $vCDAK/(v \times 6 \times 10^{20} + DAK)$, the number of analyte molecules labeled by the second, developing, antibody is given by $vCDAK[Ab^*]/[(v \times 6 \times 10^{20} + DAK)(1 + [Ab^*])]$, and the surface density of such molecules is given by $vCDK[Ab^*]/[(v \times 6 \times 10^{20} + DAK)(1 + [Ab^*])]$. Moreover, assuming that $DAK << v \times 6 \times 10^{20}$ (i.e., the amount of antibody is such that "ambient assay" conditions prevail), then the surface density (D*) of developing-antibody = $CDK[Ab^*]/[(6 \times 10^{20})(1 + [Ab^*])]$ approximately. Note that D* is independent of both v and V; also, that the ratio $D^*/D = C \times K[Ab^*]/[(6 \times 10^{20})(1 + [Ab^*])] = C \times$ const.

If the minimum detectable surface density of developing-antibody molecules (i.e., s_{Do}^*, the standard deviation of the measurement of D* when C = 0) is given by D^*_{min} (molecules per square micrometer) and C_{min} represents the minimum detectable analyte concentration in the sample, then, disregarding nonspecific binding of developing antibody within the microspot area:

$$C_{min} = D^*_{min} \times [(6 \times 10^{20})(1 + [Ab^*])]/DK[Ab^*] \quad (4)$$

For example, if $[Ab^*] = 1$, $D = 10^5$ molecules/µm², $K = 10^{11}$ l/M and $D^*_{min} = 20$ molecules per square micrometer, then $C_{min} = 2.4 \times 10^6$ molecules/ml $= 4 \times 10^{-15}$ M/l. It should be noted that, in this example, the fractional occupancy of the CAb binding sites by the minimum detectable analyte concentration is 0.04%. Figure 12 shows theoretical assay sensitivities attainable using CAb of varying affinities, plotted as a function of the minimum detectable surface density of developing antibody (D^*_{min}) within the microspot.

b. Case 2: Anti-Idiotypic Antibody ("Competitive") Assay

In this case, we assume that, following removal of the sample, the microspot is exposed to a volume V mls of a solution of (for example) a second, labeled anti-idiotypic antibody reacting with unoccupied sites on the sensor antibody. Using similar reasoning as above, we may likewise assume that the fraction of such sites which become occupied by the anti-idiotypic "developing" antibody is given by [Ab*]/(1 + [Ab*]), where [Ab*] is the developing-antibody concentration. However, the minimum detectable surface density of anti-idiotypic antibody is not, in a competitive design, the critical determinant of assay sensitivity; this parameter is essentially governed by the precision of the anti-idiotypic antibody surface density measurement.

From Equation 1, the fraction of sites *unoccupied* by analyte = 1/(1 + [An]), and the fraction occupied by anti-idiotypic antibody = [Ab*]/(1 + [An])(1 + Ab*]). Thus, if the c.v. in the measurement of anti-idiotypic antibody is ϵ, the standard deviation is ε[Ab*]/(1 + [An])(1 + Ab*]). This term also represents the s.d. of the estimate of the fraction of sites occupied by analyte. Since the total number of antibody binding sites in the spot is DA, the s.d. of the estimate of occupied sites as [An] $\rightarrow$ 0 (i.e., ^{s}Do*) approximates ϵDA[Ab*]/(1 + [Ab*]); the s.d. in the occupied site surface-density estimate is thus ϵD[Ab*]/(1 + [Ab*]). However, the s.d. of the measurement of fractional binding-site occupancy when [An] $\rightarrow$ 0 defines D_{min}, and hence the minimum detectable analyte concentration in the test samples as indicated in Equation 4. Thus:

$$C_{min} = D_{min} \times [(6 \times 10^{20})(1 + [Ab^*])]/DK[Ab^*] \tag{5}$$

$$= \varepsilon D[Ab^*]/(1 + [Ab^*]) \times [(6 \times 10^{20})(1 + [Ab^*])]/DK[Ab^*] \tag{6}$$

$$= \varepsilon/K \times (6 \times 10^{20}) \tag{7}$$

For example, if values of [Ab*] = 1, D = 10^5 molecules per square micrometer, and K = 10^{11} l/M are assumed as in the noncompetitive example discussed above, and the c.v. in the measurement of anti-idiotypic antibody density in the microspot is 1% (i.e., ε = 0.01), then D_{min} = 500 molecules per square micrometer, and C_{min} = 6 × 10^7 molecules per milliliter = 10^{-13} M/l. Fractional occupancy of the CAb binding sites by the minimum detectable analyte concentration is, in this example, 1%. It should be noted that the sensitivity limit ε/K (expressed in molar terms) is identical to that established for conventional "competitive" assays.[15]

A number of useful conclusions can be drawn from these theoretical considerations. For example, they indicate that the attainment of high microspot-immunoassay sensitivity requires close packing of CAb molecules within the microspot area, combined with the use of an instrument capable of accurately measuring a very low surface density of developing antibodies. Provided developing antibody surface densities on the order of 1 to 10 molecules per square micrometer can be detected, microspot assay sensitivities some orders of magnitude superior to those obtainable by conventional radioisotopically based immunoassays are achievable. It is also evident that it is likely to be better strategy to measure occupied rather than unoccupied capture antibody binding sites, and that sensitivities yielded by "noncompetitive" microspot assays are likely to be considerably greater than those yielded by the competitive approach.

C. MICROSPOT IMMUNOASSAY; SOME PRACTICAL CONSIDERATIONS

Though the concepts discussed above are clearly exploitable using a variety of antibody labels, the author's preliminary studies in this area[9] have relied on the use of conventional fluorescence measurement techniques using new instrumentation which has recently become

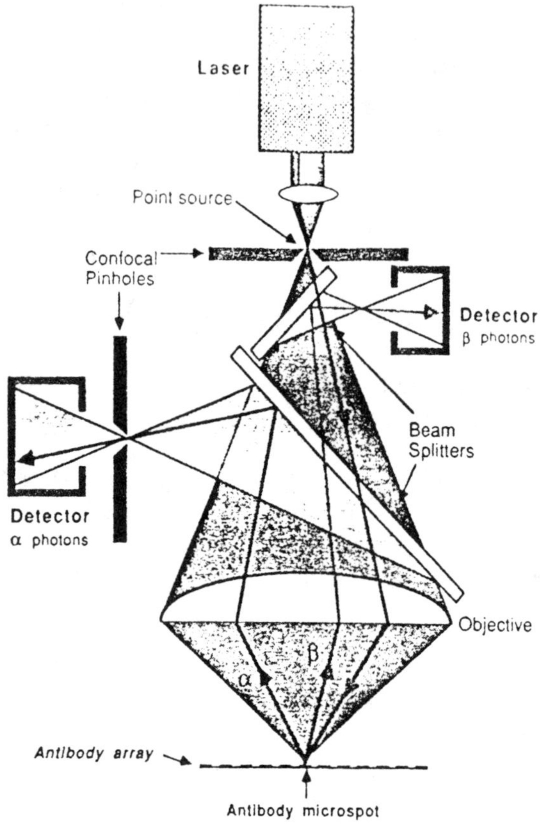

FIGURE 13. Diagrammatic representation of dual-channel confocal fluorescence microscope permitting simultaneous measurement of fluorescence signals from two fluorophors situated at the focal point. By scanning the array, the ratio of signals from each antibody microspot may be determined.

available. Though not specifically designed for the present application, and intended for use with conventional fluorophors such as Texas Red and fluorescein, it has provided the means to experimentally confirm many of the basic concepts discussed in this chapter. A brief review of the instrument itself — a laser scanning confocal microscope — may be of value.

In laser scanning confocal fluorescence microscopy, a small area of the specimen is illuminated by a focused laser beam; the fluorescence photons from this area are, in turn, focused onto a detector (see Figure 13). Both illumination intensity and light collection efficiency diminish rapidly with distance from the focal plane. At the "confocal" point, the projection of the illumination pinhole and the back projection of the detector pinhole coincide. Such systems contrast with conventional epifluorescence methods, where the specimen is exposed to an essentially uniform flux of illumination.[40]

Typically, fluorescence photons emanating from the laser-illuminated area are detected by a low dark-current photomultiplier. Electrons spontaneously emitted by the photomultiplier photocathode contribute to the background signal of the instrument, and must, for highest sensitivity, be minimized. Fortunately, the design of such instruments permits the photocathode to be small in area, so that this source of background can be expected to diminish with future improvement in photomultiplier design. Meanwhile, current instruments already display high sensitivity of detection of fluorescent signals. For example, the Zeiss microscope is claimed to display a detection limit for fluorescein of about 10 molecules per square micrometer. Most commercially available FITC-labeled IgG attains a fluorophor/protein ratio of ≈4; thus, the detection limit (D^*_{min}) of the Zeiss microscope is ~2 to 3 FITC-labeled IgG molecules per square micrometer. This implies theoretical microspot immunoassay sensitivity of ~2.4×10^5 molecules per milliliter for a noncompetitive two-site assay, assuming identical parameter values as above, or 2.4×10^4 molecules per milliliter using a "sensing" antibody of affinity 10^{12} l/M. A similar instrument is the Bio-Rad/Lasersharp microscope, on which the author's current studies of "ratiometric" assay have been based. Its argon laser emits two excitation lines at 488 and 514 nm. It is thus particularly efficient in exciting blue/green emitting fluorophores such as FITC (excitation maximum 492 nm), but is less efficient in exciting fluorophores such as Texas Red (excitation maximum 596 nm). However, the ratiometric assay principle permits considerable variation in detection efficiencies of the two labels since, *inter alia,* the specific activities of the labeled antibodies forming the antibody "couplets" can be chosen to yield signal ratios approximating unity. Inefficiency of the argon laser in exciting Texas Red is thus not a major handicap in the present context.

Though studies using this particular instrument have been encouraging, the immunoassay sensitivities attained, though of the same order as achieved by conventional RIA, have not been as high as the theoretical considerations discussed above suggest may be ultimately attainable. This has in part been due to a high fluorescence background generated within the instrument itself, a phenomenon which can nevertheless be largely eliminated by improvements in instrument design, coupled, if necessary, with the use of time resolution techniques of fluorescence measurement of the kind discussed above. A second reason has been some initial difficulty in reliably coating antibody microspots in such a way that the ratio of fluorescent signals from capture and developing antibodies across the microspot area remains constant.

It is clearly inappropriate to examine this point in detail here; nevertheless, brief discussion of the particular requirements that the solid support must fulfill in the context of microspot immunoassay is illustrative of the differing nature of the problems that may be encountered when solid phasing immunoassay reagents. As indicated earlier, constancy of the amount of CAb within the microspot area is not essential provided a dual-label approach of the kind described above is adopted, but, to achieve high sensitivity, it is important that antibody should deposited as a monolayer at high surface density and in such a manner that antibody binding sites are minimally impeded in their reaction with antigen. Thus, solid supports should display a capacity to adsorb a high surface density of antibody combined with low intrinsic signal-generating properties (for example, low intrinsic fluorescence), thus minimizing background. The author and colleagues have examined a number of materials, including polypropylene, Teflon®, cellulose and nitrocellulose membranes, and microtiter plates with these criteria in mind. White Dynatech Microfluor microtiter plates — formulated specifically for the detection of low fluorescence signals — yield high signal-to-noise ratios and have therefore been used in preliminary studies. These may be coated with antibody (by simple adsorption) at a surface density of at least 5×10^4 IgG molecules per square micrometer, a high proportion of antibody molecules so deposited appearing to retain immunological activity. Other coating procedures, e.g., linking CAb either directly (using glutaraldehyde) or via protein A as an intermediary, to silane coatings of glass and other supports have also been examined. Antibody surface densities have been found to be considerably lower using protein A albeit the functionally active coating density was essentially identical (about 1×10^3 IgG molecules per square micrometer), whichever of these approaches was adopted.

Such studies are illustrative of an increasing interest in the nature and properties of solid materials used as reagent supports for use in an immunoassay context, albeit the particular properties required inevitably differ widely depending on the particular design of the immunoassay, and the function of the solid-phased reagent in the system. Immunoassay methodology ranges in its use from simple pregnancy tests for use in the home to the possible development of the biochemical equivalent of the compact disk as described above, and the nature of the solid supports used will inevitably vary widely.

IV. SUMMARY AND CONCLUSION

Immunoassay is a ubiquitous methodology used for the sensitive measurement of thousands of substances of biological importance. Its widespread popularity has stemmed from the structural specificity deriving from its exploitation of the specificity of antibody/antigen reactions, and the sensitivity imparted to it by the use of high specific-activity labels to monitor the binding reactions between small numbers of antigen and antibody molecules. Paradoxically, controversy has long centered on the manner in which such assays should be designed in order to maximize their sensitivity, and some of the accepted ideas in the field are both unsound and misleading.

At attempt has been made in this chapter to clarify the basic concepts underlying immunoassay methodology, and to dispel some of the confusion relating to optimal assay design. In particular, it is suggested that all immunoassays rely on measurement of antibody binding site occupancy by analyte (antigen) following reaction between the two; furthermore, that there exist two basic strategies whereby such measurement can be effected: (1) direct measurement of occupied sites and (2) measurement of unoccupied sites, from which the extent of binding-site occupancy is inferred. Direct measurement of occupied sites ("noncompetitive immunoassay") is generally the better strategy, yielding higher assay sensitivity and precision.

"Solid phasing" one of the principal reagents in an immunoassay system enables the products of the immunological reaction to be efficiently and conveniently separated. "Misclassification errors" (caused by inefficient separation of reaction products) reduce immunoassay sensitivity, precision, and working range. For these reasons, the use of solid phase reagents in immunoassay systems can lead to important sensitivity improvements, albeit the reduction in reaction speed consequent upon solid phasing one reagent can be disadvantageous. Another advantage of solid phasing the CAb in such systems is that this permits the development of highly sensitive "microspot" immunoassays and the construction of microspot arrays, thus permitting, in principle, the simultaneous measurement of an unlimited number of analytes in the same sample.

REFERENCES

1. **Köhler, G. and Milstein, C.,** Continuous culture of fused cells secreting specific antibody, *Nature,* 256, 495, 1975.
2. **Yalow, R. S. and Berson, S. A.,** Immunoassay of endogenous plasma insulin in man, *J. Clin. Invest.,* 39, 1157, 1960.
3. **Ekins, R. P.,** The estimation of thyroxine in human plasma by an electrophoretic technique, *Clin. Chim. Acta,* 5, 453, 1960.
4. **Barakat, R. M. and Ekins, R. P.,** The assay of Vitamin B12 in blood: a simple method, *Lancet, ii,* 25, 1961.
5. **Wide, L., Bennich, H., and Johannson, S. G. O.,** Diagnosis of allergy by an in vitro test for allergen antibodies, *Lancet, ii,* 1105, 1967.
6. **Miles, L. E. M. and Hales, C. N.,** Labelled antibodies and immunological assay systems, *Nature,* 29, 186, 1968.
7. **Rodbard, D. and Weiss, G. H.,** Mathematical theory of immunometric (labeled antibody) assay, *Anal. Biochem.,* 52, 10, 1973.
8. **Dakubu, S., Ekins, R., Jackson, T., and Marshall, N. J.,** High sensitivity, pulsed light, time-resolved fluoroimmunoassay, in *Practical Immunoassay. The State of the Art,* Butt, W. R., Ed., Marcel Dekker, New York, 1984, 71.
9. **Ekins, R., Chu, F., and Biggart, E.,** Development of microspot multi-analyte ratiometric immunoassay using dual fluorescent-labelled antibodies, *Anal. Chim. Acta,* 227, 73, 1989.
10. **Ekins, R. P.,** The precision profile: its use in assay design, assessment and quality control, in *Immunoassays for Clinical Chemistry,* Hunter, W. M. and Corrie, J. E. T., Eds., Churchill Livingstone, Edinburgh, 1983, 76.

11. **Yalow, R. S. and Berson, S. A.,** Radioimmunoassays, in *Statistics in Endocrinology,* McArthur, J. W. and Cotton, T., Eds., MIT Press, Cambridge, MA, 1970, 327.
12. **Berson, S. A. and Yalow, R. S.,** Measurement of hormones — Radioimmunoassay in *Methods in Investigative and Diagnostic Endocrinology,* Vol. 2A, Berson, S. A. and Yalow, R. S., Eds., North-Holland/Elsevier, Amsterdam, 1973, 84.
13. **Ekins, R. P., Newman, B., and O'Riordan, J. L. H.,** Theoretical aspects of "saturation" and radioimmunoassay, in Radioisotopes in Medicine; *In Vitro* Studies, Hayes, R. L., Goswitz, F. A., and Murphy, B. E. P., Eds., U. S. Atomic Energy Commission, Oak Ridge, TN, 1968, 59.
14. **Ekins, R. P., Newman, B., and O'Riordan, J. L. H.,** Saturation assays, in *Statistics in Endocrinology,* McArthur, J. W. and Cotton, T., Eds., MIT Press, Cambridge, MA, 1970, 345.
15. **Ekins, R. and Newman, B.,** Theoretical aspects of saturation analysis, in *Steroid Assay by Protein Binding, 2nd Symposium, Karolinska Symposia on Research Methods in Reproductive Endocrinology,* Diczfalusy, E., Reproductive Endocrinology Research Unit, Karolinska sjukhuset, Stockholm, 1970, 11.
16. **Woodhead, J. S., Addison, G. M., Hales, C. N., and O'Riordan, J. L. H.,** Discussion following Addison, G. M. and Hales, C. N., The immunoradiometric assay, in *Radioimmunoassay Methods,* Kirkham, K. E. and Hunter, W. M., Eds., Churchill Livingstone, Edinburgh, 1971, 467.
17. **Yorde, D. E., Sasse, E. A., Wang, T. Y., Hussa, R. O., and Garancis, J. C.,** Competitive enzyme-linked immunoassay with use of soluble enzyme antibody immune complexes for labelling. I. Measurement of human choriogonadotrophin. *Clin. Chem.,* 22, 1372, 1976.
18. **Rodbard, D.,** Appendix III and discussion following computer analysis of radioimmunoassay and competitive protein binding assay data, in *Steroid Assay by Protein Binding, 2nd Symposium, Karolinska Symposia on Research Methods in Reproductive Endocrinology,* Diczfalusy, E., Reproductive Endocrinology Research Unit, Karolinska sjukhuset, Stockholm, 1970, 96.
19. **Jackson, T. M., Marshall, N. J., and Ekins, R. P.,** Optimization of immunoradiometric (labelled antibody) assays, in *Immunoassays for Clinical Chemistry,* Hunter, W. M. and Corrie, J. E. T., Eds., Churchill Livingstone, Edinburgh, 1983, 557.
20. **Ekins, R. P.,** Measurement of analyte concentration, UK Patent Application 8224600, 1983.
21. **Maggio, E. T., Ed.,** *Enzyme Immunoassay,* CRC Press, Boca Raton, FL, 1980.
22. **Harris, C. C., Yolken, R. H., Krokan, H., and Hsu, J. C.,** Ultrasensitive enzymatic radioimmunoassay: application to detection of cholera toxin and rota virus, *Proc. Natl. Acad. Sci. U.S.A.,* 76, 5336, 1979.
23. **Shalev, A., Greenberg, G. H., and McAlpine, P. J.,** Detection of attograms of antigen by a high sensitivity enzyme-linked immunosorbent assay (HS-ELISA) using a fluorogenic substrate, *J. Immunol. Methods,* 38, 125, 1980.
24. **Stanley, C. J., Paris, F., Plumb, A., Webb, A., and Johansson, A.,** Enzyme amplification: a new technique for enhancing the speed and sensitivity of enzyme immunoassays, *Int. Clin. Products Rev.,* 8, 44, 1985.
25. **Johannson, A., Stanley, C. J., and Self, C. H.,** A fast highly sensitive colorimetric enzyme immunoassay system demonstrating benefits of enzyme amplification in clinical chemistry, *Clin. Chim. Acta,* 148, 119, 1985.
26. **McCapra, F., Tutt, D. E., and Topping, R. M.,** Assay method utilizing chemiluminescence, UK Patent, 1461877, 1977.
27. **Schoeder, H. R., Bogulaski, R. C., Carrico, R. J., and Buckler, R. T.,** Monitoring specific protein binding reactions with chemiluminescence, *Methods in Enzymology,* de Luco, M., Ed., Academic Press, New York, 57, 424, 1978.
28. **Weeks, I., Campbell, A. K., Woodhead, S., and McCapra, F.,** Immunoassays using chemiluminescent labels, in *Practical Immunoassay. The State of the Art,* Butt, W. R., Ed., Marcel Dekker, New York, 1984, 103.
29. **Kemp, H. A., Woodhead, S., and John, R.,** Labeled antibody immunoassays, in *Practical Immunoassay. The State of the Art,* Butt, W. R., Ed., Marcel Dekker, New York, 1984, 179.
30. **Weeks, I., Sturgess, M., Siddle, K., Jones, M. K., and Woodhead, J. S.,** A high sensitivity immunochemiluminometric assay for human thyrotrophin, *Clin. Endocrinol.,* 20, 489, 1984.
31. **Whitehead, T. P., Thorpe, G. H., Carter, T. J., Groucutt, C., and Kricka, L. J.,** Enhanced luminescence procedure for sensitive determination of peroxidase-labelled conjugates in immunoassay, *Nature,* 305, 158, 1983.
32. **Thorpe, G. H., Kricka, L. J., Moseley, S. B., and Whitehead, T. P.,** Phenols as enhancers of the chemiluminescence horseradish peroxidase-luminol-hydrogen peroxide reaction; applications in luminescence monitored enzyme immunoassays, *Clin. Chem.,* 31, 1335, 1985.
33. **Haverson, F., Breinen, J. S. and Leto, J. R.,** Photoluminescence of lanthanide complexes. II. Enhancement by an insulating sheath, *J. Chem. Phys.,* 41, 157, 1964.
34. **Soini, E. and Hemmila, I.,** Fluoroimmunoassay: present status and key problems, *Clin. Chem.,* 25, 65, 1979.
35. **Hemmilä, I. and Dakubu, S.,** Methods for separating the lanthanide marker from a lanthanide marked immunochemical component, European Patent Application 0 159 066, 1982.

36. **Hemmilä, I., Dakubu, S., Mukkala, V. M., Siiteri, H., and Lovgren, T.,** Europium as a label in time-resolved immunofluorometric assays, *Anal. Biochem.,* 137, 335, 1983.
37. **Rodriguez-Ubis, J.-C., Alpha B., Plancherel, D., and Lehn, J.-M.,** Synthesis of the sodium cryptates of macrobicyclic ligands containing bipyridine and phenanthroline groups, *Helvetica Chimica. Acta,* 67, 2264, 1984.
38. **Abraham, G. E.,** Discussion following solid-phase radioimmunoassay of estrogens in biological fluids, in *Steroid Assay by Protein Binding, 2nd Symposium, Karolinska Symposia on Research Methods in Reproductive Endocrinology,* Diczfalusy, E., Reproductive Endocrinology Research Unit, Karolinska sjukhuset, Stockholm, 1970, 343.
39. **Stenberg, M. and Nygren, H.,** Kinetics of antigen-antibody reactions at solid-liquid interfaces, *J. Immunol. Methods,* 113, 3, 1988.
40. **White, J. G., Amos, W. B., and Fordham, M.,** An evaluation of confocal versus conventional imaging of biological structures by fluorescence light microscopy, *J. Cell Biol.,* 105, 41, 1987.

Chapter 7

ANTIGEN QUANTIFICATION: MEASUREMENT OF MULTIVALENT ANTIGENS BY SOLID-PHASE IMMUNOASSAY

Robert G. Hamilton

TABLE OF CONTENTS

I. INTRODUCTION

Many biologically important macromolecules can be recognized and quantified using antibodies prepared either in xenogenic species or by cell fusion hybridoma and transfectoma technologies. These antibodies bind their respective molecules, i.e., the antigen, in a specific and avid manner. Protein molecules of larger than 5000 Da are considered complete antigens because they can elicit an immune response without being coupled to a larger immunogenic carrier protein. These macromolecules are often composed of multiple polypeptide chains, held together by interchain disulfide bonds. They may occur in the presence of other biomolecules which are structurally similar, display conformationally similar epitopes and are thus potentially cross reactive. The theory and design of immunoassays for the detection and quantitation of these multiepitope biomolecules in a mixture of cross-reactive molecules can differ from those discussed in Chapter 6 for smaller molecules such as hormones and drugs. This chapter discusses the properties of macromolecular antigens and reviews current solid-phase immunoassays used in their measurement.

II. ANTIGENIC MACROMOLECULES

Table 1 lists exemplary antigens in micro- and macroenvironments that are measured in clinical immunology laboratories using solid-phase immunoassay. Antigens can be grouped into *in vivo* and *ex vivo* molecules that are either soluble (unattached) or cell associated and synthetic molecules.

A. *IN VIVO* ANTIGENS

Animal and human serum proteins (e.g., immunoglobulins, complement components, and enzymes) constitute the group of *in vivo* antigens that are most frequently measured by immunoassay. Lipoproteins (e.g., cell membranes), glycoproteins (e.g., blood group substances), and nucleoproteins (e.g., autoantigens) can be measured by immunoassay; however, antibody reagents to these are generally directed against their protein conponents. Less immunogenic molecules such as lipids (e.g., mediators), carbohydrates (e.g., bacterial polysaccharides), and nucleic acids are least frequently quantitated by immunoassay because it is difficult to produce high quality antibody reagents for the capture and detection of these molecules.

Clinically important proteins are rarely measured using tedious methods that determine their physical and chemical properties (e.g., molecular weight, net charge, hydrophobicity).[7] Rather, a variety of immunological techniques are a used to quantitate multiepitope protein antigens. These include solid- and liquid-phase immunoassays, nephelometry, and turbimetry, and immunoprecipitation (e.g., radial immunodiffusion). Solid-phase immunoassays in general have exhibited the best sensitivity and working range of these methods for the quantitation of multiepitope antigens. *In vivo* antigens circulating in the blood are readily collected by venipuncture. Of the biological specimens collected, blood may be the most uniform throughout the various populations. In contrast, other body fluids such as tears, saliva, peritoneal fluid, anniotic fluid, seminal fluid, and urine can vary in their consistency and inherent protein matrix. Tissue-associated antigens are possibly the most difficult to detect because they must be solubilized in a form which maintains their native structure.

B. *EX VIVO* ANTIGENS

Although humans generally consider their internal antigens and those of their pets and livestock to be the most diagnostically important, microbial, fungal, parasitic, and environmental antigens are frequently measured because of their role in inflammation and infectious and allergic diseases. The presence of marker antigens from infectious virus, bacteria, and

TABLE 1
Select Multivalent Antigens Measured by Solid Phase Immunometric Assays in the Clinical Allergy/Immunology Laboratory

Analyte	Method	Units	Ref.
	In Vivo Antigens		
Immunoglobulins	IRMA[a], IEMA[b]	g/l	1,2
IgG subclasses			
IgA subclasses			
Total serum IgE			
Complement	IEMA[c]	ng/ml	3
C4 allotypes (C4A, C4B)			
	Ex Vivo Antigens		
House dust allergens[d]	IEMA[b]	ng/g dust	4,5
House dust mite			
Cat epithelium			
Dog epithelium			
Cockroach			
Allergen potency	(RAST[a]	ng/ml	2,6
Short ragweed (Ambrosia elator)[e]	inhibition, IEMA[b]		

[a] IRMA = Noncompetitive binding sandwich immunoradiometric assay; RAST = radioallerosorbent test.
[b] IEMA = Noncompetitive binding sandwich immunoenzymetric assay.
[c] Activation assay is a special form of the immunoenzymetric assay.
[d] Specific aeroallergens found in house dust include house dust mite, *D. farinae* (Der f I) and *D. pteronyssinus* (Der p I); Cat epithelium, *Felis domesticus* (Fel d I); Dog epithelium, *Canis domesticus* (Can d I); and Cockroach, *Balatella germanica* (Bla g I and II).
[e] RAST inhibition is performed to standardize the allergenic content of short ragweed that is comprised of six known major allergens, Amb a I-VI.

parasites are used as indicators of infection.[8] In the field of allergy alone, more than 200 discrete extracts are routinely prepared to different species of weeds, grass and tree pollens, animal epithelium and dander, molds, dust mite excretions, insect venoms, drugs, and foods for use in allergen skin testing and as diagnostic immunoassay reagents. Immunoassays are used to identify and quantitate the potency of these antigen extracts. Moreover, antibodies have been used in conjunction with microscopy to define the species of pollens collected in communities and thus help chart the course of the pollen season.[9] Increased awareness of our personal environments has promoted the analysis of aqueous extracts of house dust by immunoassay for the presence of common allergens (Table 1).[4,5] Information from these analyses can help the allergist plan an appropriate avoidance and immunotherapy program for the asthmatic and allergic patient.

C. SYNTHETIC AND RECOMBINANT ANTIGENS

The majority of the macromolecular antigens occur in a biological millieu containing other molecules such that obtaining them in purified form is labor intensive. This problem can be circumvented in some cases by direct protein synthesis, especially for small quantities of biologically active peptides (e.g., hormones). The larger proteins which are difficult to extract from tissues have been good candidates for production by recombinant DNA technology. This technology can provide large quantities of a single protein that is free of other cross-reactive contaminants. Solid-phase immunoassays can be useful tools for manufacturing and research in the detection and quantitation of synthetic and recombinant proteins secreted from transfected cells. Hormones, like human insulin, and allergens, such as ragweed, are two examples of molecules that are being produced by recombinant DNA techniques and monitored by immunosssay.

D. ANTIGEN MULTIVALENCY

Multivalency is an inherent property of large molecular weight biomolecules and refers to the presence of more than one antigenic determinant or antibody-binding epitope on a single molecule. The size of an antigenic epitope has been estimated by determining the size of a molecule that is required to inhibit the reaction of antibody with its appropriate antigen. Classically, antibodies to polypeptides have been optimally inhibited by oligopeptides of four to five amino acids (mol wt 650), while those to dextran polysaccharide have been inhibited by six-unit saccharides (mol wt 990).[10] Thus, a 6000-mol wt protein or polysaccharide may express 6 to 10 antigenic epitopes if none are hidden or folded internally.

Steric considerations, hydrophilicity, the degree of amino acid stretch, repetition, and overlapping all influence the extent to which various regions of a molecule protrude into the aqueous environment and are accessible as binding sites for antibodies. Epitopes that are secluded within large molecules may become exposed only upon denaturation, enzymatic digestion, or antigen processing. While total denaturation of a protein can destroy its antigenicity, partial denaturation may create so-called neoantigens by minor alterations and unfolding of the molecule's tertiary structure (see Chapter 10).

The property of multivalency, possibly more than any other single characteristic of a biomolecule, allows solid-phase immunometric assays to capture and detect antigens using antibodies which bind to different immunogenic regions of the same molecule. Design of multisite immunoassays requires careful matching of capture and detection antibodies to maximize the sensitivity and working range of the assay.

III. QUANTITATIVE SOLID-PHASE IMMUNOASSAYS

Solid-phase immunoassays that use competitive and noncompetitive formats have been developed to quantitate multivalent biomolecules. Two general formats have been used: (1) the competitive, labeled-antigen immunoassay and (2) the noncompetitive immunometric assay, also known as the two-site labeled-antibody immunoassay. Format selection depends on the availablity and cost of the antigen in purified form.

A. COMPETITIVE, LABELED-ANTIGEN IMMUNOASSAY

This assay design is used when the antigen to be measured is available in an inexpensive, purified form and adaptable to direct labeling with a variety of signal generators such as radioisotopes, enzymes, fluorophors, and biotin. Antigenic determinants must not be sterically blocked or physically damaged by the label. In its generic form, the competitive, labeled-antigen immunoassay is performed by mixing a known amount of labeled antigen to a test specimen containing a variable amount of the same antigen. The mixture is then added to a limited amount of "capture" antibody (CAb) that is insolubilized on a solid phase (Figure 1). In the absence of unlabeled antigen, labeled antigen binds antibody. Increasing the amount of unlabeled antigen in the test specimen competitively reduces the amount of bound labeled antigen. This assay configuration gradually has been replaced by solid-phase labeled-antibody immunoassays.

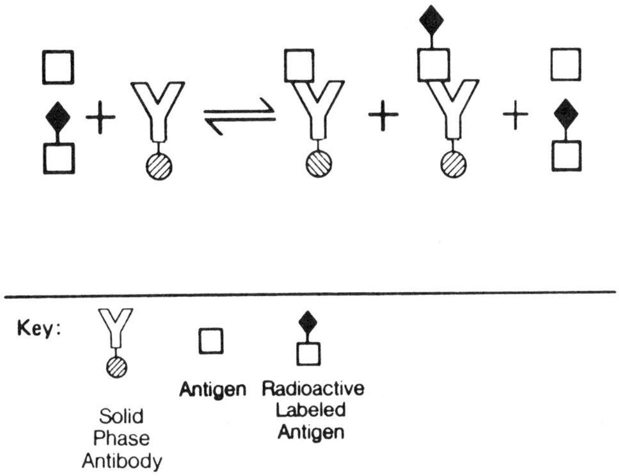

FIGURE 1. Schematic diagram of a competitive binding solid phase immunoassay. (Reproduced with permission from Reference 21.)

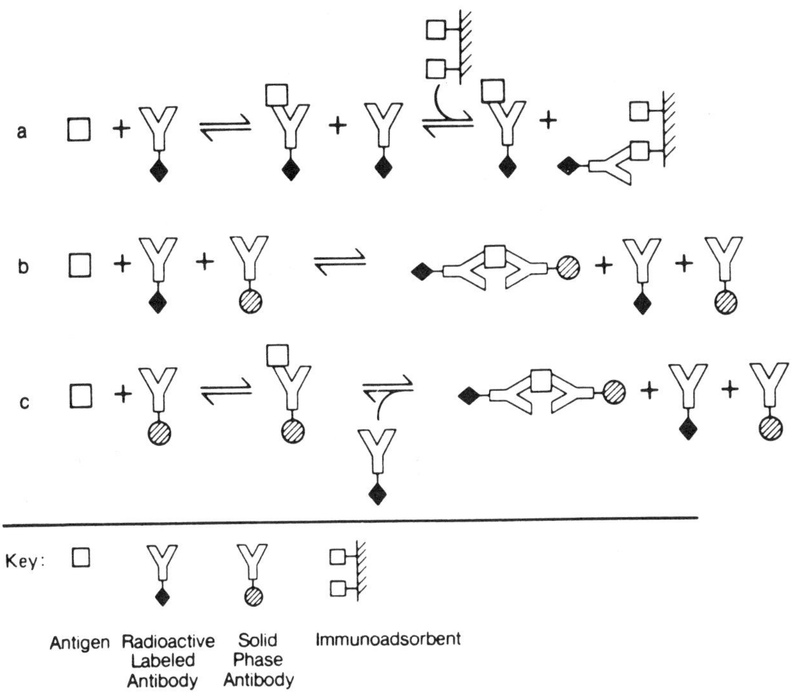

FIGURE 2. Schematic diagram of noncompetitive solid phase immunoassays. Panel a: single labeled antibody technique; b: simultaneous two-site immunoradiometric assay; c: sequential two-site immunoradiometric assay. (Reproduced with permission from Reference 21.)

B. COMPETITIVE, LABELED-ANTIBODY (IMMUNOMETRIC) ASSAY

An alternative competitive immunoassay involves the addition of the test antigen to a solution containing labeled antibody (Figure 2a). Solid-phase antigen is then added in a second step to bind free labeled antibody. The more free antigen added to the reaction mixture, the less labeled antibody will bind to solid-phase antigen. This competitive inhibition format of the solid-phase immunometric assay is used extensively to determine the "potency" or relative content of antigens in unpurified extracts. Specificity of the assay for a single antigen species is provided by a characterized polyclonal or monoclonal antibody that is directly labeled with a radioisotope, enzyme, fluorophor, or biotin.

The radioallergosorbent test or RAST is an immunometric assay that uses solid-phase allergen to detect allergen-specific IgE antibodies.[2] The RAST has been used in this competitive inhibition format to quantitate the amount of allergens in aqueous extracts of pulverized and defatted weeds, grasses, trees, molds, parasites, insects, drugs, foods, and epidermals from animals. Subsequently, the allergen mixture has been covalently attached to a cellulose or agarose particle or disk. Most recently, allergens have been adsorbed onto microtiter plate wells. In the RAST inhibition assay, polyclonal human IgE antibodies with defined specificity for the allergens of interest are combined with solid- and solution-phase allergen. The more allergen in the solution phase extract, the less IgE antibody is available for binding onto the solid phase allergen. Once this binding reaction has gone to completion, unbound antibody is removed with buffer washes and bound IgE antibody is quantitatively detected with a molar excess of labeled anti-human IgE antibody. The slope of the test and reference antigen inhibition curves are often parallel over many dilutions. The degree of inhibition at a particular dilution of test extract is compared to the inhibition observed with a standard or reference extract that contains a known concentration of allergen.[11]

Another version of the competitive inhibition solid-phase immunometric assay has been useful in the clinical assessment of cross-reactive human antibody. Since the late 1960s, IgE antibody has been an important diagnostic marker of IgE-mediated (Type I) hypersensitivity reactions. In the case of Hymenoptera venom allergens, there is well-documented cross-reactivity between vespid (e.g., Yellow Jacket, Yellow Hornet, White-Faced Hornet) and Polistes wasp venoms. It has often been difficult to determine the exact species of the offending insect which stung an allergic patient in the field. Accurate identification, however, becomes important in the selection of the proper venom for use in immunotherapy. By interchanging different soluble- and solid-phase venoms in the RAST inhibition immunoassay, one can study the extent to which the different venoms cross inhibit a patient's IgE antibodies from binding onto the venom solid phase. In this manner, one can determine whether the patient has cross reactive and/or unique venom-specific antibodies in the blood.[12]

The quality of allergen potency measurements using solid-phase immunoassays depends on several factors. Foremost is the accuracy with which allergen has been quantitated in the reference material. The International Union of Immunological Societies (IUIS) has begun a systematic preparation of allergen standards.[13] International allergen standards are available to timothy grass, short ragweed, and house dust mite. Reference allergen extracts have been characterized by physical, chemical, and immunochemical methods to ascertain the quantity of the major antigenic (and allergenic) components in the extract. Accepted standard allergens must contain >5 mg of protein per ml, >20 bands in isoelectric focusing, >3 detectable allergens, >40 μg of group I or primary allergen and the ability to bind specific IgE from human serum in the RAST. IUIS reference materials are used as primary standards against which the quantity of predominant allergens are determined in new extracts. Aqueous ragweed extracts, for instance, contain at least six allergens that differ in their molecular weights and pIs (*Amb a* I-VI)[6] (Table 1). Current ragweed extracts are also standardized in terms of their content of the group I or predominant allergen, *Amb* a I. Additional variables affecting the quality of RAST inhibition measurements include the polyclonality and variable affinity of the reference human antibodies and a multivalency of the complex allergen solid phase.

C. NONCOMPETITIVE IMMUNOMETRIC ASSAYS

The most widely performed immunoassay for the detection of multivalent antigens is the noncompetitive immunometric (labeled antibody) two-site assay. This is also called a "sandwich" assay. In its most generic form, the immunometric assay uses a solid-phase CAb to extract or bind antigen from a solution containing a mixture of cross-reactive and antigenically distinct molecules. Following the physical separation of free and bound antigen, a second immunochemical reaction is performed using a labeled "detection" antibody which binds CAb. The three primary reagents (test or reference antigen, solid-phase CAb, and labeled detection antibody) may be added together in a simultaneous (Figure 2b) or sequential mode (Figure 2c). Successful quantitation with these assays requires antigen multivalency or at least bivalency.

This noncompetitive immunometric assay has been extensively used to measure antigens in most areas of biology and medicine. Newer trends include the use of monoclonal antibodies as both capture and detection antibodies and nonisotopic labels and substrates which enhance assay sensitivity and widen its working range. Factors which influence its performance (e.g., sensitivity, specificity, precision, parallelism) include the specificity and avidity of the capture and detection antibodies, the specific activity of the labeled antibody (amount of label per unit mass of antibody), and the number of detectable epitopes on the antigen.[14]

D. ACTIVATION IMMUNOMETRIC ASSAY

Historically, a number of proteins have been used as alternative receptors to antibody to bind a specific component from a biological specimen. Intrinsic factor, for instance, has been used on a solid phase to bind vitamin B12 from serum. These assays have been called protein-binding assays; the protein must be specific and have an affinity for the particular antigen of interest.

More recently, insolublized receptor proteins have been used to not only bind but to expose unique internal antigenic sites that can be subsequently detected by labeled antibody. Aggregated human immunoglobulin G has been used as a solid-phase receptor to bind early components of human complement. A two-site immunoenzymetric assay has been established on this principle called the "activation assay". In this assay, the C1 complex (C1q/r/s) binds to aggregated human IgG on a solid phase such as a microtiter plate well. When C4 is enzymatically cleaved by the immobilized C1 complex, unique epitopes near the thioester bond are exposed in the C4 molecule. This event allows the allotypic A form of C4 to be distinquished from C4B. Enzyme-conjugated mAb specific for these unique internal epitopes in C4A and C4B are used to quantitatively detect the amount of bound C4A or C4B protein. Without activation of the C4 molecule, the allotype-unique epitopes remain hidden to labeled C4A- or C4B-specific mAbs.[3]

Quantitation of the A allotype of human C4 is accomplished with the use of a standard human serum that has been collected from a patient who is genetically predisposed to producing only C4A. By quantitating the total amount of C4 in this serum with an independent assay (e.g., nephelometry or immunodiffusion), an absolute human C4A standard is established. A human C4B standard is similarly prepared from the serum of individuals who produce only C4B. By analyzing multiple dilutions of these reference sera in each assay, and designing the assay so that all C4A or C4B is bound and activated, quantitative mass per milliliter levels of C4A and C4B are determined. Clinically, C4A and C4B levels identify individuals who are predisposed to produce lower amounts of one or both of the allotypic forms of C4 as a result of the absence of C4A and C4B genes or an abnormality in transcription or translation of C4 protein. C4A and C4B allotypes differ in their ability to bind to proteins (antibody complexes) and carbohydrates. It has been postulated that differential levels of C4A and C4B in blood of patients with autoimmune diseases may contribute to the variable rate at which immune complexes are cleared from the blood of these patients.

E. ASSAY FORMAT SELECTION

The first immunoassays reported in the 1960s were of the competitive, labeled-antigen type (Chapter 1). Today, noncompetitive immunoassays are attractive because of their simplicity and robust nature. While direct comparison of different assay formats is complicated by the many assay variables, the overall performance of the noncompetitive and competitive solid-phase immunoassays can be compared using the same antibodies. We have examined the use of a panel of IUIS-documented monoclonal anti-human IgG1, IgG2, IgG3, and IgG4 mAbs in competitive and noncompetitive immunoassays. In the measurement of total serum IgG1, the competitive immunoassay used enzyme-labeled IgG1 myeloma protein that competed with unlabeled IgG1 in the test serum for a limited amount of anti-human IgG1 mAb insolubilized on a solid phase. The noncompetitive solid phase immunoenzymetric assay in this study used the same anti-IgG1 mAb as a CAb on the microtiter plate and an enzyme-conjugated anti-human IgG Fc to complete the two-site assay. Both assays were able to achieve acceptable precision and reproducibility (e.g., intra- and interassay coefficients of variation <5 and <10%, respectively). Providing that the test serum did not contain paraproteins, both assay formats displayed comparable parallelism (e.g., interdilutional % CVs <15%). The working range of the noncompetitive immunoassay was broader. Both assay formats displayed a dependence of the working range on the time of substrate development and the level of nonspecific binding.[15,16] The decision to use a noncompetitive two-site assay in routine clinical testing was ultimately based on the low cost of monoclonal CAbs and detection antibodies and difficulty in obtaining consistent lots of purified human IgG1 paraprotein that can be labeled.

IV. SPECIMEN CONSIDERATIONS

Optimum assay results are obtained when the protein composition of the test sample is comparable to that of reference and control specimens. An imbalance in the protein content or composition between the standard and test specimens contributes to nonparallelism and random inaccuracy (see Chapter 9). Parallelism or the ability of a test specimen to titrate in a manner analogous to the reference specimen, is an important general indicator of the performance of a quantitative immunoassay. Biological fluids to be assayed can also include tears, saliva, nasal washes, bile, aqueous extracts of tissues and cells, tissue culture supernatants, and extracts of environmental specimens such as house dust. Each of these is likely to exhibit variation in their protein composition and therefore produce nonparallelism when compared to a serum or an extract standard that has been used to construct a dose-response curve. One strategy for dealing with variation in protein content of biological specimens is the use of an internal normalizing variable such as total protein or albumin in the specimen.

Serum can contain a variety of substances such as rheumatoid factors and autoantibody-antigen complexes which interfere in immunoassays.[17] Rheumatoid factors are a heterogeneous population of autoantibodies to immunoglobulin that circulate in the blood of individuals with a variety of inflammatory diseases including rheumatoid arthritis. They produce variable enhancement of antibody-binding and immune complex formation in immunoassays depending on their specificity for antibodies of different species. RF interference has been cited as the most unpredictable and poorly appreciated form of serum-factor interference in immunoassays that employ antibodies as capture and/or detection reagents. While various procedures have been proposed to minimize RF interference, the only sure method to eliminate RF interference is extraction of the RF from the specimen before analysis.

V. REAGENT CONSIDERATIONS

General procedures for reagent preparation and optimization that are important to all immunoassays naturally apply to those assays that measure multivalent antigens. In addition, certain issues should be emphasized.

A. SOLID-PHASE ANTIBODY REAGENTS

Methods for immobilizing antibody onto the solid phase vary from simple adsorption to covalent coupling. Adsorption of purified antibody in a nondetergent buffer onto microtiter plate wells is the simplest method (Chapters 12 and 17). mAb in ascites, contrary to some suggestions, does not efficiently adsorb onto plastic surfaces, even when it is diluted. Ascites contains host proteins (e.g., albumin, transferrin) that vary widely in concentration between different batches of ascites. Additionally, the pI of mAbs vary between 6.0 and 8.0 pH units; thus, different mAbs may not be absorbed onto plastic surfaces with equal efficiency.

Both mobile and stationary solid phases have been used in antigen immunoassays. Mobile solid phases that are free to move in solution have included agarose beads, cellulose disks, dextran beads, ferromagnetic particles, gelatin capsules, latex particles, liposomes, methylmethacrylate particles, nylon balls, polystyrene balls, porous glass beads, red blood cells, *S. aureus* Protein A, gold particles, and shell core particles. Stationary or rigid solid phases have included capillary tubes, cellulose threads, chromatography tubes, glass rods, indium-coated slides, microtiter plates, nitrocellulose, paper-mylar-polystyrene strips, plastic tubes, and sheets of plexiglass or polymethyl methacrylates. The binding capacity of each for antibody varies widely; solid phases are discussed in Chapter 2 and elsewhere.[18]

Early solid phases such as agarose (Sepharose®) and cellulose (disks, strips) permitted the attachment of antigens by covalent bonds for use in immunoassays and affinity chromatography. Some diagnostic allergy assays still use cellulose discs as a solid phase in an immunometric assay for total and allergen-specific IgE. Studies directly comparing the binding capacity of agarose and cellulose after cyanogen bromide activation have shown that agarose possesses the higher antibody-binding capacity. One concern with these CNBr-activated solid phases has been the variable amount of antibody that leaches during storage. Solid phases such as these are generally washed to remove unbound antibody before they are used in an assay. Otherwise, released antibody will compete with solid phase antibody for antigen; the consequence is a reduction in the sensitivity and working range of the assay.

Polyclonal antibodies possess many of the ideal characteristics desired for capture reagents in two-site assays (see Chapter 12). They bind to multiple antigen sites on a molecule and display a spectrum of affinities which enhances overall assay performance. However, mAbs possess well-known advantages such as their production in large quantities, a defined specificity and a theoretical immortality. They allow long-term standardization and stable specificity that can reduce interassay variation. In cases where multiple, structurally similar molecules are measured in the same specimen (e.g., IgG and IgA subclasses), mAbs have become an absolute necessity for insuring assay specificity.

Cocktails of mAbs may eventually mimic polyclonal antibody performance. Bias, however, can occur in which some mAbs exhibit preference for a particular allotypic form of the antigen. For instance, HP6014 is a mouse IgG1 monoclonal antibody with specificity for human IgG2 Fd that has been widely used in total human IgG2 immunoassays. It displays a tenfold higher affinity for the Fd region of IgG2 antibodies containing λ-light chain as compared to those containing κ-light chain.[19] Light-chain bias produces erroneous total serum IgG2 estimates[1] and poor immunoassay accuracy if (1) HP6014 antibodies are used alone on a solid phase to capture IgG2 in an immunoenzymetric assay, (2) the standard serum contains IgG2 antibodies of one predominant light-chain type, (3) the population of IgG2 antibodies in the test sera possess different proportions of light chain when compared to the reference serum, and (4) the binding reactions are not allowed to go to completion. To eliminate these concerns, HP6002, an IUIS-documented companion mouse monoclonal anti-human IgG2 Fc antibody, has been used either alone or together with HP6014 to balance the binding of human IgG2 antibodies with different light chains. Moreover, a pooled serum, such as the World Health Organization's 67/97, that contains a normal light chain distribution of IgG2 antibodies, can be used as the reference serum to construct an IgG2 dose-response curve.

There are two general orientations of antibody solid phases that allow the quantitation of one member of a family of structurally similar molecules which is present in a specimen in unequal amounts. Measurement of total serum human IgG3 is used for illustration. One human IgG3 immunoenzymetric assay involves the elective capture of IgG3 from serum containing IgG1, IgG2, IgG3, and IgG4 using an IgG3-specific CAb. In an asymetrical two-site format, i.e., asymetrical sandwich assay, bound IgG3 is then detected with either labeled anti-human IgG3 that binds to a different determinant on IgG3 or a labeled anti-human IgG PAN reagent that binds equally to all four subclasses of IgG. A second configuration uses a solid-phase anti-human IgG PAN to bind all four human IgG subclasses from the specimen. Bound IgG3 is then detected with a labeled anti-human IgG3 reagent. The first strategy has been most successful.[1,15] The latter approach allows high levels of IgG1 and IgG2 to competitively inhibit the binding of IgG3 to the anti-IgG PAN CAb. Dilution of the specimen to overcome this CAb-limiting constraint can dilute the IgG3 beyond the sensitivity of the assay. Thus, it is best to extract the analyte of interest from a mixture of substances first and then detect it with a reagent that binds structurally similar molecules.

Figure 3 demonstrates the dependency of the immunoenzymetric two-site assay on the affinity of the monoclonal CAb. In this example, equal amounts of two human IgG3 hinge specific mAbs (clones HP6047 and HP6050) with comparable specificities and different affinities (18.0×10^7 and 2.4×10^7 l/M, respectively) were adsorbed separately and together onto microtiter plate wells. Both assays used the same reference serum dilutions, peroxidase-conjugated detection antibody, and substrate and they were developed in parallel. The assay system that used the lower-affinity mAb (HP6050) displayed a reduced sensitivity, flatter dose-response curve, and wider working range than the assay that used HP6047. Thus, if maximum sensitivity is desired, the assay should use the highest-affinity and specificity mAb reagent available.

B. LABELED REAGENTS

There are no absolute guidelines for selection of the ideal labeled reagent. General considerations relate equally to the use of labeled antigen and detection antibody. The most commonly used reporter or signal-generating labels are radioisotopes, enzymes, and fluorophors. Alternative nonisotopic labels include molecules that allow the measurement of luminesce (e.g., luminol, acridium esters) and electronic spin resonance (e.g., nitroxide radical and 2,4,dinitrophenyl). Chapters 4 and 5 treat the topic of signal generators in detail.

Methods of conjugation are beyond the scope of this chapter; they are reviewed elsewhere.[20] Once prepared, the labeled reagent must be maintained in an immunoreactive form. Accelerated stability studies which involve incubation of the label at 37°C for up to 3 weeks help identify the ideal method of storage for maintenance of immunoreactivity. Ideally, the specific activity or amount of label per mass of reagent should be as high as possible without reducing the ability of the labeled reagent to bind. Finally, convenience and practical issues need to be considered when selecting a label. These relate to what counting equipment is available, the number of distinct assay steps which affect the assay turnaround time (e.g., use of a directly labeled reagent vs. an indirect label such as a second antibody or an avidin-biotin label system), and the experience of the investigator.

C. STANDARDS

The rules for preparing a multivalent reference antigen are generic to all standards.[18,21] The analyte should be present in a protein matrix identical to the test specimen (e.g., a human IgGl standard should be a human serum pool of at least 100 individuals with different genetic backgrounds). Thus, dilutions of the standard antigen should interact with the assay reagents in a manner identical to the test antigen. This insures parallelism of the system. The standard should represent the major allotypic and isotypic forms of the antigen molecule. Assays which use reagents that are known to have a systematic bias for one isotypic form of the antigen

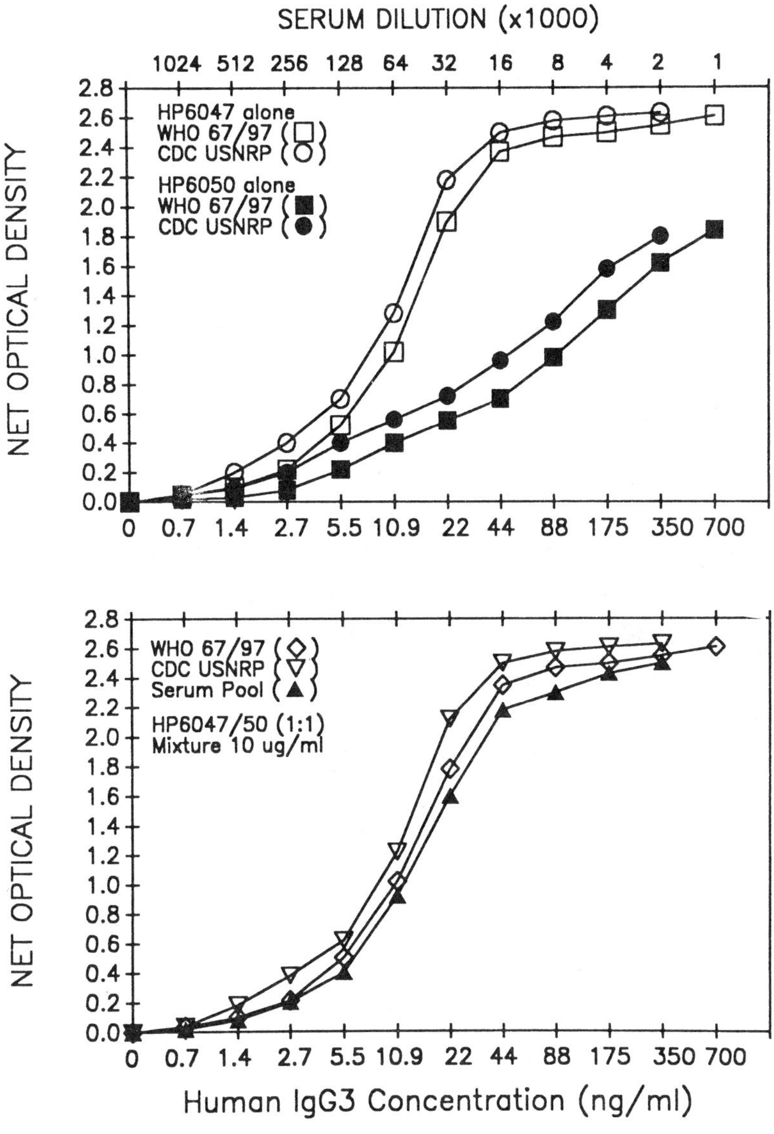

FIGURE 3. Dose response curves generated in a noncompetitive two-site IEMA for human IgG3 using the Centers for Disease Control and World Health Organization reference sera and a human serum pool (UT-86). Immulon® 2 microtiter plate removawells were coated with mAb to human IgG subclass 3 hinge (HP6050 or HP6047 alone, top panel; HP6050 + HP6047, 1:1 mixture, bottom panel) at 10 μg/ml in PBS (0.1 ml per well) overnight at 4°C. Plates were blocked with PBS-BSA, washed with PBS, and serum was added (0.1 ml per well). After 2 h at 37°C, plates were rewashed and alkaline phosphate-conjugated polyclonal mouse anti-human IgG vas added. Following 1 h at 37°C, unbound Conjugate was removed with PBS and *p*-nitrophenylphosphate substrate was added. The assay sensitivity was enhanced and working range limited with the use of the higher-affinity antibody (HP6047, ka = 18.0×10^7 l/*M* vs. HP6050, Ka = 2.4×10^7 l/*M*) and especially using both antibodies in combination. Arrows depict the IEMA's minimal detectable dose.

should be so designated. The standard should be provided in a stable, sterile, noninfectious form. Finally, the quantity of analyte assigned to the standard should be determined by cross calibration against the best-documented primary standards available.

REFERENCES

1. **Hamilton, R. G., Wilson, R. W., Spillman, T., and Roebber, M.,** Monoclonal antibody-based immunoenzymetric assays for quantification of human IgG and its four subclasses, *J. Immunoassay,* 9, 275, 1988.
2. **Hamilton, R. G. and Adkinson, N. F., Jr.,** Assessment and management of human allergic disease, in *Clinics in Laboratory Medicine,* Vol. 5, Nakamura, R., Ed., W.B. Saunders, Philadelphia, 1986, 1.
3. **Moulds, J. M., Arnett, F., Giles, C., and Hamilton, R. G.,** A novel immunosssay for the quantitation of human C4 gene products, *Complement Inflam.,* in press.
4. **Chapman, M. D., Heymann, P. W., Wilkins, S. R., Brown, M. J., and Platts-Mills,T. A. E.,** Monoclonal immunoassays for major dust mite (*Dermatophagoides*) allergens, Der p I and Der F I and quantitative analysis of allergen content of mite and house dust extracts, *J. Allergy Clin. Immunol.,* 80, 184, 1987.
5. **Chapman, M. D., Aalberse, R. C., Brown, M. J., and Platts-Mills,T. A. E.,** Monoclonal antibodies to the major feline allergen Fel d I. II. Development of a sensitive two-site immunoassay to assess Fel d I exposure, *J. Immunol.,* 140, 812, 1988.
6. **King, T. P. and Norman, P. S.,** Isolation studies of allergens from ragweed pollen, Biochemistry, 1, 709, 1964.
7. **Dunbar, B. S., Ed.,** Analysis of post-translational modification of proteins by high resolution two-dimensional PAGE electrophoresis in *Two-Dimensional Electrophoresis and Immunological Techniques,* Plenum Press, New York, 1987, 77.
8. **Hamilton, R. G.,** Application of immunoassay methods in the serodiagnosis of human filariasis, *Rev. Infect. Dis.,* 7, 837, 1985.
9. **Schumacher, M. J., Griffith R. D., and O'Rourke, M. K.,** Recognition of pollen and other particulate aeroantigens by imnunoblot microscopy, *J. Allergy Clin. Immunol.,* 82, 608, 1988.
10. **Sell, S.,** Antigens and immunogens, in *Immunology, Immunopathology and Immunity,* Elsevier, New York, 1987, 73.
11. **Yman, N., Ponterius, G., and Brandt, R.,** RAST-based allergen assay methods, *Dev. Biol. Stand.,* 29, 151, 1975.
12. **Hamilton, R. G., Wisenauer, J., Valentine, M., Golden, D., and Adkinson, N. F., Jr.,** Hymenoptera venom RAST inhibition test facilitates selection of venoms for immunotherapy, *J. Allergy Clin. Immunol.,* 85, 213, 1990.
13. **Helm, R. M., Gauerke, M. B., Baer, H., Lowenstein, H., Ford, A., Levy, D., Norman, P. S., and Yunginger, J. W.,** Production and testing of an international reference standard of short ragweed pollen extract, *J. Allergy Clin. Immunol.,* 75, 676, 1984.
14. **Raggatt, P. R. and Hales, C. N.,** Immunoassays using labelled antigens or antibodies, in *Clinical Aspects of Immunology,* Vol. 1, 4th ed., Lachmann and Peters, Blackwell, Boston, 1982, chap. 12.
15. **Hamilton, R. G. and Adkinson, N. F., Jr.,** Quantitative aspects of solid phase immunoassays, in *ELISA and Other Solid Phase Immunoassays,* Kemeny, D.M. and Challacombe,S.J., Eds., 1988, chap. 3.
16. **Butler, J. E.,** Principles and applications for the quantiative determination of immunoglobulins, in *Practical Aspects of ELISA,* Kemeny, D. M. and Challacombe, J. J., Eds., John Wiley & Sons, New York, 1987, 155.
17. **Hamilton, R. G.,** Rheumatoid factor interference in immunological methods, *Monogr. Allergy,* 26, 27, 1989.
18. **Hamilton, R. G.,** Production of immunoassay reagents, in *Immunoassay: A Practical Guide,* Chan, D. W. and Perlstein, M. T., Eds., Academic Press, New York, 1987, 25.
19. **Phillips, D. J., Wells, T. W., and Reimer, C. B.,** Estimation of association constants of monoclonal antibodies to human IgG epitopes using a fluorescent sequential-saturation assay, *J. Immunol. Methods,* 34, 315, 1980.
20. **Kabakoff, D. S.,** Chemical aspects of enzyme immunoassay, in *Enzyme Immunoassay,* Maggio, E. T., Ed., CRC Press, Boca Raton, FL, 1980, chap.4.
21. **Chan, D. W.,** General principles of immunoassay, in *Immunoassay: A Practical Guide,* Chan D. W. and Perlstein, M. T., Eds., Academic Press, New York, 1987, 1.

Chapter 8

IMMUNOBLOTTING: MEMBRANE FILTERS AS THE SOLID PHASE FOR IMMUNOASSAYS

William R. Brown, Steven E. Dierks, John E. Butler, and J. M. Gershoni

TABLE OF CONTENTS

I. INTRODUCTION

Immunoblotting techniques are ubiquitous as tools in basic and clinical research.[1-5,8] Therefore, any discussion of immunochemical aspects of solid-phase immunoassays (SPI) would be incomplete without inclusion of this methodology. Immunoblotting is an SPI because the membrane matrix involved is a particulate, insoluble, and nondiffusible reaction ingredient. The interaction of proteins with these membranes and their functional properties after immobilization are the primary focus of this chapter. Many investigators peripheral to the biotechnology of immunoblotting as well as a few who use it regularly, consider "all immunoblotting membranes to be nitrocellulose (NC)." NC has become a generic term for all materials which look like filters and which are used in immunoassays. To account for the diversity among membranes, which is described below, the generic term "blotting membrane" or "membrane" will be used throughout Chapter 8.

There are no recognized immunochemical principles that are unique to immunoblot assays, but rather are those common to most SPI. Many of these have been discussed in other chapters of this book as well as in previous reviews on immunoblotting and other SPI.[1-5,8,9] We will concentrate our discussions on two parameters which we believe are critical for understanding the immunochemical interactions associated with immunoblotting. These are (1) the binding properties of different blotting membranes and (2) the functional activity of the protein which has been bound to these membranes. This chapter will discuss factors such as matrix chemistry, method of transfer to the membrane, and binding chemistry.

Immunoblotting techniques emerged out of a need to characterize proteins after electrophoretic separation. Prior to the development of immunoblotting, the separated antigens or antibodies were detected using methods dependent on immunoprecipitation in agar gels, i.e., immunodiffusion and immunoelectrophoresis.[12] Later methods relied on the diffusion of radiolabeled or enzyme-labeled proteins into the gel.[13] These methods were lengthy and, as a consequence, the resolution of the original electrophoretic separation was often lost.

In 1975, Southern[14] described a method for the transfer of electrophoretically separated DNA from agarose gels to NC. Scientists involved in immunochemical characterization of proteins soon realized the obvious benefits of transferring proteins from a gel to an immobilizing membrane and in 1979 two groups developed and presented methods for the transfer of electrophoretically separated proteins to activated cellulose and NC membranes and their subsequent immunodetection.[15,16] Thus, the high resolution of polyacrylamide gel electrophoresis (PAGE) could be combined with the sensitivity and flexibility of SPI.

The main advantages of detecting proteins immobilized on blotting membranes include (1) short incubation times for immunochemical reactions due to the greater porosity and smaller volume of the matrix compared to agarose,* (2) the applicability of most well-known immunodetection systems and protein staining techniques, and (3) the smaller amounts of reagents (antigen, specific antibody, secondary antibodies) required.[17] Several additional characteristics have made immunoblotting a powerful tool for research and clinical analysis: (1) the transferred protein may be stored for months before further processing, (2) blots may be reprobed, and (3) specific binding may also be attained with ligands other than antibodies, e.g., hormones or toxins. A number of fairly comprehensive reviews of both experimental and clinical applications of immunoblotting are available.[1-5,8,18]

II. PRACTICAL AND IMMUNOCHEMICAL CONSIDERATIONS

A. METHOD OF TRANSFER

Immunoassays on porous membranes can be divided into two general groups, protein

* The smaller volume of the matrix could allow higher concentrations of reactants which would lead to shorter reaction times.

transfers and "dot blots".* Transfer refers to immunoblot assays which involve the transfer of electrophoretically separated proteins from a gel to an immobilizing matrix. In dot blot assays the protein sample is applied directly to the membrane. The method of application to the membrane can affect the maximal amount of protein bound to the membrane as well as its immunoreactivity.**

Dot blot assays were developed in 1982 in an effort to simplify the screening of large numbers of hybridoma clones for specific antibody production.[19,20] In practice, small aliquots (0.2 to 10 μl) of samples are applied to the surface of the membrane and allowed to penetrate into the membrane by diffusion. The membrane is then allowed to dry and the blotted substance detected immunochemically. In order to accumulate protein from dilute solutions, larger volumes can be tested by repeated application onto the same spot or with the aid of specially designed vacuum filtration manifolds (see Chapter 15). The latter procedure allows a greater accumulation of the protein on the membrane surface and provides a means to overcome the diffusion dependence of solid-phase interactions (see Chapters 18 and 19). Theoretically, this approach allows protein to be adsorbed until the maximal capacity of the membrane has been reached. The actual amount of protein which eventually binds to the membrane is variable and depends on the concentration of the protein solution, the chemistry of both the membrane and the protein, and the conditions of the binding reaction.

The method of application of proteins to membranes may also alter the ability of the immunodetection system to recognize the adsorbed molecule. Adsorption of proteins to a solid phase has been shown to alter their conformation[21,22] which leads to decreased immunochemical activity[23,24] (see Chapters 1, 2, and 10 to 13). The drying of the membrane after the protein application process may further increase the denaturation of the adsorbed molecule. This can be circumvented by storage of blots in buffered solutions or by freezing while still damp.

A variety of methods has been used to transfer protein patterns from gels to membranes. These include transfer by diffusion,[25] capillary flow,[14] capillary flow aided by vacuum,[26] and electrophoretic transfer.[16,27]

The transfer methods, which do not rely on electrophoretic forces, are not restricted to any particular buffer system and can be modified to facilitate the solubility of the protein in question. They have, however, been found to be reasonably efficient only when large pore-size gel matrices like agarose are used, as in the case of isoelectrophoresis. Diffusion transfer from acrylamide gels requires increased transfer time, has limited transfer efficiency, and may result in some loss of resolution.

Electrophoretic transfer of proteins is dependent on the applied field strength (volts per centimeter), ionic strength and pH of the transfer buffer, composition of the gel matrix, and composition of the transfer buffer.[5-7] Towbin's original procedure involved the transfer of proteins from urea-PAGE gels to nitrocellulose in a buffer containing methanol.[16] Transfer was achieved in 1 h with a field strength of 6 V/cm. A variety of other buffer systems has also been used which vary in ionic composition and may or may not contain methanol or detergent.[2,28]

Methanol and detergents have been shown to have a significant effect on the elution of proteins from the gel as well as on the binding capacity of the membrane. Transfer buffers containing SDS can facilitate the transfer of high molecular weight proteins[29] from the gel matrix but at the expense of protein adsorption to the membrane.[2,28] Transfer buffer containing methanol has been shown to increase the binding capacity of certain membranes,[28] but at the same time this causes a decrease in porosity of the gel matrix which can hinder the elution of larger proteins. The methanol used in SDS-PAGE electrotransfer buffers has been reported to facilitate binding to NC by removing SDS from SDS-protein complexes thereby favoring

* Terms such as dot immunoassay, dot immunoblot, dot spot, immunodot, dot blot, slot blot, or antigen dot test are essentially identical methods.

** In transfer blotting, the sample is dissociated into its constituents and often denatured as a result of PAGE, whereas much less protein alteration can be expected with dot blots.

hydrophobic interaction with the membrane.[30] The methanol in Towbin's original transfer buffer was added to combat swelling of the urea-containing gel during the transfer process. The continued use of methanol in alternative systems, in which gel swelling is negligible and enhanced binding is not observed, may be due more to tradition than necessity.

The ionic strength of the transfer buffer directly determines its conductance and thus the electrophoretic characteristics of the system. It may also affect the binding capacity of a membrane in addition to influencing the efficiency of transfer. Hydrophobic forces have been hypothesized to play a significant role in protein binding to membranes.[1-4] These hydrophobic forces are increased in the presence of high ionic-strength buffers although the use of such buffers can also result in protein denaturation due to excessive Joule heating during electrophoretic transfer. In general, buffers of low ionic strength, such as 25 m*M* Tris, 192 m*M* glycine, pH 8.3, are recommended even though a higher ionic-strength buffer would promote enhanced hydrophobic interactions.

The influence of pH on the binding of proteins to blotting membranes has been studied by Nyholm and Ramlau.[31] They investigated the binding of several proteins to NC over a wide pH range (2 to 12) and found that the affinity of each varied over the pH range with maximal binding occurring at pH values close to the pI value for each protein. This might be expected if hydrophobic interactions are the primary binding force and if electrostatic forces are inhibitory. However, these investigators also observed increased binding at the pH extremes. The increased binding at pH 2 and 12 was hypothesized to be caused by the exposure of internal hydrophobic residues of the protein after pH-induced denaturation. Previous studies on protein binding to polystyrene ELISA plates indicate that, although some differences can be observed, the pH of the adsorption reaction does not appear to be critical.[32,33] These results suggest that because pH can influence the binding of protein to NC, but not polystyrene, the mechanism of adsorption differs between these two surfaces. If other noncovalent binding forces, e.g., electrostatic or van der Waals, are primarily responsible for binding to membrane, one would expect the pH of the coupling reaction to have a significant effect on binding. In studies by Ramlau[34] and Gershoni and Palade,[28] no significant influence of pH on protein binding or release from charge modified nylon membranes was observed. Hence, the forces involved in the adsorption of proteins to blotting membrane may not be the same for all membranes. The binding to charge-modified nylon obviously benefits from electrostatic interactions as can be demonstrated from the comparison of "neutral" nylon 66 with the charge-modified derivative. However, binding studies at different pH illustrate that hydrophobic contacts play a significant role in protein adsorption on these membranes as well. In the case of Immobilon-AV,* a polyvinylidene difluoride (PVDF) membrane from Millipore which has been modified to block hydrophobic sites, binding has been ascribed to covalent bonding of ε-amino and arginine groups of the protein.[35]

B. BINDING MATRICES

A large number of membrane matrices are now available for use in immunoblotting assays. The most commonly used membranes are made of NC. These are available both as the pure cellulose nitrate ester and in the form of cellulose nitrate-cellulose acetate mixed esters. Other materials which have been used for immunoblotting are cationic nylon membranes ("charge modified"), uncharged nylon 66 membranes, and PVDF membranes. As discussed above, hydrophobic, electrostatic, and even covalent interactions can be involved in protein binding; membranes involving the latter, however, have not gained the popularity of those dependent on noncovalent interactions.

One of the positive features of using membrane matrices in solid-phase immunoassays is

* Immobilon-P is a PVDF membrane sold by the Millipore Corporation, not to be confused with "Immobilon" or "Immobilon-AV" which has been modified for covalent interaction.

their high protein-binding capacity compared to surfaces such as polystyrene as used in ELISA.[16,36] Since the nature of the adsorption to some membranes, e.g., nylon, may be the same as that for polystyrene, one reason for the greater capacity of membranes is their immense internal surface. It has been estimated that the internal surface can be 100 to 1000 times the frontal membrane area, depending on porosity.[31] Membranes consist of small channels (pores) which extend throughout their thickness, much like the pores found in the particles used for gel permeation chromatography (e.g., Sephadex®).

NC membranes are the most popular membranes for immunoblot assays and the pure ester membranes are preferred over the mixed ester membranes because of higher binding capacities and greater stability of protein binding.[3] NC membranes are easy to use but tend to become brittle after multiple probings or any harsh treatment. Recently, NC membranes with an inert backing have become available. This backing provides greater durability for the membrane but gives the surface an undesired texture resulting in decreased resolution during protein blotting.

Nylon membranes can also be used for immunoblot assays. Nylon membranes have an even higher protein-binding capacity than NC.[28,34] For cationic charge-modified nylon, this has been attributed to the large increase in ionic interactions[28,34] but as mentioned above, other forces apparently play a role in protein binding to this type of membrane. Higher binding capacities for unmodified nylon membranes are reportedly due to intrinsic charge groups which enable electrostatic contacts in addition to hydrophobic interactions; this fact is illustrated by the binding of RNA to these membranes as compared to NC. Although methanol was found to increase binding to NC, it has been shown to be unnecessary for high levels of protein binding to nylon membranes. The major problem associated with the use of nylon membranes in immunoblotting is increased nonspecific protein binding which often results in high background. In order to counteract this problem, solutions containing up to 10% BSA, other unrelated proteins such as hemoglobin, or mixtures like skim milk are used to quench unoccupied binding sites.

PVDF membranes are one of the newest additions on the immunoblot scene and may turn out to be the preferred membrane in the future. Conventional PVDF membranes are extremely hydrophobic when dry and must be "wetted" rendering them hydrophilic before use. Once bound, proteins are therefore retained exclusively via hydrophobic interactions. PVDF* has also been reported to have the highest retention of specific protein during subsequent blocking steps compared to NC and cationic charge modified nylon.[37] This could be particularly important when detecting proteins in very low concentrations (see below). PVDF is also very durable, resistant to many organic solvents, and is amenable to total protein staining procedures.

C. BLOCKING PROCEDURES

The immunochemical detection of proteins bound to membrane matrices requires a "blocking" (or quenching) step prior to the detection step. Unoccupied binding sites on the membrane must be blocked to prevent nonspecific binding in subsequent reaction steps. The blocking process is accomplished with detergent or protein solutions.** The actual blocking procedure and contents of the blocking solution depends on the type of membrane being used as well as the nature of the blotted material. The choice of blocking procedure is critical to the success of the assay.

* PVDF is not regarded as a high-capacity membrane which is probably the reason that Millipore has introduced Immobilon-N, a charge-modified PVDF.

** "Blocking" is a generic term, not a chemical process. Detergent and proteins may block via different mechanisms. Protein blocking on membrane binding sites may be relatively irreversible compared to detergent blocking. Furthermore, detergents may block by binding to the hydrophobic sites of proteins in solution which in turn prevents their hydrophobic interaction with membranes.

Detergent solutions are the most economical blocking reagent and are fairly effective when working with NC. Ionic detergents such as SDS and some nonionic detergents, e.g., Triton® X-100, are not very useful as blocking reagents because they can cause unacceptable losses of initially bound protein.[38] Data presented in Section III.B might suggest that the extent of such loss may be dependent on the initial incubation time. Milder nonionic detergents such as Tween® 20 and Nonidet® P-40 are the more commonly used blocking reagents. Nonidet® P-40 has also been reported to promote renaturation of adsorbed proteins as measured by the accelerated reappearance of antigenic sites on antibodies adsorbed to NC.[39] This interesting result implies that incubation of an adsorbed protein may renature to some extent by incubation with nonionic detergents probably by partial disruption of weak hydrophobic bonds between the protein and membrane. Although detergent-induced partial renaturation of solid-phase adsorbed proteins may be possible, there may be a very fine line between renaturation and total desorption of the protein from the membrane. To take advantage of this characteristic of detergent blocking solutions it may be necessary to perform detailed studies using a variety of detergents over a range of concentrations for each particular immunoblot assay in order to arrive at the correct set of conditions.

Protein blocking solutions tend to be much more effective at inhibiting nonspecific reactions between the membrane and detection systems than detergent solutions.[40] It is well known that protein solutions (up to 5 to 10%) are necessary to effectively inhibit nonspecific binding to nylon membranes. Unfortunately, protein blocking solutions can also cause desorption of substantial amounts of dotted or transferred proteins (Gershoni, unpublished results). This desorption can lead to false negative results if the levels of these proteins are reduced below the sensitivity of the assay.

III. THE ADSORPTION CHARACTERISTICS OF BLOTTING MEMBRANES FOR PROTEINS

A. THE CHARACTERISTICS OF ADSORPTION AND ITS IMMUNOCHEMICAL IMPLICATIONS

The previous section discussed the probable mechanisms by which proteins are adsorbed to blotting membranes. Unlike plastics, in which the mechanism appears to be almost exclusively hydrophobic (Chapters 1 and 2), adsorption on blotting membranes can also involve charge interactions or even covalent bonding, depending on the blotting membrane involved. As shown with studies on polystyrene, methacrylate, vinyl, and other plastic surfaces, immobilization via adsorption as well as covalent attachment can produce conformational changes in the biomolecules with a corresponding change in their activity, e.g., antigenic alteration and loss of antibody or enzymic activity (see Chapters 1, 10 to 13, and 17). The extent to which similar changes occur when proteins are bound to blotting membranes is not available in the literature. Furthermore, the generic nature of blotting membranes means that different membranes will produce different effects. Add to this the variety of conditions under which biomolecules are bound to blotting membranes: (1) electrolytic transfer, (2) direct application to wetted membranes, (3) direct application to dry membranes, (4) adsorption at different pHs, (5) adsorption at different concentrations of protein, and (6) binding in the presence or absence of ionic and nonionic detergents.

It is quite likely that the many variables that can affect the binding of biomolecules to blotting membranes yield a spectrum of results both in terms of the concentration of the bound biomolecules and their bioactivity. The latter topic will be further discussed in Section IV.

A review of the literature indicates the absence of any systematic attempt to compare the behavior of blotting membranes with plastic, e.g., polystyrene, in terms of capacity, kinetics, and mechanism. Therefore, the studies reported in this section were undertaken to gain insight into this comparative aspect of immunoblotting.

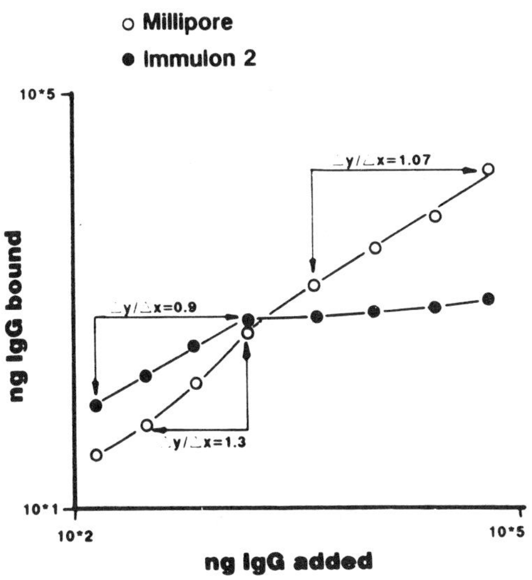

FIGURE 1. Comparison of the capacity of equal planar surface areas of 0.45 µm Millipore NC and polystyrene (Immulon® 2, Dynatech Industries) to stably adsorb ^{125}I-labeled rabbit [125]IgG. 200 µl of IgG solution were incubated either in Immulon® 2 Removawells or in covered porcelain spot plates containing a 15-mm diameter membrane disk. Data were obtained after incubation for 16 h at 4°C followed by washing in a Tween®-20-containing wash solution [PBS/Tween®-20 (0.05°%)]; microtiter wells were washed with a 200-µl volume three times each (10 sec) and NC disks, 0.45 µ, were washed with a 2-ml volume three times each (5 min) before radioactivity was measured.

B. THE ADSORPTIVE PROPERTIES OF IgG FOR NITROCELLULOSE (NC) COMPARED TO POLYSTYRENE

1. General Features and Apparent Differences in Capacity

Figure 1 illustrates the comparative capacity of an equal planar surface area of 0.45 µm Milli-pore NC* and Immulon® 2 polystyrene (Dynatech Industries, Alexandria, VA) to stably ad-sorb rabbit IgG dissolved in 0.1 *M* carbonate buffer, pH 9.6 (CB). 15-mm diameter discs were incubated with 200 µl of the IgG solution in porcelain spot plates; incubation with polystyrene was done directly in "Removawells" (Dynatech). Membranes were prewetted for 30 min. This general procedure was used throughout the studies reported in this chapter. Data were obtained after incubation for 16 h followed by washing in a Tween® 20-containing wash solution.

The adsorption of IgG on polystyrene proceeds in a log-log linear fashion (slope = 0.9 to 1.0) until approximately 600 ng per well are bound. The addition of higher concentrations results in a further increase in the adsorption of IgG but this follows a different relationship. In contrast, adsorption on NC proceeds in an approximate log-log linear fashion to a much higher concentration of added IgG. Log-log linearity is best seen after 2700 ng have been added; the highest concentration of applied IgG approaches saturation. The possibility that the changes in slope observed within the adsorption plot for NC could represent cooperative binding is discussed in Section III.E.3. The greater capacity of NC for IgG certainly resides in its three-dimensional structure. The approximate 500-fold increase in capacity demonstrated in Figure 1 for NC is consistent with an internal area which is 100 to 1000 times greater than its frontal area.[31]

* All membranes studied were 0.45-µm membranes.

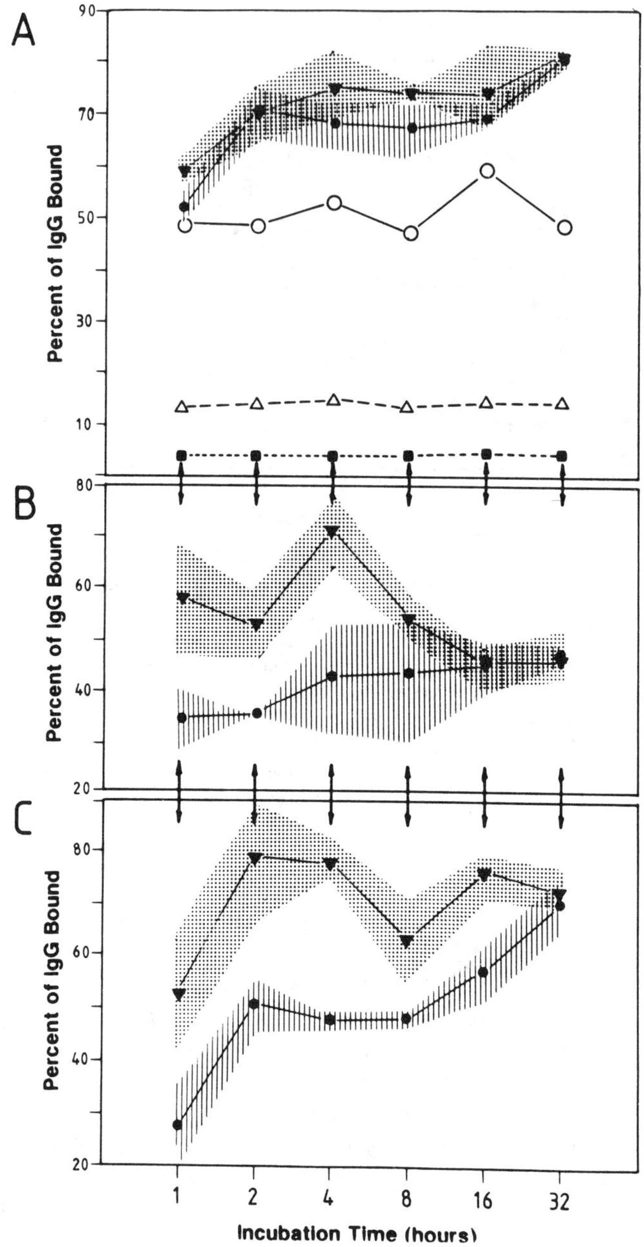

FIGURE 2. The influence of incubation time and detergent washing on the amount of rabbit IgG adsorbed to NC and polystyrene. Eight different concentrations ranging from 125 μg/ml to 0.97 μg/ml of ^{125}I-rabbit IgG were prepared in 0.1 *M* carbonate buffer, pH 9.6, and incubated in 200 μl volumes in either Immulon® 2 microtiter wells or with 15-mm diameter NC disks as described for Figure 1. After incubation at room temperature for 1 to 32 h, replicate wells or disks were subsequently washed with PBS or PBS + 0.5% Tween 20 and counted as described in Figure 1. (A) The percentage of IgG bound to polystyrene after various times of incubation. Concentrations which fall into the log-log linear binding range for plastic (Figure 1) yield a constant percent binding so that the values were pooled and expressed ± SE (shaded areas). ▼—▼ = mean percent bound in linear binding region, no detergent in wash; ●—● = same as ▼—▼, but washed with detergent; ○—○ = percentage bound after incubation with 15.2 μg/ml; values the same with or without detergent; Δ—Δ = percentage bound after incubation with 31.25 μg/ml; values the same with or without detergent; ■—■ = percentage bound after incubation with 125 μg/ml; values the same with or without detergent. (B) Percentage of added IgG over the range of 0.97 to 7.8 μg/ml which remained bound to NC after various incubation times. ▼—▼ = washed with PBS and no detergent; ●—● = washed with PBS containing 0.05% Tween®. (C) Percentage of added IgG over the range of 15.6 to 125 μg/ml which remained bound to NC after various incubation times. ▼—▼ = washed with PBS only; ●—● = washed with PBS containing detergent.

2. The Kinetics of Adsorption

The attainment of equilibrium between a solid-phase reactant on a microtiter well and its receptor or ligand in solution requires 16 h when no vigorous agitation is provided (Chapter 18).[41] Therefore, it was postulated that adsorption of protein to microtiter wells or NC membranes might also require considerable time to reach equilibrium. The data presented in Figure 2 indicate the inaccuracy of this hypothesis. Figure 2A shows that at very high concentrations of added antibody (125 μg/ml) no differences in the amount of protein adsorbed to Immulon® 2 at 1 h or any time point throughout the 32-h test period was seen. Similar results are obtained at concentrations of 7.8 and 31.25 μg/ml of added IgG. Within the log-log linear binding region, i.e., the addition of 7.8 μg/ml or less of IgG (Figure 1), the kinetics of binding are significantly different. Within this range, the mean amount bound increases about 30% between 2 and 4 h.

The kinetics of adsorption to NC at "low concentration" (0.9 to 7.8 μg/ml; Figure 2B) and after washing with detergent buffer differ from those seen over the high-concentration range (15.6 to 125 μg/ml; Figure 2C). The latter shows that progressively increasing amounts of IgG become bound as incubation proceeds from 1 h to 32 h (Figure 2B). Over the low-concentration range, differences in the amount adsorbed after 32 h were only observed during the first 2 h of incubation, but not thereafter (Figure 2B). The kinetics of adsorption to NC best corresponds to adsorption on polystyrene at concentrations within the log-log linear region (Figures 1 and 2A).

3. The Effect of a Nonionic Detergent on the Stability of Adsorbed Protein

Figure 2A shows that washing polystyrene microtiter wells incubated with concentrations of IgG in the log-log linear binding region with detergent vs. no detergent has no effect. A detergent wash also fails to affect adsorption when high concentrations of IgG are added. In contrast, detergent washing decreases the percentage of IgG which is adsorbed to NC during the first 4 h over the lower-concentration range (Figure 2B) and during the first 16 h over the high-concentration range (Figure 2C). At both low and high protein-concentration ranges, the amount of IgG which remains stably adsorbed on Tween®-washed filters tends to increase over the test period. The observation that the NC washed only with PBS shows a higher percentage of bound IgG than do Tween®-washed filters during the initial hours of incubation suggests that Tween® disrupts weak hydrophobic interactions which form during this period. The erratic nature of the binding plot for non-Tween®-washed membranes during this time may be a reflection of these labile bonds (Figures 2B and 2C). Given sufficient time, e.g., 32 h, bonds are such that detergent washing is unable to disrupt them.

In summary, Figures 1 and 2 show that polystyrene and NC behave differently toward IgG in two ways: (1) binding to plastic seems to be more stable or involves other bonds than does binding to NC from Schelicher and Schuell in that it is unaffected by washing with Tween® and (2) equilibrium is attained earlier when high concentrations of protein are incubated with plastic as compared to NC from Schleicher and Schuell.

4. The Effect of Sodium Dodecyl Sulfate in Immunoblotting

Dot-blot assays require the direct application of the antigen (or in the case of "antigen-capture", the antibody) to the NC membrane filter. The biological integrity of the immobilized protein and its concentration will both ultimately affect the success of the assay. As discussed above, hydrophobic forces are definitely involved in the binding process. However, hydrophilicity is also a factor. "Old" membranes sometime lose their ability to bind as they dry out and become completely dehydrated. The membranes can be restored to some degree by alcohol treatment enabling better wetting of the membranes.[1] Protein applied to a dry filter and left to air dry often undergoes considerable and possible irreversible denaturation. A demonstration of this can be found by simply applying alkaline phosphatase to a NC filter and letting it dry as compared to a similar filter stored in PBS. The former tends to lose its activity,

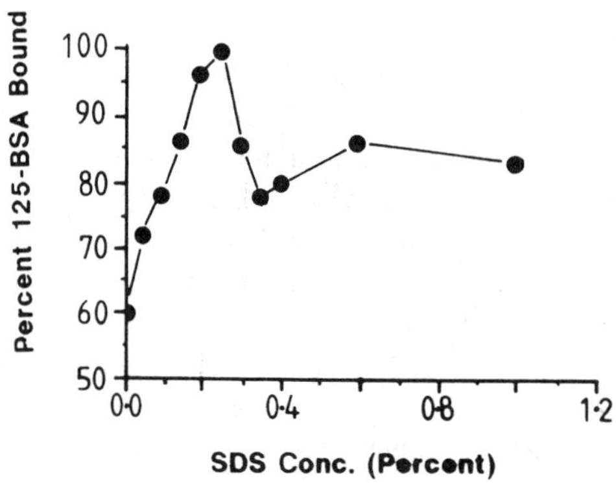

FIGURE 3. The effect of SDS on the binding of ^{125}I-labeled BSA to NC. ^{125}I-labeled BSA (0.5 μg, approximately 104 cpm) was added to PBS containing various concentrations of SDS and was applied as a 2 μl dot to dry NC filters. The filters were then rinsed in PBS and the amount of bound BSA was calculated as a percent of that applied. (Courtesy of Rachel Mosckovitz, Weizmann Institute of Science, Rehovoth, Israel.)

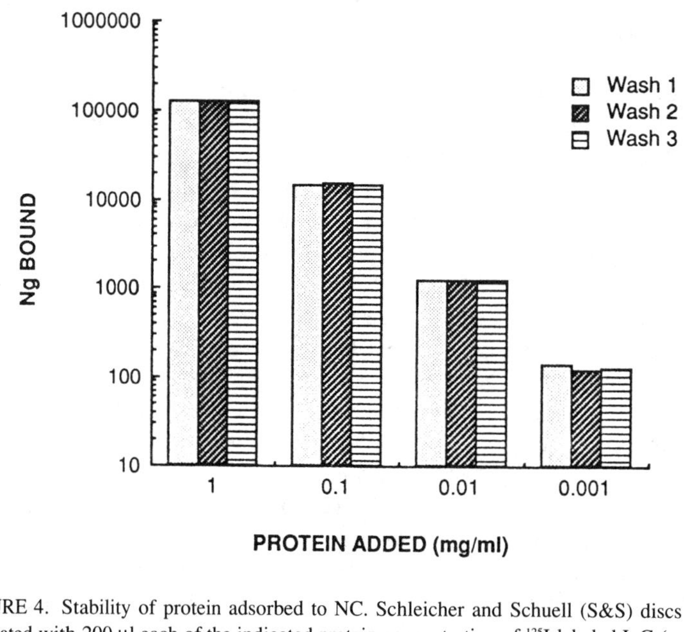

FIGURE 4. Stability of protein adsorbed to NC. Schleicher and Schuell (S&S) discs were incubated with 200 μl each of the indicated protein concentration of ^{125}I-labeled IgG (x-axis), in duplicate, for 16 h at room temperature (r.t.). Filters were counted after each 5 min wash with PBS/Tween® for a total of 3 washes. Identical results were obtained when BSA was used.

whereas the latter remains stable. Thus, this parameter should be taken into consideration in dot blots of conformation-sensitive antigens. One can always store sample-containing filters wet, in buffer, or between plastic wrap. The effect of SDS on protein binding is a bit curious. Very low concentrations of the detergent (<0.2%) tends to improve the binding of BSA to NC in dot blots (Figure 3). Increasing the concentration of SDS does not increase the amount of BSA bound but it also does not appear to markedly inhibit binding either. In experiments where higher concentrations of BSA were used, the general phenomenon was the same although the range over which maximal binding occurred was broader. Similar results were obtained with other labeled proteins, e.g., β-galactosidase and asialo-fetuin. This may suggest that a minimal amount of detergent wets the filter and thus enables more effective access to the matrix. In the case of sample application via protein transfers, e.g., electrophoretic blotting,

SDS is definitely detrimental. Here, the electromotive force drives the SDS-protein complex through the filter before it can effectively be adsorbed.

5. The Stability of Adsorbed Protein during Repeated Washing

The stability of protein adsorbed to Schleicher and Schuell NC was tested through sequential washings using two proteins of different size: IgG of 150 kDa and BSA of 66 kDa. The concentration of added protein was varied in the range from 1 μg/ml to 1 mg/ml in order to assess potential concentration-dependent differences. Figure 4 indicates that for IgG incubated with NC for 16 h, repeated washing fails to release IgG. Similar results were obtained with BSA. This stability is maintained throughout the range of added protein tested, indicating that concentration dependent differences do not exist with respect to multiple Tween® washes. Thus, in this study, one wash with PBS-Tween® is sufficient to assure stable binding to Schelicher and Schuell NC over a range of concentrations for two proteins of different molecular size.

C. THE AFFINITY AND CAPACITY OF NC FOR IgG, ALBUMIN, AND INSULIN

1. The Comparative Adsorptive Characteristics of Schleicher and Schuell NC and Immulon® 2 Polystyrene

The adsorptive characteristics of NC and Immulon® 2 polystyrene were compared with respect to the binding of three different proteins. The percent binding plot for polystyrene (Figure 5A) shows that it has only a slightly higher affinity (avidity) for IgG (approximately 70% binding in the LBR*) than for BSA and insulin (approximately 60% in the LBR).

The log-log saturation plot (Figure 5B) illustrates the inflection in the binding plot as the binding sites for protein are no longer in excess. This plot also shows that all three proteins give plots with similar Y-intercepts, indicating that affinity differences among the proteins for polystyrene are quite small. On a molar basis, insulin has the highest point of inflection followed by IgG and then BSA.

Comparative binding of the same three proteins for NC from Schleicher and Schuell is shown in Figures 5C and D. The percent binding plot (Figure 5C) reveals that approximately 90% of the BSA added to NC is bound at low concentrations but that the percentage which binds drops off rapidly with increasing concentration added. IgG also binds well; nearly 80% of the added protein is bound while insulin (mol wt 5000 kDa) binds poorly (approximately 10% binding). It should be remembered that binding studies were conducted by incubation of membrane disks until equilibrium was reached. Hence, the lower binding of insulin cannot be ascribed to rapid passage of a small protein through a highly porous membrane which could occur during electroblotting, i.e., transfer immunoblotting. Both IgG and insulin display constant percent binding over the concentration range tested. One possible interpretation of these results is that the membrane binding sites for BSA are few in number and hence become saturated at a lower concentration than in the case of IgG or insulin. This appears more evident when the data are presented in a log-log saturation plot (Figure 5D). It can be seen that addition of 10 nmol of insulin results in adsorption of approximately 1 nmol whereas BSA and IgG demonstrate nearly a 1:1 added:bound ratio, i.e., approximately 80% adsorption. This difference in affinity is equally apparent by attention to the Y-intercepts for the three proteins. These data also indicate that IgG and BSA have similar affinities for NC when only low concentrations of BSA are tested.

The comparison of the adsorptive characteristics of three proteins for polystyrene and NC reveals some major differences between these two widely utilized solid phases. Figure 5B shows that all three proteins have similar affinity for polystyrene and >60% of the protein

* LBR (linear binding region) is defined as the region of the percent binding curve involving a homogenous receptor-ligand interaction in which the receptor is in excess of the ligand; thus, the percentage of bound ligand within this region for relatively high affinity interactions remains constant. Percent binding begins to decrease progressively as higher ligand concentrations are added (see Figure 3, Chapter 1).

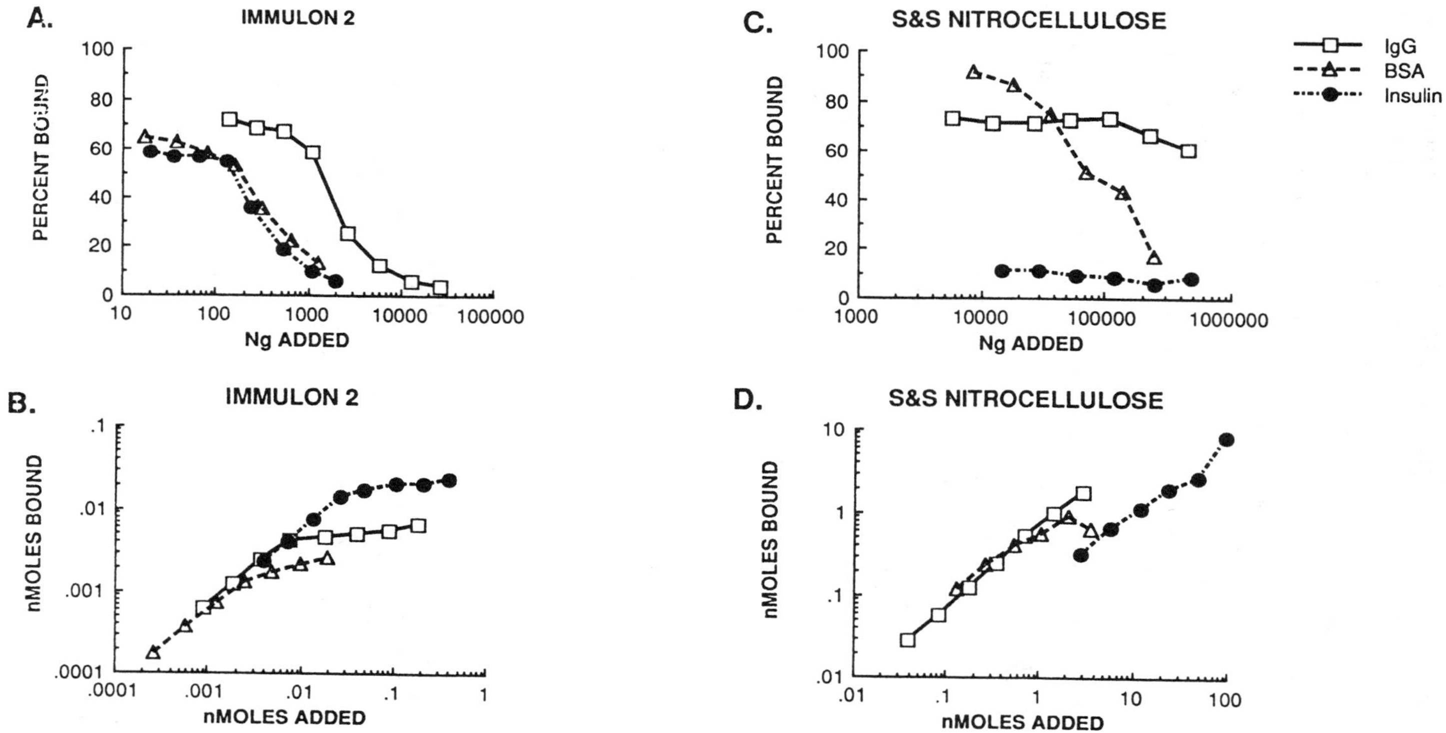

FIGURE 5. Percent binding plots (A and C) and saturation plots (B and D) comparing the adsorption of three different proteins for Schleichler and Schuell NC and polystyrene. Tests were done in duplicate using the same protocol described for Figure 1. NC discs were washed 3 times for 5 min each with 2 ml PBS/Tween® while plastic wells were washed 3 times with 200 µl before counting. The remainder of legend is on the figure.

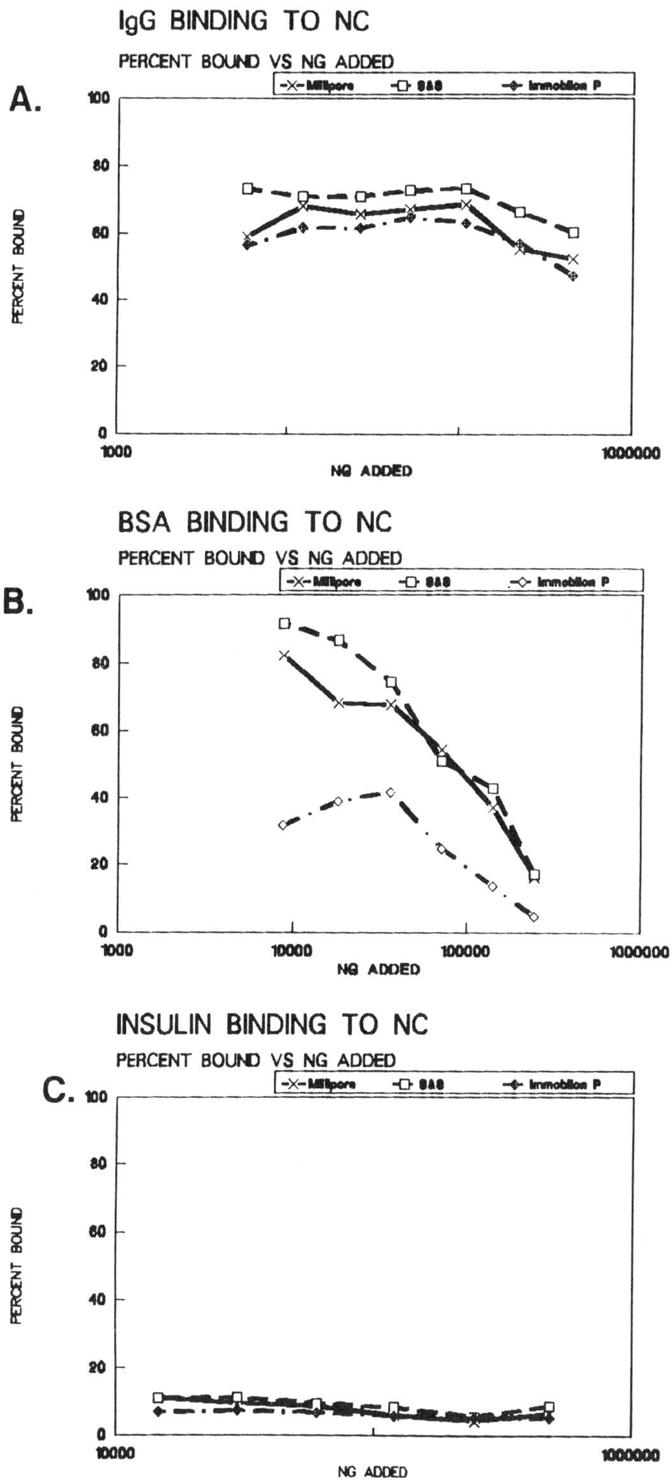

FIGURE 6. The comparative affinity of three different proteins for three different 0.45-μm membranes. Experiments were conducted as described for Figures 1 and 5. The remainder of the legend is on figure.

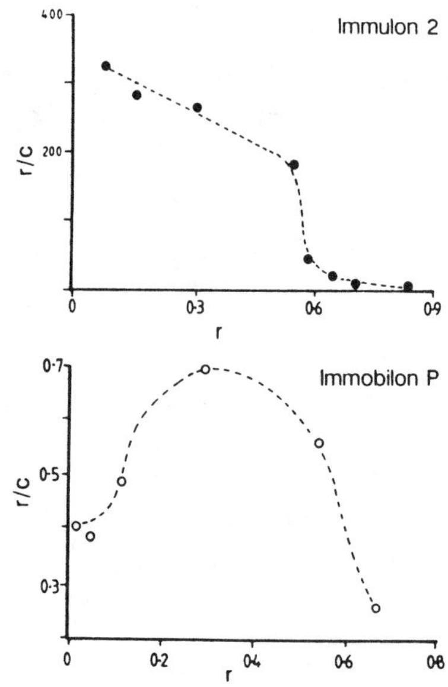

FIGURE 7. Scatchard analysis of the adsorption of ^{125}I-labeled IgG on Immulon® 2 and on Immobilin®-P. The values for functional antibody concentration [Ab_t] were calculated from Langmuir plots (data not shown). r = molar ratio of bound protein to [Ab_t]; c = molar concentration of unbound protein at equilibrium.

added in the linear binding region is bound (Figure 5A). BSA binds to membranes differently than IgG or insulin, achieving apparent saturation of high-affinity binding sites at levels where the binding sites for the other two proteins are still in excess. Overall, the membranes have a much higher capacity for protein than polystyrene.

2. Differences in the Adsorptive Characteristics of Various Blotting Membranes

The differential adsorptive capacity of three different membranes was studied: Schleicher and Schuell NC (SS), Millipore NC (MP) and Immobilon®-P (IP). Three different proteins were tested on each filter over a range of concentrations and the results are given in Figure 6.

Figure 6A shows a percent binding plot for IgG. Although there are minor differences among the three filters, they all bind IgG in a similar manner: binding, which is totally within the LBR, is between 60 and 75%. In contrast, BSA behaves quite differently (Figure 6B). SS and MI bind BSA very well at the lowest concentration tested (approximately 80%) but the percent bound drops precipitously as more is added. IP binds BSA quite differently; not only is binding much less than to the other filters (maximum = 40%), but the pattern of binding is different. Percent binding with IP generally increases up to approximately 40,000 ng of added protein while the addition of higher concentrations of protein results in a decrease in the percentage of bound BSA. Figure 6C shows that all three filters bind insulin in a consistent manner, albeit poorly (10% or less).

Thus, whereas all three filters show low but consistent binding to insulin, they also show high, consistent binding to IgG; this contrasts with the binding of BSA which is membrane dependent. Hence, the adsorption of proteins to various membranes can vary depending on the type of membrane and the source of the NC membrane, as well as the type of protein.

3. Do Proteins Bind Cooperatively to Blotting Membranes?

Figure 7 shows Scatchard plots for the adsorption of IgG to IP and polystyrene. All three blotting membranes gave the same results. Studies on plastic have shown that although the values presented for r and r/c may not be accurate due to reaction volume considerations, the form of the plot appears to be reliable[43] (Chapter 12). The Scatchard plot for IP shows that r/c initially increases with increases in r. At higher input of protein, the value of r/c decreases with increasing r as would be the case in normal noncooperative ligand-receptor interactions.

By contrast, a Scatchard analysis of the binding of IgG on polystyrene reveals a plot suggesting the presence of homogeneous binding sites on the plastic until 600 ng per well are bound; thereafter, the binding appears heterogeneous. The Scatchard plot for IP in Figure 7 is typical of cooperative binding, i.e., as the first molecule of protein is adsorbed to the membrane, its presence positively influences the binding of additional molecules. The cooperative-type interaction which we observed may also explain the changes in slope observed with IgG adsorbed to MP (Figure 1). At low concentrations of added protein, a slope greater than 1.0 is observed, whereas above an added concentration of 2.7 μg/ml the binding plot has a slope of approximately 1.0. This phenomenon may be associated with specific proteins and certain sources of NC since it is also observed with BSA on IP but not with insulin on any of the filters (data not shown).

There are several possibilities which could explain this "cooperative-type" effect. First, proteins may be forming clusters in such a manner that adsorption of the first molecule facilitates adsorption of the second because it interacts not only with the membrane binding site, but also with the previously bound protein. Second, cooperative binding may simply reflect protein aggregation in that subsequent molecules do not bind the membrane but a previously bound molecule. Finally, adsorption may be a "solubility effect" as described in Chapter 1, so that increasing the protein concentration increases aggregate formation which in turn facilitates adsorption and could yield the cooperative-type plot observed with IP (Figure 7). Whatever the basis, we observed this cooperative effect on all three immunoblotting membranes which were tested. As will be discussed in Section V, this effect may not be unique to adsorption on blotting membranes but may merely be easier to demonstrate on such surfaces because higher concentrations of protein can be used.

IV. THE FUNCTIONAL PROPERTIES OF PROTEINS ADSORBED ON NC

A. THE COMPARATIVE ACTIVITY OF ANTI-FLUORESCEIN ON NC

1. The Comparative Behavior of Polyclonal Anti-Fluorescein on NC and Polystyrene

Figure 8 illustrates the functional properties of three concentrations of polyclonal rabbit anti-fluorescein antibodies adsorbed to polystyrene or NC. Varying concentrations of a radiolabeled antigen, α-lactalbumin conjugated to ditriazinoamino fluorescein (FLU-ala) were incubated with polyclonal anti-FLU which had been previously adsorbed on NC by incubation of the NC with antibody at concentrations of 5, 50, or 150 μg/ml. All three concentrations of adsorbed capture antibody (CAb) behave similarly on polystyrene and all performed better than when 5 μg/ml of anti-FLU was tested on the various immunoblotting membranes (Figure 9). The fact that all three curves in Figure 8 (top) failed to display a LBR (section III.E.1) for the range of Ag tested indicates that incubation of polystyrene with greater than 5 μg/ml of antibody neither significantly increases the functional surface CAb concentration nor the ability of the additional CAb to capture more antigen. It should be noted that the concentration range of Ag used was chosen to emphasize the comparative differences between polystyrene and blotting membranes, not differences among plastics; the latter topic is dealt with in Chapter 12.

The lower panel of Figure 8 shows that when antibody is adsorbed on NC, the functional concentration increases with the amount of bound CAb. This is reflected in an increase in the length of the constant binding region with increasing CAb; it is longest when 150 μg/ml is added, intermediate at 50 μg/ml, and essentially absent when 5 μg/ml is used. There is also

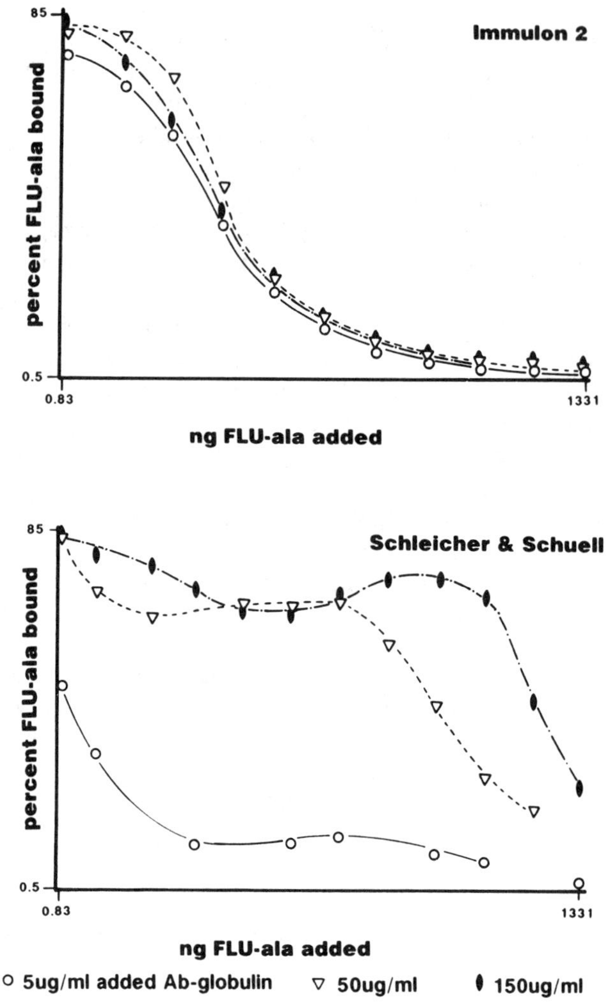

FIGURE 8. Comparison of the functional properties of capture antibodies (CAb) on NC vs. polystyrene. Varying concentrations of radiolabeled Ag, α-lactalbumin conjugated to ditriazinoamino fluorescein (FLU-ala), were incubated for 16 h at 4°C with polyclonal antifluorescein Ab that had been previously adsorbed to NC or polystyrene by incubation for 16 h at 4°C at three different concentration of anti-FLU. Membranes and wells were blocked with 2 ml or 200 μl, respectively, of 0.5% BSA in PBS/Tween and shaken for 2 h at room temperature before addition of FLU-ala. Washing and counting conditions were the same as described in Figure 1. Results are expressed as percentage of antigen captured as a function of the amount added.

a substantial increase in AgCC (antigen capture capacity) when CAb concentration is increased from 5 to 50 μg/ml. The plateau region is not a LBR per se, since it actually shows increased AgCC when more antigen is added. This increased AgCC cannot be explained as an antibody effect by the Law of Mass Action because affinity differences within the polyclonal population should favor involvement of lower affinity antibodies as higher concentrations of Ag are added. This should result in a lower percentage of bound Ag, *not* an increase! Rather, we postulate that this phenomenon is an Ag effect, i.e., at high concentrations the antigen becomes aggregated which results in a greater radioactive signal. Studies were not conducted to test this possibility.

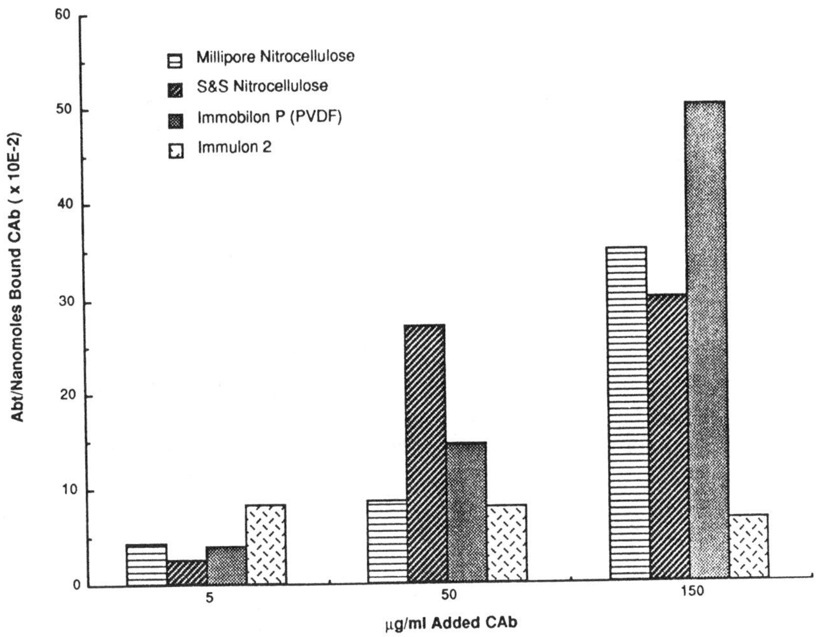

FIGURE 9. The functional activity of CAb adsorbed at three different concentrations to three different membranes and Immulon® 2 polystyrene. Experimental conditions are as described in Figure 8. The ordinate describes functional activity as the ratio of the functional antibody [Abt] or Ab sites, determined from the Y-intercept of Langmuir plot (data not shown), to the nanomoles of bound CAb. The latter was determined independently using ^{125}I-anti-FLU as described in Figure 1.

2. The Behavior of Polyclonal Anti-Fluorescein on Different Blotting Membranes

The functional activity of three concentrations of CAb, with respect to different membranes, is shown in Figure 9. All three membranes show an increase in functional activity as the concentration of CAb is increased. Activity is defined as the ratio of the functional antibody or antibody site concentration (defined from the Y-intercept of a double reciprocal, i.e., Langmuir plot), to the nanomoles of bound CAb. Increases in functional activity were not parallel among the membranes tested. Immobilon®-P, for example, performed best when 150 μg/ml was used while SS NC performed best when 50 μg/ml was tested.

B. POTENTIAL PITFALLS IN THE INTERPRETATION OF DATA OBTAINED BY IMMUNOBLOTTING

Data presented in this section are interesting, but have limited application to immunoblotting as it is most commonly employed. Rather than adsorbing CAb, immunoblotting membranes are more often used to measure antigen, either directly adsorbed through dot-blotting techniques or transferred to blotting membranes from PAGE or other sources. Hence, it is the functional activity of the antigen rather than the CAb that is at question. Data presented here may however be insightful for ELISPOT assays conducted on blotting membranes[42] or in enzyme immunofiltration assays (see Chapter 15).

It is known through the "trade" that polyclonal detection antibodies often work much better than monoclonal detector antibodies for antigens adsorbed on immunoblotting membranes even though the mAb recognized the antigen in solution. Assuming that protein antigens adsorbed on immunoblotting membranes undergo some of the same effects as reported for antigens on plastic, e.g., epitope loss (Chapters 1, 10, and 13), it would not be surprising that a MAb specific for a single epitope would no longer recognize the adsorbed antigen whereas the polyclonal, which could recognize multiple epitopes, would have a greater probability of recognition.

Unfortunately, we have presented no data to indicate that such adsorption-induced epitope loss occurs in immunoblotting membranes. Furthermore, the diversity of chemical bonding proported to be involved in adsorption to the various immunoblotting membranes suggests that different membranes may produce different results.

It is also known that certain mAb only recognize epitopes that are present on denatured biomolecules (Chapter 10). This can be due to the method used for hybridoma screening or can be a direct result of how the biomolecule was prepared. If the protein was eluted from PAGE and then used as an immunogen, it might well display a different conformation than if eluted from an antibody affinity column. Monoclonal technology should be used with some care when the potential of recognizing denature protein is present. For example, a mAb to a "native" protein might recognize the "novo epitope" of an unrelated immunoblotted protein while a mAb specific for an "adsorption-induced epitope" might recognize a native epitope in an unrelated protein.

There may also be major differences between dot blots and protein-transfer procedures. It stretches the imagination to think that a molecule separated in ionic detergent by electrophoresis and then subsequently transferred to NC or some other immunoblotting surface by electrotransfer could be expected to express as many native epitopes as the antigen that was merely dot blotted. On the other hand, electrotransfer may offer the opportunity for the electrophoretically separated biomolecule to renature to some degree. Also, for SDS gels, the ionic detergent may facilitate adsorption to the blotting membranes as shown in Figure 3.

The possible antigenic changes associated with binding to immunoblotting membranes appears to be magnified beyond that faced by investigators using plastic. The involvement of both electrostatic and hydrophobic interactions in binding to immunoblotting membranes may be paralleled by a greater diversity of conformations and consequent antigenic diversity than is seen on plastic where hydrophobic interactions are believed to be nearly the exclusive force.

Perhaps it is useful for each investigator to ask the following questions of his/her dot blot or transfer blot.

1. Is it important to the interpretation of the data that an immunoblotted biomolecule retain its native configuration?
2. Is the molecule recognized on the immunoblot the same molecule for which the antibody is presumed specific?
3. Does incubation of the specific antibody in solution with its "native" antigen block the recognition of the same or other antigens adsorbed on the immunoblotting membrane?
4. How can it be tested as to whether a negative immunoblot result means that the antigen in question is truly absent from the blotting membrane?
5. Does the antigen in question perform the same way in a dot blot as when transferred electrophoretically from a separation gel to an immunoblotting membrane?
6. Against what form of the antigen was the mAb or polyclonal antibody prepared, i.e., solution-phase "native" molecule or the antigen eluted from an immunoblot, or some other solid phase?
7. How was the specificity of the antiserum or antibody established? Was this done by adsorbing the antigen on plastic, testing it in a dot blot, or by some type of solution-phase, coprecipitation method?
8. To what extent must "nonspecific" binding of the antibody be blocked to render the blot specific for the antigen of interest? Extensive blocking can merely be a method for quenching low-affinity reactions which may be significant but do not fit the hypothesis being tested!
9. Can it be shown on transfer or dot blots, by varying either the concentration of adsorbed antigen or the detection antibody, that some cross-reactions are merely low-affinity

antigen or the detection antibody, that some cross-reactions are merely low-affinity interactions, the detection of which is easily altered by varying the concentration of the reactants?

10. How do the components of the sample buffer and/or washes, affect the antigenicity of the biomolecule? For example, a biomolecule may tolerate either SDS or a reducing agent, e.g., DTT, but not their combined action. Hence, buffer constituents should be tested in dot blots before attempting their use in protein transfer systems.

V. SUMMARY

Chapter 8 has presented an overview of biochemical and immunochemical aspects of relevance to the use of membranes in SPI. Much of the technical and procedural literature on "immunoblotting" or "Western blotting" has been previously reviewed.[1-5,8,17,18] As with other solid-phase systems, the literature is weak in providing information on the immunochemical and biochemical aspects of protein adsorption and its effect on the activity of the protein. Hence, the authors of this chapter have made an attempt to bring awareness to these issues through discussion and presentation of preliminary data.

The data presented indicate certain striking differences between membranes and polystrene. First is the issue of capacity. On the basis of planar surface area, membranes have at least a 500-fold greater capacity for IgG. However, it is believed that the internal surface area may be 100 to 1000-fold the planar area,[31] which could explain our findings. It has been argued by some[44] that certain problems associated with SPI are a result of the limited capacity to immobilize antigens or antibodies. This problem has been a favorite argument for the use of microparticles in particle concentration assays[45,46] or in the use of microparticles in general (see Chapter 14). The high functional CAb concentration on the membranes we studied compared to polystyrene, probably reflects the greater surface area making the former desirable for immunoassays in which assays with broad dynamic ranges are needed. However, such an increase in dynamic range also comes at the expense of a greater amount of CAb. Immunoassays conducted in microtiter trays with membrane bottoms such as those developed by Millipore, might perform quite differently than immunoassays conducted on ordinary plastic microtiter trays if major differences in receptor density exist. The issue of adsorption-site density is an important one which is unfortunately unanswered by the preliminary data we present because of the lack of quantitative data on the total surface area. A mere increase in surface area would only affect the dynamic range of an immunoassay whereas true differences in adsorption-site density would influence such factors as affinity dependence and dissociation rates (see Chapter 19).

The proteins adsorbed on NC membranes and Immobilon®-P appeared very stable in the short-time studies we conducted. Little or no desorption was observed after a single detergent wash for the proteins BSA and IgG, although this might be a protein-specific effect. Furthermore, we provide no data on long-term stability and we only examined the stability of proteins that had been incubated for 16 h with the membranes. The data presented in Figure 2 indicate that such long-term incubation results in more stable bonding as evidenced by the failure of detergent to displace the adsorbed protein. Hence, different results may be obtained following short-term incubation, using different protein, different membranes, or when membranes have been stored for a long time. Also, we included no protein in our wash buffers. As it is known that one protein may replace another during the adsorption process (see Chapter 1), it is possible that such treatment could result in less stable adsorption than we report here.

Detergent washing does affect adsorption to NC of protein during the first 16 h of incubation whereas such an effect is not observed with plastic. The kinetics of adsorption to plastic are very rapid when high protein concentrations are used whereas a longer time is required for equilibrium to be reached on membranes. This could be a simple consequence of the ratio of protein:binding site when comparing the two surfaces. If the true surface of NC is 100 to 1000 times larger than its planar surface with little change in binding-site density, the type of data presented in Figure 2 would be anticipated; equilibrium is concentration dependent for plastic but similar overall for NC. The adsorption plot shown in Figure 1 for NC might therefore represent a "concentration-expanded view" of the corresponding plot for polystyrene. The evidence we present, that binding of proteins at low concentrations to immunoblotting membranes behaves cooperatively, might also be seen on polystyrene if extremely low concentration of protein could be used. However, it is technically difficult to obtain highly labeled proteins which permit accurate measurement over such low concentration ranges. Hence, studies on immunoblotting membranes might provide useful insight in the binding of biomolecules on surfaces in general.

The adsorption of protein on polystyrene microtiter wells proceeds much more rapidly than do subsequent reactions involving surface-adsorbed proteins and their antibodies or antigens (see Chapter 18).[41] This could be a consequence of the plastic binding-site concentration or the nonreversibility of the adsorption interaction. The data of Figure 2A, which suggest that equilibrium is reached very quickly when high concentrations of added IgG are tested, should not be interpreted to mean that higher concentrations are better. Data presented in Chapter 12 and in Figure 9 indicate that functional activity of adsorbed antibodies progressively decreases as higher concentrations are added. The very low percentage of protein adsorbed when high concentrations are incubated with plastic no doubt forms the basis for the often-practiced procedure of repeated transfer of unbound protein from the wells of one microtiter plate to the next. While the 80 to 90% excess provides a more than adequate concentration for reuse, the assumption that concentration does not influence the quality of the adsorbed protein is probably erroneous (see Chapter 12).

The adsorption properties of the immunoblot membranes, however, cannot be simplistically viewed as merely a planar surface with a large "hidden" internal surface. This would not explain the striking differences seen when the affinity of IgG, BSA, and insulin for polystyrene and blotting membranes were compared. While polystyrene may be mono- or oligomeric as regards binding-site heterogeneity, membranes most likely possess a greater diversity of binding sites. Furthermore, types of membrane may differ from each other more than types of plastic (see Chapter 12). The differences in the binding properties of the three proteins studied for the three membranes (Figure 6) compared to the similarity of their binding on plastic (Figure 5) illustrate that each membrane should be treated as a separate adsorption media for SPI that differs from plastic not merely in surface area but in binding site heterogeneity and affinity for proteins.

The differences in affinity of the various proteins for membranes, while binding similarly to polystyrene, may also reflect differences between membranes and polystyrene in the chemistry of the bond. The observation that SDS actually promotes rather than inhibits the adsorption of protein on NC (Figure 3), while it is known to inhibit adsorption to plastic (Chapter 1), may reflect the involvement of a different spectrum of chemical bonds. As discussed in Section I, electrostatic bonds are also thought to be important for membranes, but not polystyrene. It must be emphasized that the comparative data we report here were obtained at only one pH. If electrostatic effects are more important on membranes than plastic, the differences in affinity of the three proteins we studied for the different membranes might be different as pH is changed. Evidence for such an effect has been cited.[31]

The data presented on the functional activity of adsorbed CAbs differ from those seen when CAbs are immobilized on plastic (see Chapter 12). As the amount of CAb adsorbed is increased, activity improves whereas the opposite is true on plastic. If the true comparative surface areas of the membrane and polystyrene wells we tested was, for example, 500:1, then adding 150 μg/ml of anti-FLU to a membrane is equivalent to adding 300 ng of anti-FLU to plastic. In studies with antigens, Dierks et al.[24] showed that the antigenic activity of rabbit IgG was more distorted when low amounts of protein were adsorbed than when large amounts were used. Furthermore, this distortion could be corrected by coadsorption of BSA. Therefore, the improved functional activity observed as higher concentrations of anti-FLU are added could be due to the same phenomenon. Presumably, when high concentrations are added, the antibodies serve as their own "stabilizing protein" (see Chapter 1, Figure 1). Should this be the case, the coadsorption of irrelevant stabilizing protein could be a useful economy measure which simultaneously yields CAbs with the highest activity per nanogram bound.

ACKNOWLEDGMENT

The authors gratefully acknowledge the assistance of Dr. Harry Towbin, Ciba-Geigy, Basel, Switzerland for his thoughful suggestions concerning the preparation of this chapter and for his critical review of the manuscript. We also wish to thank Dr. Kalpana Joshi for conducting the studies on polystyrene, Dr. Larry Blankstein of the Millipore Corporation for providing the membranes used in the work, and Ms. Marcia Reeve for preparation of the typescript and coordination of the mailings to the authors located at very different sites.

REFERENCES

1. **Gershoni, J. M. and Palade, G. E.,** Protein blotting: principles and applications, *Anal. Biochem.,* 131, 1, 1983.
2. **Towbin, H. and Gordon, J.,** Immunoblotting and dot immunoblotting — current status and outlook, *J. Immunol. Methods,* 73, 313, 1984.
3. **Bers, G. and Garfin, P.,** Protein and nucleic acid blotting and immunobiological detection, *Biotechniques,* 3, 276, 1985.
4. **Bjerrum, O. J. and Heegaard, N. H. H., Eds.,** *Handbook of Immunoblotting of Proteins,* Vol. 1, CRC Press, Boca Raton, FL, 1988.
5. **Gershoni, J. M.,** *Protein Blotting: A Manual in Methods of Biochemical Analysis,* Vol. 33, Glick, D., Ed., John Wiley & Sons, New York, 1988.
6. **Baldo, B. A. and Tovey, E. R., Eds.,** *Protein Blotting: Methodology, Research and Diagnostic Applications,* S. Karger, Basel, 1989.
7. **Kemeny, D. M. and Challacombe, S. J., Eds.,** *ELISAs and other Solid-Phase Immunoassays,* John Wiley & Sons, New York, 1988.
8. **Axelson, N. H., Ed.,** Handbook of immunoprecipitation in gel techniques, *Scand. J. Immunol.,* 17(Suppl. 10), Blackwell Scientific, Oxford, 1983.
9. **Heegaard, P. M. N. and Borg-Hansen, T. C.,** *Immunoelectrophoretic Methods in Analytical Gel Electrophoresis of Proteins,* Dunn, M. J. and Hilger, A., Eds., London, 1985.
10. **Kohn, J.,** An immunoelectrophoretic technique, *Nature (London),* 1980, 986, 1957.
11. **Burridge, K.,** Changes in cellular glycoproteins after transformation: identification of specific glycoproteins and antigens in sodium dodecyl sulfate gels, *Proc. Natl. Acad. Sci. U.S.A.,* 79, 4457, 1978.
12. **Southern, E. M.,** Detection of specific sequences among DNA fragments separated by gel electrophoresis, *J. Mol. Biol.,* 98, 503, 1975.
13. **Renart, J., Reiser, J., and Stark, G. R.,** Transfer of proteins from gels to diazobenzyloxymethyl-paper and detection with antisera: a method for studying antibody specificity and antigen structure, *Proc. Natl. Acad. Sci. U.S.A.,* 76, 3116, 1979.
14. **Towbin, H., Staehelin, T., and Gordon, J.,** Electrophoretic transfer of protein from polyacrylamide gels to nitrocellulose sheets: procedure and applications, *Proc. Natl. Acad. Sci. U.S.A.,* 76, 4350, 1979.
15. **Heegaard, N. H. H. and Bjerrum, O. J.,** Immunoblotting — general principles and procedures, in *Handbook of Immunoblotting of Proteins,* Vol. 1, Bjerrum, O. J. and Heegaard, N. H. H., Eds., CRC Press, Boca Raton, FL, 1988, 1.
16. **Bjerrum, O.J. and Heegaard, N. H. H., Eds.,** *Handbook of Immunoblotting of Proteins,* Vol. 2, CRC Press, Boca Raton, FL, 1988.
17. **Hawkes, R., Niday, E., and Gordon, J.,** A dot-immunobinding assay for monoclonal and other antibodies, *Anal. Biochem.,* 119, 142, 1982.
18. **Herbrink, P., VanBussel, F. J., and Warnaar, S. O.,** The antigen spot test (AST): a highly sensitive assay for detection of antibodies, *J. Immunol. Methods,* 48, 293, 1982.

19. **Bull, H. B.,** Adsorption of bovine serum albumin on glass, *Biochim. Biophys. Acta,* 19, 464, 1956.
20. **Burghardt, T. P. and Axelrod, D.,** Total internal reflection fluorescence study of energy transfer in surface-adsorbed and dissolved bovine serum albumin, *Biochemistry,* 22, 979, 1983.
21. **Kennel, S.,** Binding of monoclonal antibody to protein antigen in fluid phase and bound to solid supports, *J. Immunol. Methods,* 55, 1, 1982.
22. **Dierks, S. E., Butler, J. E., and Richerson, H. B.,** Altered recognition of surface adsorbed compared to antigen bound antibodies in the ELISA, *Mol. Immunol.,* 23, 403, 1986.
23. **Bowen, B., Steinberg, J., Laemmli, U. K., and Weintraub, H.,** The detection of DNA-binding proteins by protein blotting, *Nucleic Acid Res.,* 8, 1, 1980.
24. **Peferoen, M., Huybrects, R., and deLoof, A.,** Vacuum-blotting: a new simple and efficient transfer of proteins from sodium dodecyl sulfate-polyacrylamide gels to nitrocellulose, *FEBS Lett.,* 145, 369, 1982.
25. **Kyhse-Anderson, J.,** A simple horizontal apparatus without buffer tank for electrophoretic transfer of proteins from polyacrylamide to nitrocellulose, *J. Biochem. Biophys. Methods,* 10, 203, 1984.
26. **Gershoni, J. M. and Palade, G. E.,** Electrophoretic transfer of proteins from sodium dodecyl sulfate-polyacrylamide gels to a positively charged membrane filter, *Anal. Biochem.,* 124, 396, 1982.
27. **Burnette, W. N.,** Western blotting: electrophoretic transfer of proteins from sodium dodecyl sulfate-polyacrylamide gels to unmodified nitrocellulose and radiographic detection with antibody and radioiodinated Protein A, *Anal. Biochem.,* 112, 195, 1981.
28. **Lin, W. and Kasamatsu, H.,** On the electrotransfer of polypeptides from gels to nitrocellulose membranes, *Anal. Biochem.,* 128, 302, 1983.
29. **Nyholm, L. and Ramlau, J.,** Nitrocellulose membranes as solid phase in immunoblotting, in *Handbook of Immunoblotting of Proteins,* Bjerrum, O. J., and Heegaard, N. H. H., Eds., CRC Press, Boca Raton, FL, 1988, 101.
30. **Dierks, S. E.,** Differentiation Recognition of Solid-Phase Immunoglobulin and Antibody Bound to Solid Phase Antigen: a Method of Antibody Quantitation, M.S. thesis, University of Iowa, Iowa City, 1985.
31. **Peterman, J. H.,** Factors which Influence the Binding of Antibody to Solid-Phase Antigen: Theoretical and Experimental Investigations, Ph.D. thesis, University of Iowa, Iowa City.
32. **Ramlau, J.,** Protein binding to charge derivatized membranes in blotting procedures — a critical study, in *Handbook of Immunoblotting of Proteins,* Vol. 1, Bjerrum, O. J. and Heegaard, N. H. H., Eds., CRC Press, Boca Raton, FL, 1988, 109.
33. **Blankstein, L. A. and Dohrman, L.,** An advanced affinity membrane for immunodiagnostic tests, *Am. Clin. Products Rev.,* 33, 1985.
34. **Cantarero, L. A., Butler, J. E., and Osborne, J. W.,** The absorptive characteristics of proteins for polystyrene and their significance in solid-phase immunoassay, *Anal. Biochem.,* 105, 375, 1980.
35. **Pluskal, M. G., Przekop, M. B., Kavonian, M. R., Vecoli, C., and Hicks, D. A.,** Immobilon PVDF transfer membrane: a new membrane substrate for western blotting procedures, *Biotechniques,* 4, 272, 1986.
36. **Batteiger, B., Newhall, W. J., and Jones, R. B.,** The use of Tween 20 as a blocking agent in the immunological detection of proteins transferred to nitrocellulose membranes, *J. Immunol. Methods,* 55, 297, 1982.
37. **Petit, C., Sauron, M. E., and Theze, J.,** Direct detection of idiotypic determinants on blotted monoclonal antibodies, *Ann. Immunol (Inst. Pasteur),* 133D, 77, 1982.
38. **Batteiger, B.,** Blocking of immunoblots, in *Handbook of Immunoblotting of Proteins,* Vol. 1, Bjerrum, O. J., and Heegaard, N. H. H., Eds., CRC Press, Boca Raton, FL, 1988, 145.
39. **Koertge, T. E., Butler, J. E., and Dierks, S. E.,** Evaluation of the soluble immune complex (EIC) of the amplified ELISA and the use of this assay for quantitation by reaction stoichiometry. *J. Immunoassay,* 6, 371, 1985.
40. **Czerkinsky, C., Moldoveanu, Z., Mestecky, J., Nilsson, L. A., and Ouchterlony, O.,** A novel two colour ELISPOT assay. I. Simultaneous detection of distinct types of antibody-secreting cells, *J. Immunol. Methods,* 115, 31, 1988.
41. **Joshi, K. S. and Butler, J. E.,** The immunochemistry of sandwich ELISAs. V. The capture antibody performance of polyclonal antibody-enriched fractions prepared by various methods, *J. Immunol. Meth.,* in press.
42. **Butler, J. E., Koertge, T. E., and Peterman, J. H.,** The amplified enzyme-linked immunosorbent assay (a-ELISA), in *Enzyme Mediated Immunoassay,* Ngo, T. T. and Lenhoff, H. M., Eds., Plenum Press, New York, 1985, 241.
43. **Cleveland, P. H., Richman, D. D., Redfield, D. C., Disharoon, D. R., Binder, P. S., and Oxman, M. N.,** Enzyme immunofiltration technique diagnosis of herpes simplex virus eye infections in a rabbit model, *J. Clin. Microbiol.,* 16, 676, 1982.
44. **Jolley, M. E.,** Particle concentration fluorescence immunoassay (PCFIA) new, rapid immunoassay technique with high sensitivity, *J. Immunol. Methods,* 67, 21, 1984.

Chapter 9

QUANTITATION OF SPECIFIC ANTIBODIES: METHODS OF EXPRESSION, STANDARDS, SOLID-PHASE CONSIDERATIONS, AND SPECIFIC APPLICATIONS

John E. Butler and Robert G. Hamilton

TABLE OF CONTENTS

I. INTRODUCTION

Early immunological studies demonstrated that antibodies specific for foreign antigens or pathogens appear in serum following an accidental or natural exposure to foreign immunogens. Hence, monitoring the serum antibody activity to antigens of interest has become standard practice in clinical and experimental immunology. Antibodies to allogenic and autologous cell surface antigens and autoantibodies to native and modified self-proteins have also been measured as indicators of inflammatory disease processes.

Unlike immunoassays configured to measure antigens, specific antibody immunoassays (SpAbI) do not normally generate readouts in gravimetric units. Rather, antibody immunoassays provide measures of *antibody activity*. While serologists still refer to this as *titer,* others have expressed this activity in units relative to a reference standard (RS) or in "absolute" weight per volume units (see Section IV). Does the widespread use of relative activity mean that expression of humoral immune responses in weight per volume units (e.g., microgram of antibody per milliliter) is incorrect? *Yes, maybe, sometimes,* and *no* are all correct answers. The issue is knowing when it is immunochemically valid to do so.

Effective use of the SpAbI depends in part on a knowledge of the structure and heterogeneity of antibodies. The first section of Chapter 9 therefore reviews salient features of antibody structure and formation; those well versed in this area should proceed to Section III. The main portion of Chapter 9 (Sections III to VI) examines the immunochemistry of SpAbI, while the chapter closes with a discussion of the use of SpAbI in special situations (Section VII).

II. THE BIOCHEMISTRY AND BIOLOGY OF ANTIBODIES

A. THE STRUCTURE AND ORIGIN OF ANTIBODIES

1. Basic Structure

Antibodies are members of the immunoglobulin (Ig) supergene family which share structural features (Figure 1A). The basic unit is a domain of approximately 110 amino acids that contains an intradomain disulfide linkage (Figure 1) and is encoded by a gene segment (exon) of 330 bp. This protein subunit is folded into a compact structure composed of parallel and antiparallel β-pleated sheets, the so-called β-barrel, the ends of which are joined together by more flexible looped sequences which comprise the "complementary determining regions" (CDR). IgG, the most abundant class of serum antibody, typifies the basic structure of all antibodies (Figure 2B). IgG is composed of two identical heavy chains and light chains that are held together by noncovalent and disulfide bonds. Each heavy chain (55 kDa) is composed of four domains, the N-terminal variable (V_H) region domain, and three constant (C_H) domains. The κ or λ light chains (22 kDa) contain one V_L and one C_L domain. V-regions contain the antigen binding sites which define the antibody's specificity. C-regions determine the class (isotype) and light-chain type of the antibody.

2. V-Region Heterogeneity and Antibody Spectrotypes

The V_H and V_L regions fold to form the antibody binding site. The V_H domain originates through DNA rearrangement and joining of V_H, D_H, and J_H exons. The V_L domain originates by a similar process which involves only V_L and J_L exons. It is estimated that enough V-D-J germ-line exons exist to yield 10^5 different V_H regions and enough V-J exons to yield 10^3 different V_L regions; hence, 10^8 different binding sites appear to be encoded in the germ line. Moreover, somatic mutation and insertion of extra nucleotides at the junction between the V-D-J gene segments can further enhance variability. Each rearranged V-region domain contains conserved "framework sequences" and "hypervariable regions"; the latter give rise to the CDRs (see Section II.B).

In contrast to V-regions, the C-regions of antibodies are the gene products of a small number of closely linked, single copy, germ-line genes that so far have not been shown to be subject to somatic mutation, junctional diversity, or exon rearrangement characteristics. Thus, the charge heterogeneity which characterizes the serum antibodies of a particular isotype appears mostly to result from V-region sequence variability, e.g., bovine IgG2a.[1] Additional intra-isotypic heterogeneity may result from posttranslational modification.

3. Isotype, Allotype, and Idiotype

Igs are proteins which express epitopes that can be antigenically recognized by the immune system of other species and, in the case of autoimmune responses, the same species. Healthy mammals are tolerant to their own serum proteins and thus lack antibodies to their own Ig C-regions. However, intraspecies C-region germ-line heterogeneity is present and is distributed as Mendelian traits. Such Ig variants, called allotypes, are analogous to population blood group variants. Hence, when an individual is exposed to Ig molecules bearing non-self constant region epitopes (by blood transfusion or escape of cord blood across the human placenta), they may produce antiallotypic antibodies.

To obtain antibodies that recognize constant-region (isotype or subisotype) epitopes which are invariant within a species, it is necessary to immunize a different species with the Ig of interest, e.g., mice or rabbits with human IgG. These antiantibodies most commonly recognize isotypically unique determinants in the Fc and C_L regions of the immunizing immuno-

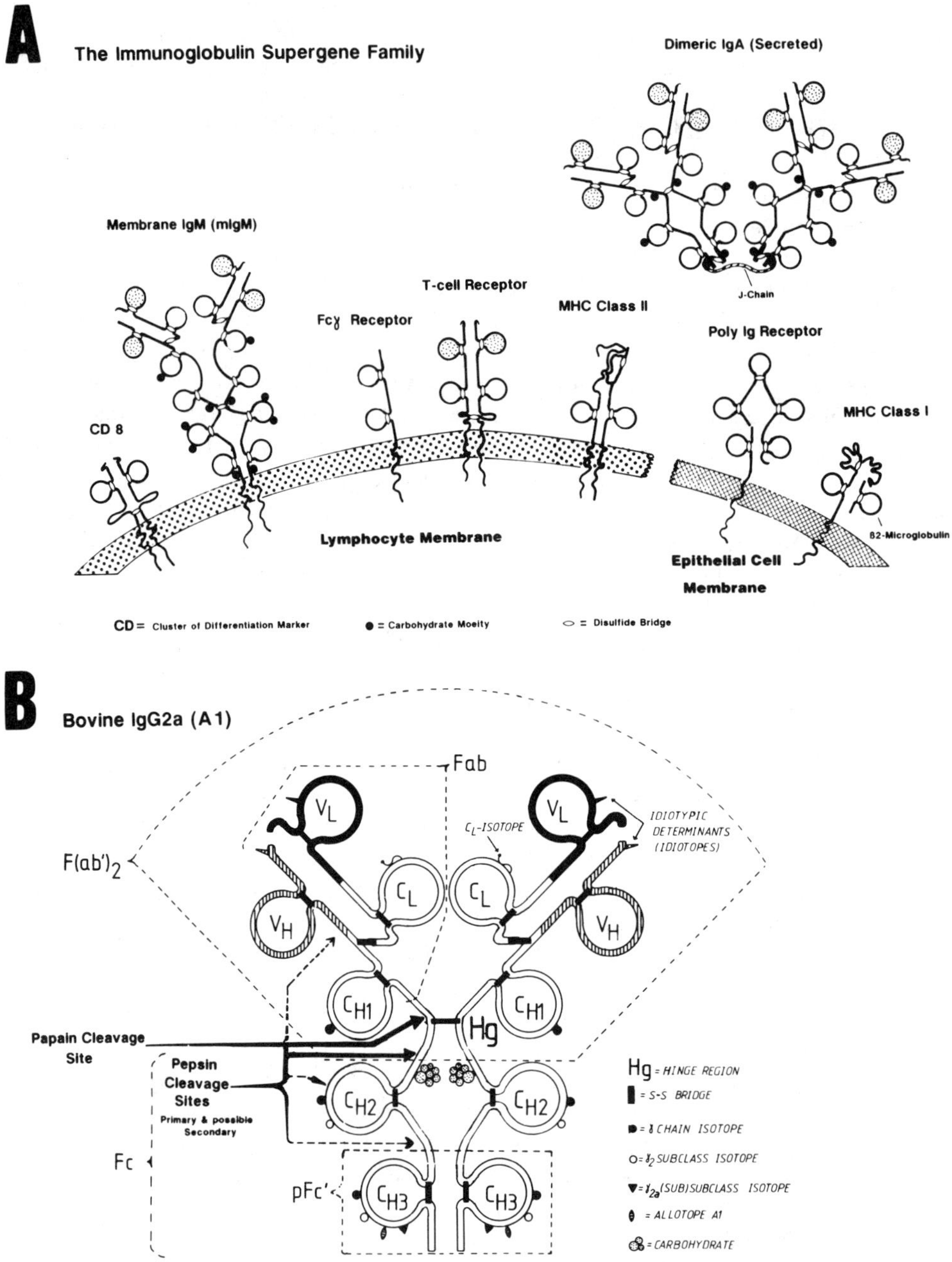

FIGURE 1. Antibodies and the immunoglobulin (Ig) supergene family. (A) Various members of the family are illustrated either as attached to the plasma membrane by their transmembrane tails or in secreted form. Each nearly closed circle represents the intrachain disulfide loop which characterizes the domains of molecules in this family. Note that the distal domains of MHC class II, which binds peptides rather than configurational determinants as do antibodies, lack this characteristic structure. Membrane IgM (mIgM) as well as membrane forms of other Igs, serve as B-cell antigen receptors while T-cells possess a different antigenic receptor but which is a member of the supergene family. IgA is shown as an example of a secreted polymeric Ig. Secreted IgM is a pentameric polymer and like the IgA dimer, also contain a J-chain. Variable region domains, which are responsible for specific antigen recognition, are stippled. All members of the family are glycosylated, although sites are only shown for Igs. (B) The planar chain structure of IgG. The peptide structure of the most common class of antibodies is shown by this model of the A1 allotype of a bovine IgG2a. The various enzyme cleavage sites, the major fragments generated by proteolytic digestion, the various domains and the various types of antigenic epitopes are indicated. (Modified from Butler[46] using recent data from Heyermann et al.[64] With permission.)

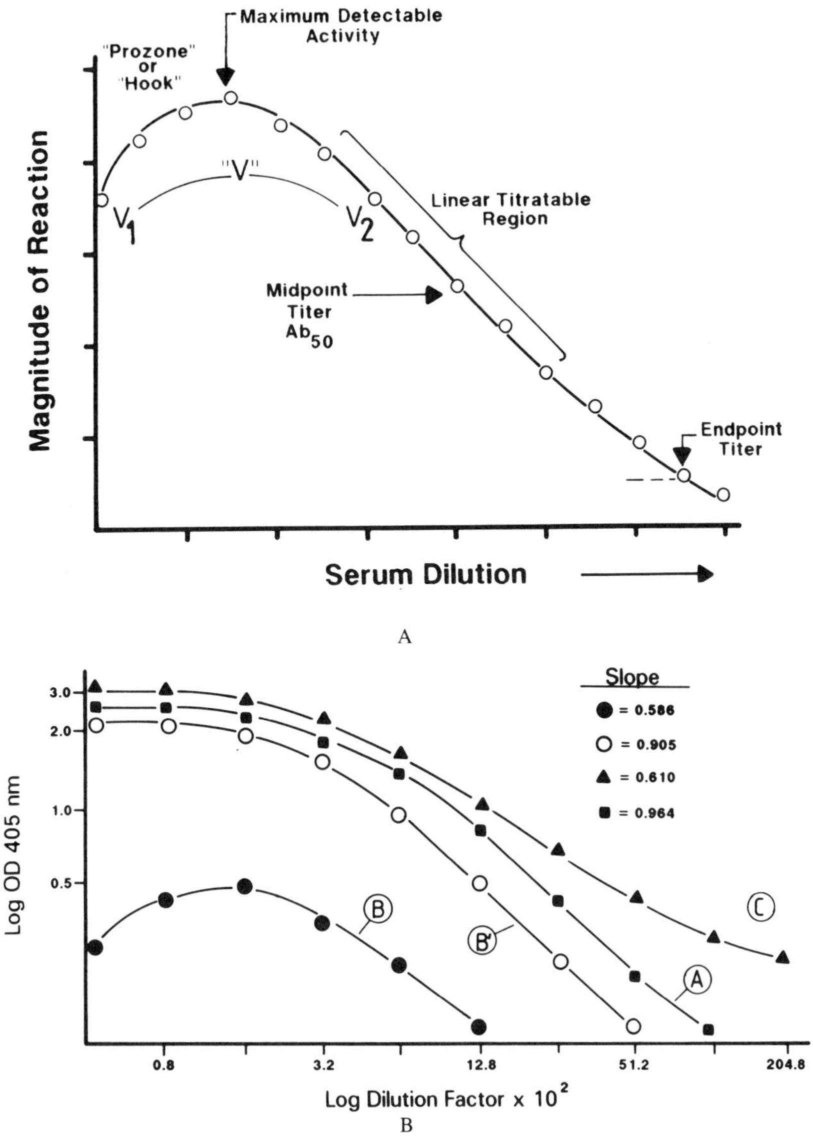

FIGURE 2. The anatomy of a titration plot. (A) A typical titration plot, showing such salient features as prozone, the titratable region, maximal activity, and endpoint titer. V_1 and V_2 illustrate two points which generate an equal signal but cannot be distinguished by single point determination. (B) Factors which influence parallelism in titration plots. Plot A, proper titration of an SpAbI which expressed in log-log fashion has a negative slope of approximately 1.0; plot B, the titration of IgA anti-OA in rabbit serum in which extreme prozone inhibition is observed as well as a titration slope of significantly <1.0; plot B', = The titration of IgA anti-OA in rabbit serum after partial depletion of IgG anti-OA;[44] plot C, a titration plot with decreased slope caused by nonspecific binding at high sample dilution (see Chapter 3).

globulins. Epitopes that define the isotype and subisotype of antibodies are therefore recognized by xenogeneic antisera while allotypic differences are most often recognized by antibodies made in other members of the same species.* The capacity of the immune system to recognize foreign immunoglobulin epitopes has been a cornerstone in the development of diagnostic reagents for the measurement of human and animal antibodies according to isotype and allotype.[2]

* Antiallotypic antibodies can also be raised in heterologous (xenogenic) species and using monoclonal technology.

Extreme variability in the V-domains of antibodies would suggest that within one individual any particular set of V_H and V_L sequences will be present in very low amounts. Hence, tolerance is not developed to this broad V-region epitope repertoire; this has no pathological consequences because any one expressed V-region epitope is in low concentration. Antibodies which recognize V-region epitopes in the individual's own antibodies are called antiidiotypic antibodies. The epitopes they bind are called idiotopes. Any V_H/V_L combination can display a variety of such idiotopes and collectively they define the idiotype, derived from the Latin for *idiot* or that which is unusual in occurrence.

4. Isotypic Diversity and Physiological Distribution

Five different C-region isotypes, IgM, IgG, IgA, IgD, and IgE have been identified in nearly all mammals; some examples appear in Figure 1. Many species have additional C-region exons that have evolved to code for subisotypes of IgG and IgA. Four IgG subclasses have been identified in both man and mouse which display >90% homology between their C-region domains. Three IgG subisotypes have been defined in cattle,[3,4] one in the rabbit,[5] and an unknown number in most other species. While humans possess C-region genes for only two IgA subisotypes, rabbits produce 13 subisotypes of IgA![6] IgG is quantitatively the predominant antibody in the serum and IgA antibodies are predominant in most exocrine body fluids. Most antibodies exist as monomeric units as in the case IgG, but secreted IgM is pentameric and secretory IgA is a dimer (Figure 1A).

B. THE ANTIBODY BINDING SITE

The site in the antibody that binds antigen exists as a CDR-lined cleft formed from the combination of the V_L and V_H domains. The "lips" of the cleft are formed in part by looped sequences at the end of the β-barrel. The six CDRs, three from V_L and three from V_H in these loops, become clustered through the juxtapositioning of the V_L and V_H domains. Early studies suggested the binding site could accommodate epitopes the size of a pentasaccharide or octapeptide[7,8] and more recently fluorescein has been regarded as a site-filling hapten.[9] Larger epitopes, perhaps comprising 15 amino acids, can be accommodated by adapting the V-region ("induced complementarity") through antigen-induced repositioning of CDRs to achieve optimal complementarity.[10] Structurally related epitopes may fit the same binding site, however, the reaction is normally one of lower affinity than that which occurs with the homologous epitope, thus permitting antibodies to discriminate between similar epitopes by subtle energetic differences resulting from differences in electrostatic forces, hydrogen bonding, van der Waals forces, and hydrophobic interactions. The net binding energy may be supplied by a few critical interactions within the binding site or may be the cumulative effect of multiple interactions. It is also possible that one antibody binding site can accommodate two unrelated epitopes, especially if they are small and can interact with amino acid side chains in different subregions of the binding site[11] or can recognize determinants in otherwise unrelated cells, organs, and tissues.[12] Stereochemically, a binding site could accommodate two epitopes the size of the hapten, dinitrophenyl. Antibody valency, or the number of binding sites, is an important factor which varies among isotypes and affects the degree of interaction of an antibody with multivalent antigens. The polymeric immunoglobulins IgM and dimeric IgA, have valences of ten and four, respectively (Figure 1A).

C. FUNCTIONS OF THE ANTIBODY HINGE REGION AND C-DOMAINS

Unique structural features of the C-regions of the different antibody isotypes endow them with unique biological activities. Structural differences in the C-domains of IgA and IgM permit their ordered polymerization yielding multivalent antibodies of high avidity.[13,14] Certain human isotypes (IgM, IgG1, or Ig3) upon binding antigen can activate the classic complement cascade while others cannot. Certain isotypes are specifically recognized by Fc receptors (Figure 1A) on the surfaces of phagocytic cells, regulatory T-cells, and epithelial cells involved in their transport. Dimeric IgA and IgM antibodies are recognized by the poly-Ig receptor (Figure 1A) which initiates their transport across various epithelial cells and secretion into exocrine body fluids. IgE antibodies bound to the high-affinity ε-receptors on mast cells and basophils can trigger the release of vasoactive mediators in the presence of antigen, producing immediate-type hypersensitivity.

The hinge is a sequence of usually 15 to 22* amino acids found between the C_H1 and C_H2 domains of IgG and IgA antibodies (Figure 1B). It permits flexibility of movement for the two Fab arms. Longer hinge regions present in human IgG1 and IgG3 antibodies enhance their flexibility and may favor their binding to multiple antigenic determinants in close proximity on a solid phase. The extended hinge, however, makes them more prone to enzymatic degradation than isotypes with more compact hinges (IgG2, IgG4, and IgA2). Moreover, the heavily glycosylated IgA2 hinge appears to enhance *in vivo* and *in vitro* stability by blocking degradative proteases. Thus, exocrine body fluids may contain measurable, intact secretory IgA antibodies while IgG may be degraded.[15,16] If antibody is digested by enzymes, Fab fragments may still bind to solid-phase antigens; however, they will go undetected in SpAbIs that employ antiglobulin conjugates specific for Fc determinants.

D. THE CLONALITY OF ANTIBODIES

Antibody specificity at the B-cell level involves the rearrangement of V-region genes to produce mRNA which encodes "composite exons" which become translated into the binding-site specificity of the antibody. During B-cell differentiation, a productive rearrangement which leads to the synthesis of functional H- and L-chain initiates a feed-back mechanism so that the B-cell does not rearrange remaining DNA and thus locks into the synthesis of that one V-region module. Such B-cells give rise to offspring clones which express the same V-region for the remainder of their lives. A so-called "switch mechanism" however can be activated to produce B-cell progeny which secrete antibody with the same V-region and a different C-region, i.e., a different antibody isotype.

Furthermore, a period exists in the life of a B-cell when it is subject to V-region somatic mutations, an event that can result in modified V-region specificity. Selection of B-cell clones by antigen depends on the affinity of surface antibody which differs from secreted antibody only by a transmembrane tail (Figure 1A). Hence, B-cells which have somatically mutated binding sites with higher antigen affinity are preferentially activated. This "affinity maturation" process occurs throughout the lifetime of an individual which can cause antibody affinity to increase over the course of continual antigen exposure. Normally, antigen-binding selects multiple clones for each epitope that differs in affinity. The clonal heterogeneity of the response is correspondingly increased when complex antigens bearing multiple epitopes are involved. Therefore, the polyclonal sera or secretions most often analyzed in solid-phase SpAbI contain a spectrum of antibodies, the distribution, affinity, and isotype of which may change during a course of immunization and over the lifetime of the individual. The assayist must contend with this heterogeneous population of antibodies of different isotypes, subisotypes, affinities, specificities, and allotypes. The design of immunoassays to quantitatively measure clinically relevant antibodies specific for a complex antigen mixture is a challenge.

III. MEASUREMENT OF ANTIBODY ACTIVITY

A. MASS ACTION PRINCIPLES IN ANTIBODY IMMUNOASSAYS

1. Special Considerations in Solid Phase Systems

The Law of Mass Action governs solid-phase as well as liquid-phase immunoassays, although kinetics (Chapter 18), absolute affinity constants (Chapter 12), and dissociation rates[17] (Chapter 19) may be quite different. Users of solid-phase antibody immunoassays

* Human IgG3 is an exceptional Ig with 62 residues in the hinge.

rarely need to know the exact concentration of their reactants. This is fortunate, since reaction volumes are interfacial and difficult to define (Chapters 1 and 12).

The use of mixtures of multivalent antigens creates additional problems: (1) the concentration of any one epitope is difficult to determine, (2) the proportion of functional determinants in immobilized mixtures may differ from that in solution due to denaturation and steric hindrance, and (3) limitation on the antigen density of certain solid phases can mean that infrequent epitopes are present at very low concentration. Therefore, investigators must optimize their assay not only by manipulation of the test antibody dilution and antigen concentration but also by utilizing procedures that minimize immobilization-induced antigen alteration and/or epitope expression.

2. Antibody Heterogeneity

The major application of SpAbI has been in the analysis of polyclonal antibody mixtures in human and animal serum and secretions. Mass Action principles dictate that at equilibrium a greater proportion of high-affinity antibodies will be bound than low-affinity antibodies. Therefore, when equal amounts of monoclonal antibodies with differing affinities are mixed together and analyzed in a solid-phase antibody immunoassay, the antibody that contributes most to the titer (Figures 2A and 3), or units per milliliter, is the high-affinity population.[18] This population can represent >80% of the bound antibody while comprising <20% of the mixture. The antibody detected is that which remains stably bound and *not all* the antibody in the specimen. "High antibody activity" is therefore the consequence of both affinity and concentration. With rare exceptions, such as human rheumatoid factors, a high concentration of a low-affinity antibody is rarely as significant clinically as a low level of high-affinity antibody. After all, "specific" antibody binding is typically distinguished from nonspecifically bound antibody, i.e., background, by the criterion of affinity.

B. ANALYTICAL FACTORS AFFECTING QUANTITATIVE ANTIBODY MEASUREMENTS

Various interrelated performance parameters of SpAbI are used in assay evaluation. These are summarized below but are also treated in Chapter 3 and especially in Chapter 6.

1. Sensitivity

The minimum detectable dose or sensitivity of an antibody immunoassay depends on the signal-to-noise ratio and the precision at the zero dose. Unless the immunochemistry of the assay is altered to enhance site occupancy and high-affinity antibody interaction, amplification of the signal from the test specimen will generally be paralleled by amplification of "noise", i.e., background signal.

2. Precision

The agreement of replicate measurements or precision within or between assays within a laboratory or between laboratories can be used as a measure of assay quality. Precision can be described by the coefficient of variation (CV) which is computed as a measure of the assay variation (standard deviation or SD) that has been normalized for the mean (e.g., CV = SD/mean). A plot of the variation in the measured response (e.g., SD or CV in counts per minute bound or optical density) for each concentration of a standard preparation over the working range of the assay (i.e., titratable region, Figure 2) defines the response-error relationship of the assay. The CV of the interpolated dose that results from imprecision in the response measurement can be plotted over a range of doses to produce a precision profile of the assay. The precision profile is useful in defining the sensitivity and working range of an assay that meets a predetermined level of variation. Moreover, intra- and interassay and interlaboratory precision profiles provide an objective basis upon which to optimize assay reagents and perform-

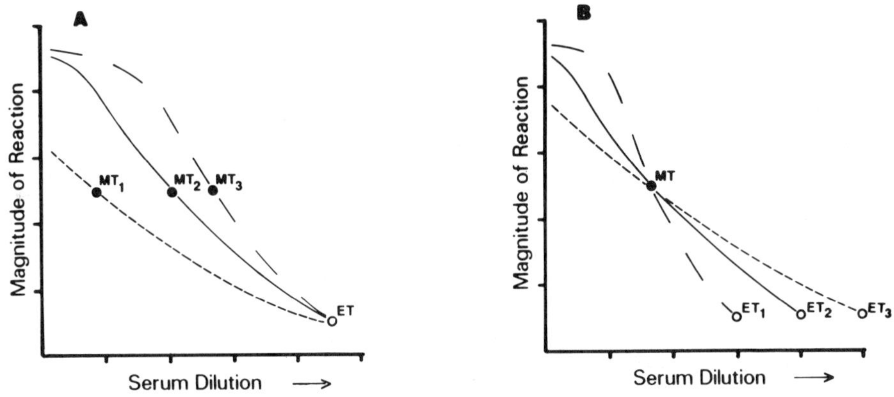

FIGURE 3. A graphic comparison of midpoint and endpoint titration. (A) The titration of three different samples which all give the same endpoint titer (ET) but due to nonparallelism have different midpoint titers (MT_1-MT_3). (B) The titration of three different samples which have the same midpoint titer (MT) but due to nonparallelism, give different endpoint titers (ET_1-ET_3).

ance. The response-error relationship and precision profile are illustrated in detail in Chapter 6.

3. Parallelism

Parallelism refers to the ability of test and reference antibody populations to produce dilution curves in an antibody assay that are equidistant and nonintersecting[20] (Figures 2B and 3). It is shown by interpolating response values obtained for various dilutions of a test antibody from a reference dose-response curve and/or by comparing the slope in the linear titration region. Correction of the interpolated dose for dilution should generate the same concentration of antibody no matter which dilution was analyzed. In general, two types of deviations are commonly encountered: flattened slopes due to nonspecific binding at high dilution (plot C, Figure 2B) or inhibition at low dilution (plot B, Figure 2B).

Parallelism is a sensitive indicator of imbalances between the nonanalyte factors, i.e., ionic strength, protein composition, and lipid content in a patient's serum when compared to the RS. Physiological buffers containing irrelevant proteins such as bovine serum albumin are commonly used in antibody immunoassays to minimize nonanalyte differences and to block nonspecific interactions of antibody with the antigen or the solid phase. However, the test serum itself may contain an abnormally high amount of the analyte being measured which, through nonspecific effects, will result in nonparallelism. Nonparallelism can be seen in the radioallergosorbent test (RAST) for allergen-specific IgE when a serum contains a total IgE level >1000 ng/ml. An interdilutional percent CV of the dose <15% has been used in the RAST and other antibody assays as a practical indicator of parallelism for test specimens.

C. UNITS OF ANTIBODY MEASUREMENT

1. Titer, Activity, and Immunobiology

An antibody level in a specimen can be described in a variety of ways. Titer is an historical unitage that equals (1) the reciprocal of the highest dilution of the specimen at which the antibody activity can still be detected (end-point titer) or (2) a midpoint in the titration plot where antibody activity is a percentage of the maximum detectable activity, e.g., 50% = Ab_{50} (Figures 2A and 3). The titration method is applicable to most assays, but it can be misleading and may reflect the technical bias of a laboratory. A titer of 1024 for instance may be equivalent to a 2048 titer from another laboratory, although it represents merely a single twofold dilution difference! Expression of the antibody quantity as a percentage of functional antibody in a RS is an alternative method of expressing activity. In enzyme-based solid phase immunoassays, "ELISA Units per milliliter" has been used by directly comparing the observed response level from the test antibody to that produced by a RS antibody that has been assigned a quantity of 100 or 1000 units per milliliter. When antibody levels are reported in arbitrary units based on interpolation from a multipoint dose-response curve (see Chapter 3), they tend to be more comparable between laboratories than titer, provided everyone uses the same RS.

2. Absolute (Microgram Per Milliliter) Units of Antibody

Expression of antibody in absolute gravimetric units, e.g., μg/ml, is often difficult and controversial. A major problem stems from the lack of common methods for determining the weight per volume amount of antibody in the RS. Moreover, the assigned absolute value represents the amount of antibody in the standard and *not* the amount of functional antibody that binds to antigen, so that values for test samples interpolated from the RS are accurate only when the proportion bound in both RS and test specimen are the same. The topic is treated in detail in Section IV.

3. Immunoreactive vs. Biologically/Clinically Relevant Antibody

Exposure of a population of individuals to a multideterminant antigen mixture could presumably elicit a polyclonal antibody response against "all" of the immunogenic epitopes. As the specificity and nature (isotype, allotype, idiotype) of human immune responses to allergenic and pathogenic antigens is increasingly characterized, it appears that not all the antibodies produced in these immune responses are equivalent in their ability to elicit biologically relevant or clinically important effector functions.* Preferably, an immunoassay should be designed to quantitate antibodies in a biological specimen that are not only immunoreactive with "all" immunogenic epitopes on the antigen, but are isotypes that have *in vivo* or diagnostic relevance (see Section VII.F).

D. CONVENTIONAL TITRATION OF ANTIBODY ACTIVITY

1. Serological Titration

Titration may be defined as "changing the concentration of the variable of interest until the result of a reaction achieves a pre-determined magnitude." In antibody immunoassays, this technically means preparing a dilution sequence of the test specimen and assaying these dilutions for their ability to achieve the antigen-antibody reaction of the desired magnitude, i.e., endpoint. As antibody quantity is progressively reduced by dilution, the titration enters a region of the dose-response curve, i.e., the titratable region or working range, where solid-phase antigen is in relative excess to antibody and the immunoassay reaction becomes antibody dependent (Figure 2A).

The need to rapidly analyze many samples has caused analysts to make several changes in their titration methods. First, dilution sequences have been broadened, e.g., log dilutions can replace twofold dilutions to make sure that a point is reached where the concentration of antibody falls in the titratable range of the assay. This process frequently avoids the need for repeat testing of specimens. Second, the number of dilutions that are being tested is being reduced in some cases even to a single dilution. The lack of multiple dilutions provides no opportunity to test parallelism (Section III.B.3).

Various algorithms have been used to plot titration data: arithmetic, semi-log, log-log, and logit-log transformations of dose-response data (see Chapters 3 and 20). For each of these,

* Certain antigens and epitopes are immunodominant. A small contamination of a protein preparation with α_2-macroglobulin can result in a more pronounced response to α_2-macroglobulin than the major protein in the preparation. Similarly, certain epitopes behave immunodominantly.[21]

parameters such as affinity, heterogeneity and nonspecific binding have different meaning. This has been illustrated theoretically[18] and empirically[22,23] (Chapter 3). For example, when antigen-antibody titrations are expressed in a log-log manner, the linear titratable region has a log-log slope of 1.0. Deviations from 1.0 reflect immunochemical perturbations within the assay. The interpretation of a dose-response titration plot depends on the algorithm used; when ideal response data are plotted log-log, the Y-intercept defines the affinity of the reaction whereas in a semi-log presentation, slope reflects affinity.

2. Endpoint vs. Ab50 Titration

Endpoint and midpoint (Ab_{50}) titrations are both used, each subject to certain inherent errors. Figure 3A illustrates three dilution plots that titrate to the same endpoint although they produce different midpoint titers. Such nonparallelism can also produce the opposite effect (Figure 3B). The potential criticism of the end- or midpoint titrations can be overcome by (1) documenting that the titration plots obtained with multiple sera are parallel or (2) performing a multipoint titration analysis on every specimen. Establishing parallelism for routine assays is preferable because it not only shows that each specimen reacts immunochemically in a similar manner, but permits complete titration plots to be abbreviated to save time and materials.

Figures 2A and B both illustrate the prozone problem which have special significance in single point measurements. Without the use of a dilution sequence, the low dilution (prozone) V_1 value cannot be distinguished from the V_2 value attained at higher dilution (Figure 2A).

3. Multipoint Analyses

The errors inherent in mid-, end-, and single-point analyses can be overcome by multipoint analysis. Multipoint analysis involves the comparison of complete titration plots from multiple specimens. Documentation of a new antibody assay should involve the analysis of multiple dilutions of at least ten sera containing different levels of antibody. Once parallelism has been shown, two high-titered antisera can be selected. One should be analyzed in six to eight dilutions to generate a dose-response or standard curve and the second should be used as a control specimen which is tested at only three dilutions.

E. ANTIBODY STANDARDS AND DATA EXPRESSION

1. Direct Expression of Antibody Activity

Titer, either as endpoint or Ab_{50} (Figure 3), is still a commonly used expression of antibody. This mode of antibody expression is useful for comparative purposes within an experiment or laboratory but it remains of little value for workers in other laboratories.

2. The Reference Standard Method

RSs are typically pools of serum or body fluid containing high levels of antibody that meet certain requirements of performance (see III.E.3 below). An RS may be assigned units (U) of antibody activity, e.g., 100 U/ml, or an absolute, gravimetric quantity of antibody. The RS method is useful for comparing data among different laboratories and among experiments in the same laboratory. RS given in units per milliliter have the disadvantage of being quantitatively misleading because, for example, a serum might be assigned a value of 100 ELISA Units (EU) of IgE anti-ragweed and also 100 EU of IgG anti-ragweed. Tables and figures describing patients as having 150 EU of IgE/ml and 95 EU/ml of IgG could be misinterpreted to mean that more ragweed-specific IgE antibody activity is present than IgG antibody! One method of arbitrarily preventing such misinterpretation is to assign EU in relation to the total amount of the IgG and IgE in serum. For a serum with 10,000 μg/ml of IgG antibody and 0.1 μg/ml of IgE antibody, the RS could be assigned a value of 100,000 EU/ml of IgG and 1 EU/ml of IgE; IgA in the same serum would be assigned a value of 20,000 EU/ml. RS expressed

3. Types of Reference Standards

A standard is a substance that has been established by an authority as a rule for the measurement of the quantity, weight, extent, value, or quality of that substance in a test specimen. An ideal RS for SpAbI should possess (1) well-documented *potency* (e.g., units per milliliter) or *amount* of antibody in easily understood international units such as weight or mass per volume rather than arbitrary units, (2) defined *specificity* with antibodies that bind to all the immunoreactive and clinically or biologically relevant epitopes on the antigen of interest, and (3) a distribution of *affinities* that approximates those possessed by the population of antibodies in the test specimens.

Unlike most hormone and drug standards that are prepared with purified compounds, antibody standards are either individual or pooled sera that have been characterized empirically against a defined antigen preparation for the quantity, quality (isotype, affinity), and antibody specificity. The most well-characterized standards are international or primary reference preparations that have been collected, tested, and aliquoted under the guidance of the World Health Organization (WHO) International Laboratory for Biological Standards. For instance, the WHO 67/97 is the present WHO standard for human serum proteins such as total serum IgG, IgA, and IgM.[24] Few antibody standards with defined potency and specificity for known antigens have ever achieved the level of WHO documentation, in part because the "gold standard" detection methods have been in a constant state of evolution and because reference and test sera are composed of a heterogeneously reactive population of antibodies that have been derived from various clones of cells (Section II.D).

Secondary standards are not as extensively tested but they do have a defined potency and/or specificity based on replicate measurements by the producer [e.g., National Institutes of Health or Centers for Disease Control (CDC)]. Secondary reference standards, for example, are available from the CDC for autoantibody assays. Working or tertiary standards are prepared in house, typically using a patient's serum or a pool representing >100 individuals and then standardized against the secondary or primary standards.

4. Expression of Antibody as Specific Activity

A useful method in the field of mucosal immunity is to express antibody activity in exocrine body fluids as a ratio of the titer (either as reciprocal of dilution or units per milliliter) to the total amount of that particular isotype or subisotype measured by a sandwich ELISA (see Section VII.C).

5. Homologous vs. Heterologous Interpolation

Values generated in SpAbI by dilution of the test specimen may be quantitatively interpolated from a standard curve that has been constructed with dilutions of a characterized antibody specific for the same or "homologous" antigen, or a different, "heterologous" antigen (Section IV, Figure 4D). For illustration, the amount of IgE antibody detected in an immunoradiometric assay that uses yellow hornet venom (YHV) antigen may be quantitated by the interpolation of bound cpm from a "homologous" IgE anti-YHV standard or a "heterologous" IgE anti-yellow jacket venom standard curve. Either standard has a defined amount of antibody as determined in weight per volume units by the methods discussed in Section IV.

Heterologous interpolation methods are attractive when (1) sera containing antibody specific for less immunogenic antigens are not available in large amounts, (2) two antigen systems are highly cross reactive as exemplified by the Hymenoptera venoms from the yellow jacket, yellow hornet, white-faced hornet, and Polistes wasp, or (3) one wishes to compare the relative amounts of antibodies of different isotypes within one serum that bind to the same solid-phase antigen, but for which homologous antibody standards are not available (e.g., IgG1, IgG2, IgG3, and IgG4 anti-ragweed). One major requirement for successful use of heterologous interpolation methods is parallelism within the assay system.

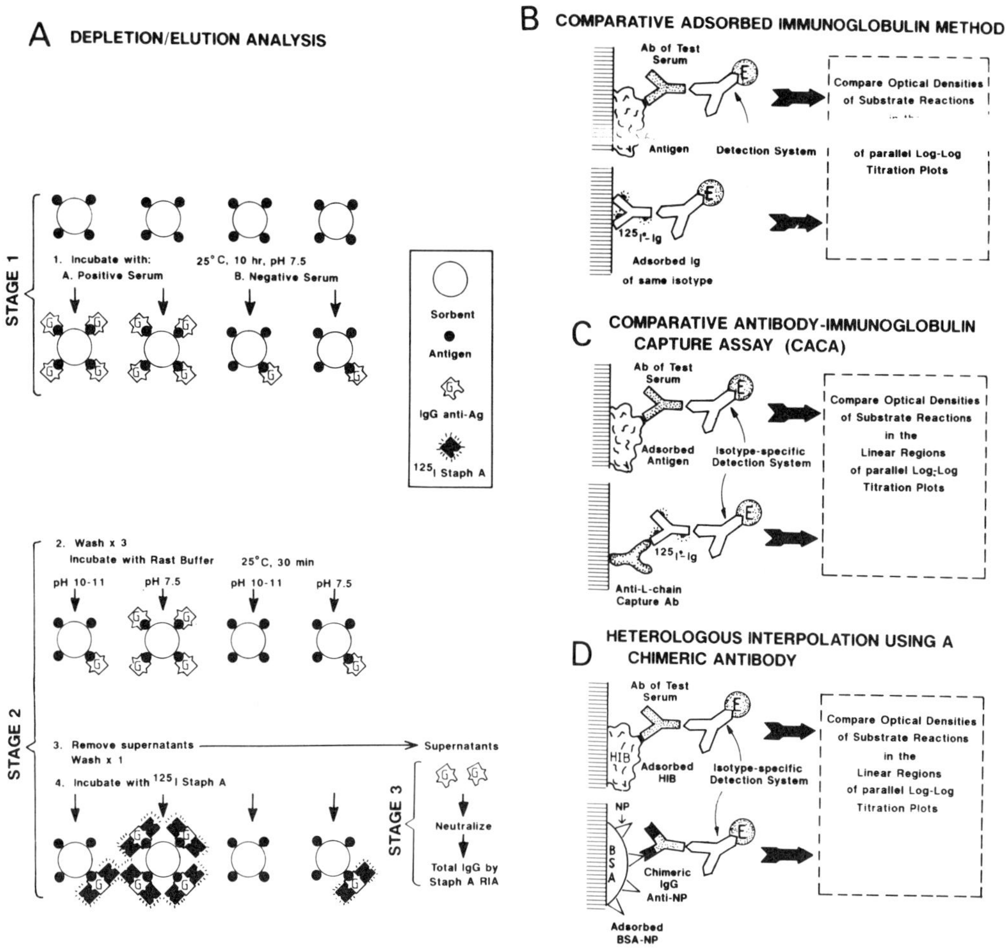

FIGURE 4. Methods for the determination of the absolute quantitation of antibody. (A) The depletion/elution analysis. (B) The comparative adsorbed immunoglobulin method. (C) The comparative antibody-immunoglobulin capture assay (CACA). (D) Heterologous interpolation using chimeric antibody. The steps involved in conducting each of these assays is described in the text.

IV. GRAVIMETRIC QUANTITATION OF ANTIBODIES

A. METHODS OF GRAVIMETRIC QUANTITATION

The accuracy of the various methods described for the absolute measurement of antibody depends on the assumptions made when determining the antibody concentration of the RS; different methods employ different assumptions.

1. Saturation Analysis

Scatchard proposed a method that relates the amount of bound labeled antigen at saturation to the number of antibody binding sites and ultimately to the mass of the antibody present in a reaction mixture.[25] Following the separation of free and bound antigens after equilibrium, the unitless ratio of bound/free antigen is plotted as a function of the amount of antigen bound. For homogeneous interactions, the slope of the Scatchard plot equals the equilibrium constant of the antibody while the X intercept (achieved at saturation of the antibody) equals the moles/ per liter of antigen bound. Assuming the absence of steric hindrance and an antigen/antibody ratio of two for bivalent antibody, the molar concentration of antibody present should equal one half of the molar concentration of antigen at the X intercept. Scatchard analysis is only practical for small molecular weight antigens (no steric hindrance) and for antibodies with limited heterogeneity.

2. Depletion Analyses

A RS serum can be standardized by depletion of their specific antibodies on solid phase antigen affinity columns. Depletion analysis works best for minor isotypes such as IgE *where the amount of antibody specific for a particular antigen mixture can represent a large percentage of the total immunoglobulin protein (e.g., 10 to 40%).*[26] After depletion, the unabsorbed (sham control) and adsorbed serum are analyzed for total IgE protein by immunoassay; the difference before and after in IgE content (Chapter 7) is considered the amount of specific antibody in the serum. When the percentage of specific antibody is low, a small difference between before and after Ig content can introduce a large error in the estimation of antibody. The depletion method may also yield misleading values because the amount of specific antibodies removed from the specimen by the affinity column may be proportionally greater than the amount that will bind solid phase antigen in an SpAbI.

3. Affinity Purification

The speciflc antibody eluted from an antigen affinity column can be used as a "purified" RS antibody. Cited disadvantages of this approach include the partial denaturation of antibody after elution, the presence of blocked binding sites when antibody is eluted with soluble antigen, and the presence of nonantibody proteins in the eluant which makes the determination of antibody content difficult. When polyclonal sera are involved, the antibody population bound and subsequently affinity purified, may not represent the typically wide affinity spectrum of the starting material; high-affinity antibodies may remain on the column. Finally, antigen covalently coupled to a support may express a different set of epitopes than antigen insolubilized on microtiter plate wells.

4. Depletion/Elution Analyses

This solid-phase elution technique combines the solid-phase depletion and affinity chromatography methods.[27] The sensitivity of the depletion method is enhanced by allowing small amounts of specific antibody to be adsorbed, eluted, and quantified apart from the large background of antibodies with the same isotype but other specificities. A schematic of the method is presented in Figure 4A. In an initial step, an amount of antigen-coated particles is incubated with an antiserum, such that >90% of the specific antibody is bound. After washing, 30 to 60% of the bound antibody is eluted by pH change and its Ig content determined as described in Chapter 7. The antibody remaining on the solid phase is also quantitated *in situ,* permitting the calculation of a better estimate of the total bindable antibody in the specimen. Some of the errors inherent in affinity purification methods are also relevant to this method.

5. The Comparative Adsorbed Immunoglobulin Method

Purified Ig, over a range of concentrations, is bound directly to a desired solid phase. The technique assumes that direct adsorption of, for example, a myeloma protein of a particular isotype, subclass, light-chain type, and allotype to the solid phase mimics antibody bound to antigen.[28] Parallel titrations are then performed using this purified protein standard and a RS and its specific antigen (Figure 4B).

Several problems limit the usefulness of this method. First, a purified Ig protein of the same isotype, subclass, and allotype spectrum identical to the antibody that is being detected in the assay must be available; for minor subisotypes this is a difficult matter. Second, the method assumes incorrectly that all the Ig pipetted into a well binds to the plastic in a manner that is sterically similar to the antibody in the sample being studied (e.g., with its Fab oriented toward the solid phase and its Fc extended into the fluid phase). Under optimal conditions, approximately 50% of the added antibody becomes adsorbed to plastic[29] (see Chapter 12) but it may only be recognized $^{1}/_{2}$ to $^{1}/_{25}$ as well as when the same antibody is bound to antigen.[30]

6. Heterologous Interpolation using a Chimeric Antibody Standard

A heterologous dose-response curve can be constructed using a genetically engineered chimeric human-mouse antibody with mouse V-regions that define the antigen specificity (not necessarily the same as the test antibody) and constant region domains of the same isotype or allotype of the species antibodies being studied.[31] Figure 4D illustrates how the amount of human IgG2 *H. influenzae* type B (HIB) in a proposed RS is estimated with heterologous interpolation from a dose-response curve generated with IgG2 human/mouse chimeric antibody specific for 4 hydroxy, 3 nitrophenyl (NP) coupled to bovine serum albumin (BSA). Simultaneous titrations are performed and the amount of bound human IgG2 anti-HIB and human/mouse IgC2 anti-NP are detected with the same monoclonal anti-human IgG2Fc (HP6002) enzyme conjugate. Provided the two dilution curves are parallel, the amount of bound IgG2 anti-HIB can be interpolated from the heterologous chimeric anti-NP titration plot. Use of radiolabelled chimeric antibody can be used to standardize the system for the actual amount of chimeric IgG2 anti-NP bound to NP-BSA. As with the above methods, heterologous interpolation from a chimeric antibody dose-response curve only provides a measure of the amount of test antibody which binds the solid-phase antigen. The efficiency of extraction of antibody from the sample must be determined, possibly by sequential transfer methods (see below) to define the actual content of specific antibody in the test specimen.

7. The Comparative Antibody-Immunoglobulin Capture Assay (CACA)

The CACA system[32] is somewhat similar to the chimeric antibody standard system except that it requires only purified Ig of the appropriate isotype without regard for specificity (Figure 4C). Simultaneous titration plots are prepared, one in which iodinated, purified Ig is captured by an anti-L chain CAb and the other in which the antibody of a RS binds to the solid-phase antigen. Detection of the captured Ig or bound antibody is then detected with the same isotype-specific enzyme-antibody conjugate. The relationship between the amount of bound immunoglobulin and the extent of the enzymatic reaction (optical density) can be standardized with radioiodinated Ig of known specific activity (counts per minute per microgram protein). Moreover, the relationship between bound and free antibody at equilibrium and actual percentage of bindable antibody (Z) is determined by the equation $Z = X–Y/X*100$, where $X =$ amount of antibody added and $Y =$ amount of antibody which cannot bind under any circumstanccs. The latter is determined by sequential transfer of the supernatant to fresh, antigen-coated wells and values calculated by the equation $Y = X(1–b1)(1-b2)(1-b3)$ where b1 is the fractional amount of antibody bound after the first incubation, b2 the fractional amount bound after the second, etc. Additional details are provided elsewhere.[32,33]

B. COMPLICATIONS WITH ABSOLUTE ANTIBODY QUANTITATION

1. The Meaning of Absolute Units of Antibody; Bound vs. Free

The absolute amount of antibody in a RS has different meanings depending upon how the determination was made. The basic differences among methods reside in whether the value reflects the total amount of immunoreactive antibody present in the RS or merely the amount that has bound to solid-phase antigen in the assay. Determination of absolute quantity by saturation analyses, using the comparative adsorbed immunoglobulin method or by heterologous interpolation using chimeric antibody, measures only antibody which has bound antigen. On the other hand, use of an affinity purified RS or the depletion/elution method is a measure of total immunoreactive antibody in the RS. In some cases only 50% of added specific antibody binds to antigen adsorbed on microtiter wells at equilibrium. Hence the concentration antibody binds to antigen adsorbed on microtiter wells at equilibrium. Hence the concentration of antibody determined in test samples by interpolation from the dose response curve of such a RS, could actually be twofold higher. Multiple transfer of reaction supernatants as described for the CACA system[32,33] has been used to correct this discrepancy.

2. Differential Affinity of RS and Test Populations

According to the Law of Mass Action, a polyclonal serum containing antibodies of multiple affinities to even a single epitope reacts with antigen such that those of higher affinity will be preferentially represented in the bound fraction. Hence, differences in bound:unbound ratios among polyclonal antibody mixtures when complex solid-phase antigens* are used could reflect affinity differences as well as differential epitope exposure or concentration. Therefore, the differential binding of polyclonal antibodies must be considered in relation to the *affinity of the RS and* the *method used to quantify the antibodies in the RS.* If the affinity of the RS is *considerably greater* than the average affinity of the polyclonal antibody mixture, the total antibody concentration in the specimen will be *underestimated* when the concentration of antibody in the RS is based *on total,* rather than the bindable antibody. This occurs when the antibody content in the RS is established by, for example, affinity purification. If on the other hand chimeric antibody technology is used, affinity differences should be automatically corrected because this method depends on the *bound antibody* in the RS. However, unless bound-to-free ratios are determined, as in the CACA, *bound* antibody is reported instead of total antibody.

3. Additional Factors Complicating Absolute Quantitation

The detection of bound primary antibody in SpAbI is indirect; the final signal is influenced not only by the magnitude of binding of the primary antibody, but also the binding of the secondary anti-antibody. For example, IgM bound adjacent to IgG could sterically inhibit the anti-IgG detection system from binding. The compatibility and quality of the RS is also important. A RS purified and standardized by affinity chromatography may (1) contain antibodies which are unable to recognize antigen due to elution-dependent denaturation or blocked binding sites, and (2) reflect a population of antibody which binds antigen on the affinity column but which may not recognize the solid-phase antigen (see below).

V. PROPERTIES OF THE SOLID-PHASE ANTIGEN

A. ANTIGEN CONCENTRATION AND EPITOPE DENSITY

The Law of Mass Action predicts that an increase in the antigen concentration will increase the proportion of antibody in a specimen that is bound at equilibrium (Chapter 3). Beyond a certain excess of antigen, any further increases will have a negligible effect. Increasing the epitope density on hapten-protein conjugates or naturally occurring antigens with repeating sequences will have its greatest effect on the detectability of low affinity antibody.

Solid phases like microtiter wells are generally more antigen limiting than microparticles, nitrocellulose, or some surfaces to which antigens are covalently attached. Large surface areas and high antigen concentrations are desirable when assays with wide dynamic ranges are needed.

B. ACTUAL VS. FUNCTIONAL ANTIGEN CONCENTRATIONS

Chapters 10, to 12, 14, and 17 support the view that both adsorbed and covalently attached proteins undergo alterations during the immobilization process which effect their functional concentration or immunoreactivity. The degree of such alteration may vary among proteins.

* Complex antigens are those displaying multiple, chemically distinct epitopes.

plastic compared to when bound to its antigen which has been adsorbed on plastic. IgG shows differences of 25- to 60-fold when compared in this manner.[30] Hence, investigators should select a solid phase and method of immobilization that optimizes functional as well as total antigen concentration. Although the actual amount of antigen on a solid phase may seem low, the reaction volume is believed to be largely confined to the solid interface so the interfacial antigen concentration may be high. Furthermore, antigen may not always be randomly distributed on a solid phase. At sites where antigen clusters, its concentration may be very high.

C. OTHER COMPLICATIONS WITH SOLID-PHASE SpAbI

1. Specificity and Isotype Bias

There are circumstances during the measurement of humoral immune responses when all antibody that can react with a solid-phase antigen is in fact not detected. The assay may be selective for those antibodies of highest affinity, i.e., "most specific" and the highest in concentration, e.g., IgG vs. IgE in serum. Occasionally, antibodies of the same isotype go undetected by a process called specificity or affinity* bias. Isotype bias is commonly experienced in clinical allergy laboratories where nanogram per milliliter levels of IgE antibody must be detected in the presence of microgram per milliliter levels of IgG antibody (see VII.B). Specificity bias can result from large differences in the affinity of antibodies of the same isotype in a specimen. Most bias of the latter type goes unnoticed because the antibodies inhibited from binding are of the same isotype as those detected.

2. Bias Caused by Epitope Presentation

Antibodies to complex antigens may be prevented from binding to their respective antigen epitopes for a variety of reasons. Immunodominant epitopes, frequently only a minor proportion of epitopes displayed on multivalent antigens, may elicit the formation of large amounts of antibodies. Due to the limited expression of the epitope on the antigen, these antibodies may be underestimated in SpAbI because the individual antigen epitopes are not in molar excess. Disproportionate epitope presentation can also result from adsorption-induced conformational changes (Chapter 10) or nonrandom adsorption. We have observed that when ovalbumin is adsorbed onto plastic, <50% of the antibodies which recognize covalently bound ovalbumin are detected. Hence, careful studies are needed to insure that the epitopes of clinical and biological interest are expressed on the solid-phase antigen.[33] Saturation analysis with increasing amounts of defined monoclonal antibodies can be useful in documenting epitope expression. Ideally, the best solid-phase antigen is one that can bind antibodies with specificities for all of its antigenic determinants. The controversial use of "whole HIV" antigen (U.S. system) or a preformed mixture of recombinant HIV proteins (German system) as solid-phase antigen in acquired immune deficiency syndrome (AIDS) serology, exemplifies this issue.

3. Diffusion-Related Effects

The diffusion dependence of solid-phase immunoassay (Chapters 1, 18, and 19) can favor the binding of certain antibodies over others due to concentration or molecular weight differences especially when short incubation times are used. The use of agitation or two-step systems in which the initial reaction occurs in semisolution on microparticles followed by particle concentration have been introduced to overcome this problem.

VI. DETECTION SYSTEMS FOR SPECIFIC ANTIBODY IMMUNOASSAYS

Detection systems for solid-phase immunoassays have been reviewed in Chapter 1 and discussed in Chapters 4 to 6 and 17. The signal-to-noise ratio, labeled reagent stability,

* Specificity is defined by affinity.

amplication potential, and the efficiency of detection equipment are among the important considerations for use in SpAbI. Several aspects of SpAbIs may immunochemically favor a particular detection system over another. One factor is the size of the detection system. It has been shown[22] that the amount of primary antibody bound to antigen and the amount of the secondary antibody-enzyme conjugate bound do not bear a constant relationship to each other over the titration range. This phenomenon was shown to result from steric hindrance as demonstrated by using antibody-enzyme conjugates of different sizes. The smallest detection system, e.g., the radioiodinated Fab of the antiglobulin, will produce the least steric hindrance and the widest assay working range. However, a variety of logistic and safety issues do not make the latter the label of choice. The decision to select a particular label and detection system may therefore not be solely dictated by immunochemical criteria, but rather by convenience, safety, and expediency.

VII. PRACTICAL ASPECTS IN THE MEASUREMENT OF ANTIBODIES

A. IgG SUBCLASS ANTIBODY STANDARDIZATION

Human IgG antibodies, once bound to antigen, vary as a function of their subclass in terms of the degree to which they activate complement and bind onto cell surface Fc receptors. The subclass distribution of IgG antibodies ultimately determines the degree of antibody-mediated immunopathology or protection. Restriction of an IgG antibody response to a particular subisotype changes its biological effect. IgG1 and IgG3 antibodies are associated with destructive processes because they efficiently cross link antigen, bind onto monocytes Fc receptors, and readily fix Clq. IgG4 antibodies, due to their compact hinge, do none of these functions well. Rather, they appear to circulate as functionally monovalent antibodies and they are associated with protective processes.

In the field of allergy, protein allergens are known to elicit a predominantly IgG1 antibody response upon acute allergen exposure. The subclass distribution of allergen-specific IgG antibodies gradually shifts to IgG4 upon chronic allergen challenge.[34] Rarely have allergen-specific IgG2 and IgG3 antibodies been detected in allergy patients on immunotherapy. The consequence of the IgG1/IgG4 antibody shift during immunotherapy and the role of these subclass in modulating or blocking allergen-induced IgE-mediated immediate-type hypersensitivity is still the subject of investigation.

In microbiology, human IgG antibody responses to bacterial carbohydrate antigens such as the polysaccharides on HIB can be restricted to IgG2 in many adults and children. The injection of a protein-conjugated polysaccharide vaccine appears to shift the subisotype of the antibody response toward IgG1.[35] The shift of the immune response from IgG2 to IgG1 is assumed to enhance the destructive and opsonizing activity of the antibody. The relative ability of human IgG1 and IgG2 anti-HIB antibodies with the same affinity and epitope specificity to clear a *H. influenzae* infection is under investigation.

IgG1 and IgG3 autoantibodies against cell surface antigens (e.g., red blood cells and intradermal cells) have been shown to bind and lyse cells, causing inflammation of surrounding tissue by complement activation. Chronic autoantigen exposure can in cases such as the skin disease pemphigus reportedly induce IgG4 autoantibodies that may bind to the cell surface and "protect" it from the more destructive IgG1 and IgG3 antibody binding.[36]

The subclass distribution of these and other human IgG immune responses has been studied with the SpAbI. Insolubilized antigen has been used to bind antibodies of all isotypes from a test specimen. Once bound, a particular isotype is commonly detected with labeled isotype-specific antibody. The major problem with these studies has been the lack of a suitable method to relate the relative response levels across the subclasses (e.g., comparison of data from IgG1 SpAbI with those obtained in IgG2, IgG3, and IgG4 SpAbIs). Each assay is performed in an identical manner, except that they use different enzyme-conjugated mouse monoclonal antibodies specific for human IgG1, IgG2, IgG3, and IgG4 and different specimen dilutions. No simple method has been developed to compare the assay response signals (e.g., optical density = OD) and determine the relative distribution of IgG antibody among the bound subisotypes. One early approach called equipotency[37] involved optimization of the IgG suclass enzyme-SpAbIs so that an OD of 1.0 in the IgG1, IgG2, IgG3, and IgG4 antibody assays would roughly correspond to the same amount of bound antibody. The measured response signal in the different subclass SpAbIs, however, has been shown to be only an indicator and not a direct measure of the concentration of antibody bound. The OD measured in a human IgG1 enzyme-SpAbI with 10 ng of human IgG1 antibody bound to a solid-phase antigen may not be similar to the OD that is detected in the IgG2 SpAbI with 10 ng of bound IgG2 antibody. This is mainly attributed by differences in the affinity and specific activity (e.g., units of enzymatic activity per unit weight of antibody) of the enzyme-conjugated monoclonal anti-IgG1, IgG2, IgG3, and IgG4 detection antibodies.

The only rigorous approach to determine the subclass distribution of a human IgG response is to directly quantitate the amount of IgG antibody of each subclass in the specimen with four individual human IgG subclass assays, each with their unique standard and control sera. Ideally, each IgG subclass antibody standard should be specific for the antigen used on the solid phase. However, this has been complicated when one considers the more than 100 different antigen specificities of clinical importance and the difficulty of making homogeneous human antibodies standards of defined specificity and restricted subclass.[38] An existing panel of human-mouse chimeric antibodies with specificity for a hapten are currently being used to bind a defined amount of human IgG subclass protein onto a solid-phase antigen. Due to their high affinity for NP (see Section IV.A.6), >90% of this antibody binds to its antigen on microtiter plates during a first incubation as shown by sequential transfer experiments. Once a known abount of chimeric human antibody of a single IgG subclass has been bound, one can relate the assay response signal detected to amount of bound IgG protein and construct a heterologous dose-response curve from which an estimation of bound antibody can be performed on the test sample (Figure 4D).[39] Recombinant DNA techniques will be increasingly used to prepare matched panels of engineered human IgG1, IgG2, IgG3, and IgG4 antibody standards with specificity for clinically relevant antigens.

B. ISOTYPE BIAS IN THE ALLERGY LABORATORY

The antigen-limiting nature of microtiter well-based SpAbIs causes problems when large differences exist between the concentration of antibodies of different isotypes or subisotypes, e.g., IgG and IgE, in serum. A sufficient excess of solid-phase antigen is rarely present to permit both isotypes to bind noncompetitively. If one assumes that antibodies of different isotypes have the same affinity, then this competitive inhibition problem should be resolvable by (1) dilution of the serum so that the antigen on the solid-phase antigen achieves relative molar excess in relation to antibody, (2) increasing the concentration of the antigen on the solid phase, or (3) removing the major antibody, i.e., IgG, from the test sample to permit accurate measurement of antibodies in isotypes which are present at low concentration. In the first case, low levels of a minor isotype antibody such as IgE can be diluted beyond the sensitivity of the assay. The second approach may not be feasible with microtiter wells due to the limited antigen-binding capacity of these solid phases. Furthermore, increasing the concentration of antigen may not resolve the problem unless the surface area of the solid phase is also increased because steric hindrance alone can inhibit the binding and/or detection of antibodies of a minor isotype or low affinity.

A microtiter well can absorb 200 to 1000 ng of antigen depending on its molecular weight and affinity for the receptors on plastic. At these levels, IgE antibodies can be inhibited from binding antigen by the more abundant IgG antibodies.[40] Serum IgG has been depleted by the

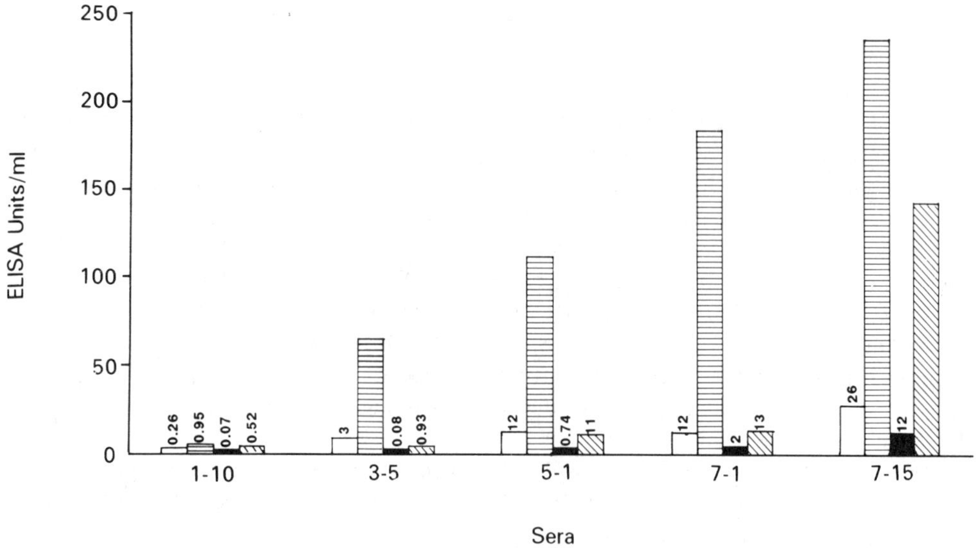

FIGURE 5. Improvement in the detection of IgA anti-BSA and anti-OA in rabbit sera after partial depletion of IgG by *S. aureus* absorption. □ = anti-OA before treatment, ⊟ = anti-OA after treatment, ■ = anti-BSA before treatment, and ▨ = anti-BSA after treatment. Anti-IgA is expressed in ELISA/ml compared to a pooled RS. Very low values are directly indicated at the top of each bar. Although data are presented here for IgA antibodies, differences in detectability of IgE of similar magnitude, before and after depletion of competing IgG, can also be seen. (From Butler, J. E., Peterman, J. H., and Koertge, T. E., *Enzyme-Mediated Immunoassays*, Ngo, T. T. and Lenhoff, H. M., Eds., Plenum Press, New York, 1985, 241. With permission.)

addition of 35% ammonium sulfate, leaving IgE in solution.[41-43] A similar phenomenon may occur in the measurement of serum IgA in species where levels are <0.5 mg/ml.[44] Figure 2B, plot B illustrates the effect of such inhibition on the titration plot. A flattened slope and prozone inhibition region are notable. Partial depletion of IgG with *S. aureus* yields a normal IgA titration plot (Figure 2B, plot B′). Figure 5 shows how the detectability of rabbit serum IgA anti-OA and anti-BSA is improved after partial depletion of competing rabbit serum IgG anti-OA and anti-BSA. Isotype bias can be best identified by performing a complete titration plot and testing for parallelism (see Section III.B.3).

C. THE MEASUREMENT OF SECRETORY IMMUNE RESPONSES

Secretory or mucosal immunity is often evaluated by the measurement of antibodies in exocrine body fluids, e.g., milk, saliva, intestinal fluids, and tears. Antibodies of the IgA class predominate in most secretions of mammals, the exceptions include milk of those species in which IgG antibodies is transferred exclusively via colostrum rather than across the placenta.[45] It is possible that competitive inhibition can be encountered in extreme cases such as in bovine colostrum where IgG1 may reach levels >70 mg/ml while IgA and IgM levels may be <5 mg/ml.[46]

The loss of antibody activity to proteases and other substances found in these secretions is a major problem in this field of research. Storage of human saliva at –20°C for 18 months results in complete loss of IgA antibody activity[47] while sera can be stored for years under the same conditions with only minimal loses in IgA antibody levels. Most remedies involve addition of antiproteases[48] but lyophilization has also been used.[49] In comparative studies with bile, intestinal fluid, and lung lavage fluids from rabbits, storage at –70°C in 50% glycerol resulted in the least loss of antibody activity. Observed losses after such treatment were negligible after 30 d but ranged up to 30% when storage was continued for 90 d.[50]

Another factor is the unequal recovery of antibodies from exocrine body fluids because they associate with other constituents of exocrine secretions. For instance, Laven et al.[51] and Honkanen-Buzalkski and Sandholm,[52] have shown that IgA in milk is preferentially associated with fat globules in human and bovine milk, respectively. When iodinated immunoglobulins were incubated with bovine milk, the IgA and IgM preferentially associated with the fat layer while IgG1 and IgG2 remained in the whey fraction.[53] Others have shown that IgA[54] and IgA, IgM, and secretory component[55] are preferentially associated with the mucous or the detergent-soluble portion of respiratory fluids. This raises the question of how exocrine body fluids should be collected and processed. Are the antibodies in saline lavage fluids representative of the antibodies present on the lavaged mucosa or do they represent nonmucous-associated antibodies? Similarly, when antibody levels in milk are compared using milk whey, are these representative of total milk antibodies?

A third issue is that of specimen dilution. Contrary to serum, exocrine body fluids can possess different compositions as a function of the glandular secretion rate, physiological disturbances, and methods of collection. For example, the total immunoglobulin concentration in the first 50 ml of rabbit lung lavage fluid is normally 10-fold higher than in fluid collected with subsequent lavages.[56] Hence, it can be meaningless to compare the titer of antibodies in exocrine fluids because they vary so widely due to collection processes and physiological conditions. The concentration of antibodies in secretions can be normalized to their protein content or antibody activity (Equation 1). Specific antibody activity has been expressed as

$$\frac{\text{ELISA Units of Activity/ml}}{\mu\text{g of the appropriate Ig/ml}} \times 100 \qquad (1)$$

This method allows the specific activity of immunoglobulins of a particular isotype in a secretion to be compared to the level of immunoglobulins of the same isotype in serum. For instance, the specific activity of IgA to the oral pathogen *Streptococcus mutans* is greater in parotid saliva than in whole saliva which is in turn greater than in serum.[32]

A fourth issue concerns contamination of such secretions by serum. Mastitis[57,58] and alveolitis[59] result in an increase in the level of albumin and IgG in the body fluid adjacent to the site of inflammation. Thus, antibody detected in exocrine body fluids are not necessarily a consequence of local antibody production, but also a result of inflammation-associated transudation from serum. A popular approach to show local production is to measure albumin and IgG levels in serum and then to calculate "local" IgG based on albumin:IgG ratios in serum (Equation 2).

$$[\text{IgG}]_{\text{local}} = \frac{[\text{IgG}]_{\text{secretion observed}} - [\text{Alb}]_{\text{secretion}}}{\text{Alb: IgG}_{\text{serum ratio}}} \qquad (2)$$

The factors discussed above should be considered in the design of experiments and interpretation of data in the field of mucosal and secretory immunity.

D. MEASUREMENT OF SPECIFIC ANTIBODIES IN TISSUES

Tissue-embedded antigens and free-floating cells are extensively used together with immunohistochemistry and flow-cytometry methods to detect antibodies specific for internal and surface antigens. Such measurements have been useful in the diagnosis of autoimmune diseases and the detection of cytotrophic antibodies. Quantitation of the actual amount of antibody in a serum or on a population of cells is generally less important in these assays than accurate and sensitive detection of the antibody. Moreover, the actual pattern (e.g., spindle) or location (e.g., nuclear vs. cytoplasmic) of the antibody binding can be associated with a defined disease. The fluorescence intensity can be used as an indicator of a weak or strong antibody response and it relates to the combined effects of antibody concentration, affinity, and specificity.

When antibody is not directly pathogenic but rather a marker associated with a disease process, it is of limited value to quantitate the absolute amount of the antibody in the patient's serum or tissue. If the antibody is directly involved in the disease (e.g., pemphigus), changes in the antibody concentration can be valuable in monitoring the effect of therapy. Dilutions of the specimen (typically serum) are assayed on tissue sections or fixed cells until an endpoint titer is determined. With certain exceptions, the principles governing the standardization of these cell-based SpAbIs are analogous to those used for microtiter plate-based SpAbIs. Differences relate to the variable nature and amount of antigen present in tissues or on cells from different donors and species. Insolubilized antigen cannot be assumed to be in excess unless it is shown with multiple dilutions of the specimen. Moreover, dose-response curves with standard antibody preparations are not generally performed. Antibody-containing sera are most commonly used only as positive quality control reagents.

E. SPECIFIC ANTIBODIES AND VETERINARY IMMUNOLOGY

The principles of SpAbI discussed in this chapter apply equally to the measurement of serum and secretory antibodies in animals. Accurate measurement of antibodies in cattle, swine, dogs, poultry, and fish is constrained by the general unavailability of reliable reagents with defined specificity for the immunoglobulin of interest. The best reagents are those prepared by individual investigators and thus available from private sources. Standardization has become the major problem with the use of private reagents. This problem has been compounded by the incomplete characterization of most animal immunoglobulins, especially the subclasses and minor immunoglobulin isotypes. This is partly due to limited funds that have been allocated for research in veterinary immunology. Moreover, there are few training programs in immunology and immunochemistry at schools of agriculture and veterinary medicine in the U.S. These problems have resulted in unreliable and nonstandardized antibody measurements throughout veterinary immunology. Perhaps when such animals become used as models for medical science, the situation will improve.

F. MEASUREMENT OF IgE ANTIBODIES TO PURIFIED AND COMPLEX ALLERGENS

Human IgE-mediated diseases result from the exposure of a sensitized individual to the sensitizing allergen. The allergic symptoms are induced when vasoactive mediators are released from the basophils and mast cells immediately following the cross linking of surface-bound IgE antibody. A diagnostic SpAbI called RAST was developed in 1968 to detect specific IgE antibody in serum to over 100 allergens in weeds, grasses, trees, epidermals, molds, insects, and foods.[60] In its simplest form, allergen is coupled to a solid phase (e.g., paper disk) and incubated with a dilution of serum. IgE antibodies that bind to the solid phase allergen are detected with labeled anti-human IgE.

The goal of the RAST and its nonisotopic counterparts (e.g., the EAST, MAST, FAST) is to detect the presence of allergen-specific IgE in serum of patients who are suspected, based on their symptoms and history, of having an allergy. RAST results are also used to clarify equivocal skin tests. Maximal sensitivity, reproducibility, and minimal nonspecific binding are considered paramount. Specificity is afforded by the quality of the allergen on the solid phase. Parallelism tests and quantitative interpolation are not commonly performed in RAST assays except when IgE is detected to well-characterized Hymenoptera venom (bee, vespid, and fire ant) and drugs (insulin, penicillin, protamine). The American Academy of Allergy's position paper on standardizing technologies used in the diagnosis of allergies recommends 0.1 ml of serum and an overnight serum incubation with the allergosorbent to maximize RAST sensitivity.[61]

Most clinically important allergens are aqueous extracts composed of mixtures of glycoproteins >5000 mol wt, each with multiple antigenic (and allergenic) determinants.

Drugs are among the few allergens which are available in purified form. Allergosorbents are prepared by either covalently coupling or adsorbing protein mixtures to a solid phase. Allergosorbents vary between lots by virtue of inherent variation in the efficiency of allergen attachment. Reproducibility and quality control of the allergosorbent is best performed by saturation analysis using well-characterized human sera that contain a spectrum of IgE antibodies with multiple allergen specificities.[62]

RAST results are reported in semiquantitative units (e.g., positive/negative, net counts per minute, test/buffer ratio) and units interpolated from a dose-response curve that has been assigned arbitrary values (e.g., Pharmacia relative units per milliliter, (PRU/ml) and fluorescent relative units per milliliter (FSU/ml). Moreover, arbitrary class schemes have been proposed based on PRU/ml or normalized counts per minute bound. The level of specific IgE antibody in some reference sera have been standardized by depletion (see Section IV.A.2) in weight/volume units, however, it is difficult to prepare homologous primary IgE antibody standards for all the allergen specificities. Ideally, these "reference" sera should contain antibodies specific for all the major allergen specificities and the antibodies should be in comparable levels with comparable affinities to those in the test sera. Alternatively, a serum pool is used to approximate all the major allergen specificities. Allergy laboratories have mostly relied upon heterologous interpolation from a dose-response curve of a different allergen specificity to quantitatively estimate the amount of specific IgE in the specimen.[63] Problems with isotype and specificity bias (Section V.C) can be identified in RAST assays when there is extensive nonparallelism.

Contrary to common laboratory practice, many diagnostic allergy laboratories have adopted a policy of analyzing test specimens in singlicate and at one dilution. Inherent assay variation due to random errors predestines the IgE antibody results produced in these laboratories to be inaccurate, especially when one is detecting levels of IgE that are near the threshold of sensitivity. Good allergy laboratory practice is to analyze test specimens at least in duplicate and to analyze a negative control serum on each solid-phase allergen specificity in each assay. Nonspecific binding levels of the different allergosorbents vary as a function of the amount and quality of allergen on the solid phase and the age of the anti-IgE label. RAST sensitivity depends on the level and random variation of the NSB of each allergosorbent (see Section III.B).

VIII. SUMMARY

Humoral immune responses to environmental and self antigens are composed of antibodies of many isotypes, specificities, and affinities. In its ideal form, the solid-phase SpAbI is configured to have a functional molar excess of solid-phase antigen and an isotype-restricted detection reagent that permits sensitive and specific quantitation of specific antibodies in the diagnosis and monitoring of immunologically based diseases. This chapter has focused on immunochemical considerations in the optimization and reagent selection in SpAbI. When the presence of an antibody provides sufficient information to diagnose a disease, sensitivity, reproducibility at zero antibody dose, and nonspecific binding become critical properties of the assay used to measure it. Alternatively, the immune system can be challenged with antigen or regulatory molecules. In these situations, accurate quantitation of the antibody level and changes in the antibody quality (specificity, affinity) can be valuable. Parallelism, antigen quality (valency, epitope density), antibody binding capacity, and extraction efficiency become important considerations in these immunoassay applications. The measurement of antibody is in a constant evolution toward more accurate quantitation.

REFERENCES

1. **Butler, J. E., Borca, M. V., Heyermann, H., Dillender, M., and Bielecka, M.,** The heterogeneity of bovine IgG2. III. The basis of the ion-exchange heterogeneity of bovine IgG2a is the result of V_H-region variation, *Mol. Immunol.*, 24, 1317, 1987.
2. **Jefferis, R., Reimer, C. B., Skvaril, F., de Lange, G., Ling, N. R., et al.,** Evaulation of monoclonal antibodies having specificity for human IgG subclasses: results of an IUIS/WHO collaborative study, *Immunol. Lett.*, 10, 233, 1985.
3. **Butler, J. E., Heyermann, H., Frenyo, V. L., and Kiernan, J.,** The heterogeneity of bovine IgG2. II. The identification of IgG2b, *Immunol. Lett.*, 16, 31, 1987.
4. **Knight, K. L., Suter, M., and Becker, R. S.,** Genetic engineering of bovine Ig. Construction and characterization of hapten-binding bovine/murine chimeric IgE, IgA, IgG1, IgG2, and IgG3 molecules, *J. Immunol.*, 140, 3654, 1988.
5. **Knight, K. L., Burnett, R. C., and McNicholas, J. M.,** Organization and polymorphism of rabbit immunoglobulin heavy chain genes, *J. Immunol.*, 134, 1245, 1985.
6. **Scheiderman, R. D., Hamly, W. C., and Knight, K. L.,** Expression of rabbit IgA heavy chain genes in *E. coli* and in murine myeloma cells, *Vet. Immunol. Immunopathol.*, 17, 279, 1987.
7. **Kabat, E. A.,** The upper limit for the size of the human antidextran combining site, *J. Immunol.*, 84, 82, 1960.
8. **Cebra, J. J.,** Studies on the combining site of the protein antigen silk fibroin. III. Inhibition of the sild fibroin-antifibroin system by peptides from the antigen, *J. Immunol.*, 86, 205, 1961.
9. **Voss, E. W., Jr., Eschenfeldt, W., and Root, R. T.,** Fluorescein: a complete antigenic group, Immunochemistry, 13, 447, 1976.
10. **Getzoff, E. D., Tainer, J. A., Lerner, R. A., and Geysen, H. M.,** The chemistry and mechanism of antibody binding to protein antigens, *Adv. Immunol.*, 43, 1, 1988.
11. **Givol, D.,** Structural analysis of the antibody combining site, in *Contemporary Topics in Medical Immunology*, Reisfeld, R. A. and Mandy, W. J., Eds., Plenum Press, New York, 1973, 2.
12. **Srinivasappa, J., Saegusa, J., Prabhakar, B. S., Gentry, M. K., Buchmeier, M. J., Wiktor, T. J., Koprowski, H., Oldstone, M. B. A., and Notkins, A. L.,** Molecular mimicry: frequency of reactivity of monoclonal antiviral antibodies with normal tissue, *J. Virol.*, 57, 397, 1986.
13. **Greenbury, C. L., Moore, D. H., and Nunn, L. A. C.,** Reaction of 7S and 19S components of immune rabbit antisera with human group A and AB red cells, *Immunology*, 6, 421, 1963.
14. **Adinolfi, M., Mollison, P. L., Polley, M. J., and Rose, J. M.,** γA blood group antibodies, *J. Exp. Med.*, 123, 951, 1966.
15. **Tax, A. and Korngold, L.,** Comparison of the effect of elastase on human secretory IgA and serum IgA, *J. Immunol.*, 107, 1189, 1971.
16. **Shim, B.-S., Kang, Y.-S., and Kim, W.-J.,** Self-protective activity of colostral IgA against tryptic digestion, *Nature*, 222, 787, 1969.
17. **Nygren, H., Czerkinsky, C., and Stenberg, M.,** Dissociation of antibodies bound to surface-immobilized antigen, *J. Immunol. Methods*, 85, 87, 1985.
18. **Peterman, J. H.,** Factors which Influence the Binding of Antibody to Solid-Phase Antigens: Theoretical and Experimental Investigations, Ph.D. thesis, Univeristy of Iowa, Iowa City, 1986.
19. **Ekins, R. P.,** The precision profile and its use in RIA assessment and design, *J. Clin. Immunoassay*, 4, 33, 1981.
20. **Hamilton, R. G. and Adkinson, N. F., Jr.,** Quantitation of antigen specflc IgG in human serum. II. Comparison of radioimmunoprecipitation and solid phase RIA techniques for measurement of IgG specific for a complex antigen mixture (yellow jacket venom), *J. Allerg. Clin. Immunol.*, 67, 14, 1981.
21. **Butler, J. E. and Heyermann, H.,** The heterogeneity of bovine IgG2. I. The A1 allotype is a major determinant recognized by rabbit antibodies to bovine IgG2, *Mol. Immunol.*, 23, 291, 1986.
22. **Koertge, T. E. and Butler, J. E.,** The relationship between the binding of primary antibody to solid-phase antigens in microtiter plates and its detection by the ELISA, *J. Immunol. Methods*, 83, 283, 1985.
23. **Rodbard, D., Lenox, R. H., Wray, H. L., and Ramseth, D.,** Statistical characterization of the random errors in the radioimmunoassay dose-response variable, *Clin. Chem.*, 22, 350, 1976.
24. **Rowe, D. S., Anderson, S. G., and Grab, B. A.,** A research standard for human serum immunoglobulin IgG, IgA and IgM, *Bull. WHO*, 42, 535, 1970.
25. **Scatchard, G.,** The attraction of protein for small molecules and ions, *Ann. N.Y. Acad. Sci.*, 51, 660, 1949.
26. **Gleich, G. J., Jacob, G. L., Yunginger, J. W., and Henderson, L. L.,** Measurement of the absolute levels of IgE antibodies in patients with ragweed hay fever, *J. Allerg. Clin. Immunol.*, 60, 188, 1977.
27. **Hamilton, R. G. and Adkinson, N. F., Jr.,** Quantitation of antigen-specific IgG in human serum. I. Standardization by a Staph A-solid phase radioimmunoassay elution technique, *J. Immunol.*, 124, 2966, 1980.
28. **Butler, J. E., Cantarero, L. A., and McGivern, P. L.,** Measurement of bovine subclass antibodies (IgG1and IgG2) using the amplified enzyme-linked immunosorbent assay, *Mol. Immunol.*, 17, 645, 1980.
29. **Cantarero, L. A., Butler, J. E., and Osborne, J. W.,** The binding characteristics of various proteins to polystyrene and their significance in solid-phase immunoassay, *Anal. Biochem.*, 105, 375, 1980.

30. **Dierks, S. E., Butler, J. E., and Richerson, H. B.,** Altered recognition of surface-adsorbed compared to antigen-bound antibodies in the ELISA, *Mol. Immun.,* 23, 403, 1986.
31. **Hamilton, R. G.,** Engineered antibodies as immunologic reagents, *Ann. Bio. Clin.,* 48, 473, 1990.
32. **Butler, J. E., Spradling, J. E., Peterman, J. H., Joshi, K. S., and Challacombe, S. J.,** Humoral immunity in root caries in an elderly population. I. Development of reliable solid-phase immunoassays for routine measurement of antibody to oral bacteria and quantitation of secretory immunoglobulin, *Oral Microbial Immunol.,* 5, 98, 1990.
33. **Robinson, J. E., Holton, D., Liu, J., McMurdo, H., Murciano, A., and Gohd, R.,** A novel enzyme-linked immunosorbent assay (ELISA) for the detection of antibodies to HIV-1 envelope glycoproteins based on immobilization of viral glycoproteins in microliter wells coated with concanavalin A, *J. Immunol. Methods,* 132, 63, 1990.
34. **Hamilton, R. G.,** Human IgG subclass measurements in the clinical laboratory, *Clin. Chem.,* 33, 1707, 1987.
35. **Makela, O., Mattila, P., Rautonen, N., Seppala, I., Eskola, J., and Kayhty, H.,** Isotype concentrations of human antibodies to Hemophils influenzae type B polycsccharide (Hib) in young adults immunized with the polysaccharide as such or conjugated to a protein (Diphtheria toxoid), *J. Immunol.,* 139, 1999, 1987.
36. **Jones, C. C., Hamilton, R. G., and Jordon, R. E.,** Subclass distribution of human IgG autoantibodies in pemphigus, *J. Clin. Lab. Immunol.,* 8, 43, 1988.
37. **Djurup, R., Sondergaard, I., Magnusson, C. G. M., and Weeke, B.,** A three layered immunoradiometric assay for determination of IgG subclass antibodies in human sera. *Allergy,* 39, 51, 1984.
38. **Hamilton, R. G. and Adkinson, N. F., Jr.,** Serological methods in the diagnosis and management of human allergic disease, *Crit. Rev. Clin. Lab. Sci.,* 21, 1, 1984.
39. **Hamilton, R. G.,** Engineered human antibodies as immunologic quality control reagents, *Ann. Biol. Clin.,* in press.
40. **Butler, J. E., Peterman, J. H., and Koertge, T. E.,** The amplified enzyme-linked immunosorbent assay (a-ELISA), in *Enzyme-Mediated Immunoassays,* Ngo, T. T. and Lenhoff, H. M., Eds., Plenum Press, New York, 1985, 241.
41. **Haba, S. and Nisonoff, A.,** Quantitation of IgE antibodies by radioimmunoassay in the presence of high concentrations of nonIgE antibodies of the same specificity, *J. Immunol. Methods,* 85, 39, 1985.
42. **Dierks, S. E.,** Construction and Characterization of Hapten-Binding Chimeric IgE Molecules, Ph.D. Thesis, University of Iowa, Iowa City, 1988.
43. **Gershwin, T. J. and Friebertshauser, K.,** Demonstration of pasteurella specific immunoglobulin E in bovine serum, *Am. J. Vet. Res.,* 48, 169, 1987.
44. **Butler, J. E., Heo, Y., Adams, P., and Richerson, H. B.,** The antigen-limited nature of microtiter ELISA requires partial depletion of IgG to permit reliable determination of rabbit serum IgA antibody activity, *Mol. Immunol.,* 27, 319, 1990.
45. **Butler, J. E.,** Transmission of immunity from mother to young, in, *Fertility and Sterility,* Hasegawa, T., Hayashi, M., Ebling, F. J. G., Henderson, I. W., Eds., *Excerpta Medica,* Amsterdam, 1971, 92.
46. **Butler, J. E.,** Biochemistry and biology of ruminent immunoglobulins, *Prog. Vet. Microbiol. Immunol.,* Pandey, R., Ed., S. Karger, Basel, 1985, 1.
47. **Butler, J. E., Spradling, J. E., Rowat, J., Ekstrand, J., and Challacombe, S. J.,** Humoral immunity in root caries in an elderly population. II. Collection, processing, preservation and antibody activity and influence of diurnal and monthly variation on collection procedures, *Oral Microbiol. Immunol.,* 5, 113, 1990.
48. **Elson, C. O., Ealding, W., and Lefkowitz, J.,** A lavage technique allowing repeated measurement of IgA antibody in mouse intestinal secretions, *J. Immunol. Methods,* 67, 101, 1984.
49. **Svennerholm, A.-M.,** University of Göteborg, personal communication.
50. **Butler, J. E., Heo, Y., and Richerson, H. B.,** The stability of rabbit IgA, IgG and specific antibodies of these isotypes in bile, intestinal and bronchial lavage fluids, pending.
51. **Laven, G. T., Crago, S. S., Kutch, W. H., and Mestecky, J.,** Hemolytic plaque formation by cellular and non-cellular elements in human colostrum, *J. Immunol.,* 127, 1967, 1981.
52. **Honkanen-Buzalkski, T. and Sandholm, M.,** Association of bovine secretory immunoglobulins with milk fat globule membranes, *Comp. Immunol. Infect. Dis.,* 4, 329, 1981.
53. **Frenyo, L. V., Butler, J. E., and Guidry, A. J.,** The association of extrinsic bovine IgG1, IgG2, SIgA and IgM with the major fractions and cells of milk, *Vet. Immunol. Immunopathol.,* 13, 239, 1986.
54. **Clamp, J. R.,** The relationship between secretory immunoglobulin A and mucus, *Biochem. Soc. Transactions,* 5, 1579, 1977.
55. **Butler, J. E.,** A concept of humoral immunity among ruminants and an approach to its investigation, in *The Ruminant Immune System,* Butler, J. E., Ed., Plenum Press, New York, 1981, 137, 3.
56. **Richerson, H. B., Heo, Y., and Butler, J. E.,** unpublished data.
57. **Butler, J. E., Kiddy, C. A., Pierce, C. S., and Rock, C. A.,** Quantitative changes associated with calving in the levels of bovine immunoglobulins in selected body fluids. I. Changes in the levels of IgA, IgG1 and total protein, *Can. J. Comp. Med.,* 36, 234, 1972.

58. **Ost, M., Guidry, A. J., and Shainline, W. E.,** Sequential response of milk leukocytes, serum albumin, immunoglobulins and electrical conductivity following infusion of *E. coli* endotoxin into the mammary gland, *J. Dairy Sci.,* 61(Suppl. 1), 159, 1978.
59. **Butler, J. E., Richerson, H. B., Kopp, W. C., Suelzer, M. J., and Barfknecht, C.,** The influence of muramyl dipeptide on secretory imrnunity, in *The Secretory Immune System,* Kunkel, H., Mestecky, J., and McGhee, J., Eds., *Proc. N.Y. Acad Sci.,* 1983, 409, 669.
60. **Hamilton, R. G. and Adkinson, N. F., Jr.,** Serum IgE antibodies by the radioallergosorbent test (RAST): diagnosis of IgE-mediated diseases, *Immunopathology,* 6, 1, 1982.
61. **Bernstein, L.,** Proceedings of the task force on guidelines for standardizing old and new techniques used for the diagnosis and treatment of allergic diseases, *J. Allergy Clin. Immunol.,* 82, 487, 1988.
62. **Hamilton, R. G. and Adkinson, N. F., Jr.,** Quantitation of venom specific human IgE and IgG antibodies, *Folia Allergae. Immunol. Clin.,* 33, 31, 1986.
63. **Hamilton, R. G. and Adkinson, N. F., Jr.,** Quantitative aspects of solid phase immunoassays, in *Theoretical and Technical Aspects of ELISA and other Solid Phase Assays,* Kemeny, M. and Challacombe, S. J., Eds., John Wiley & Sons, New York, 1987.
64. **Heyermann, H., Butler, J. E., and Frangione, B.,** The heterogeneity of bovine IgG2. V. Differences in the primary structure of bovine IgG2 allotypes, pending.

Critical Issues and Special Topics

Chapter 10

THE SPECIFICITY OF MONOCLONAL ANTIBODIES FOR ENZYMES IN SOLUTION VS. IMMOBILIZED ON SOLID PHASES

Lisa Djavadi-Ohaniance and Bertrand Friguet

TABLE OF CONTENTS

I. INTRODUCTION

With the recent development and improvement of quantitative and highly sensitive techniques in immunochemistry, monoclonal antibodies (mAbs) are becoming powerful tools in analyzing the conformational changes that may occur upon the acquisition of functional properties of enzymes. mAbs are also becoming precise tools to dissect the structural events which up to now have only been described in terms of kinetics during the folding of the proteins as well as to monitor quantitatively the appearance of immunoreactive folding intermediates. Such analysis require that the mAbs chosen as conformational probes recognize the native form of the enzyme.

It is also important that the native conformation of the antigen be preserved when antipeptide antibodies are screened in order to select those that cross react with the whole native protein as is the case in experiments designed to develop a synthetic peptide vaccine. Antibodies that recognize a denatured form of the viral protein, but not the native conformation, are unlikely to possess neutralizing or protective potential. Conversely, mAbs specifc for the "nonnative" form of a protein can yield precious information about the spontaneous unfolding of the protein in solution.

Therefore, a prerequisite for studying antibody-antigen interaction, either qualitatively or quantitatively, is to discriminate between antibodies recognizing the native and the nonnative forms of the antigen. We will underline this important point by referring to some results of our work carried out with mAbs directed against an enzyme of *Escherichia coli*: the β_2 subunit of tryptophan synthase.

II. RECOGNITION BY MONOCLONAL ANTIBODIES OF SURFACE-ADSORBED OR SOLUTION-PHASE ENZYME

mAbs directed against the β_2 subunit of *E. coli* tryptophan synthase (EC 4.2.1.20) were produced from hybridoma clones as previously described.[1,2] The selection of hybridoma clones secreting specific immunoglobulins was done by indirect ELISA test in which the protein is first adsorbed onto a microtitration plate.

The effects of the selected mAbs on the enzymatic activity of β_2 were studied.[2] After the incubation (for 15 min) of the mAb with β_2, some mAbs inhibited the enzymatic activity of β_2 while no immediate enzyme inactivation was observed with others. We then investigated whether extending the incubation time to 24 h for the latter group of mAbs could effect the enzymatic activity of β_2. The results showed that after a 24-h incubation, mAbs in this group inhibited the enzymatic activity of β_2 to different degrees. These observations raised the question of whether mAbs in this category actually could bind the enzyme after a 15-min incubation in solution. To test the binding of the mAbs with the enzyme in solution, the β_2/mAb mixture was incubated for 15 min, analyzed by exclusion chromatography and the enzymatic activity of β_2 tested in the eluted fractions. In all cases, the enzymatic activity in the elution profile for β_2/mAb mixtures was superimposable with that obtained for β_2 alone. When the same experiment was repeated after a 24-h incubation of the β_2/mAb mixture, the remaining enzymatic activity was confined to the remaining free β_2. Thus, the inhibitory effect of mAbs that was only observed after a long incubation time seemed to be a consequence of the slow rate of formation of the β_2/mAb complex in solution.

When the same "slow-binding" mAbs were tested in an ELISA against enzyme adsorbed on plastic, they reacted within 15 min. These results prompted us to propose that the slow-binding mAbs recognized antigenic determinants normally hidden in the native enzyme but which become exposed when the enzyme is adsorbed on the solid phase. The eventual binding of these mAbs to the enzyme even in solution could be explained by the transient exposure of epitopes due to the slow spontaneous unfolding-refolding equilibrium of the protein in solution.

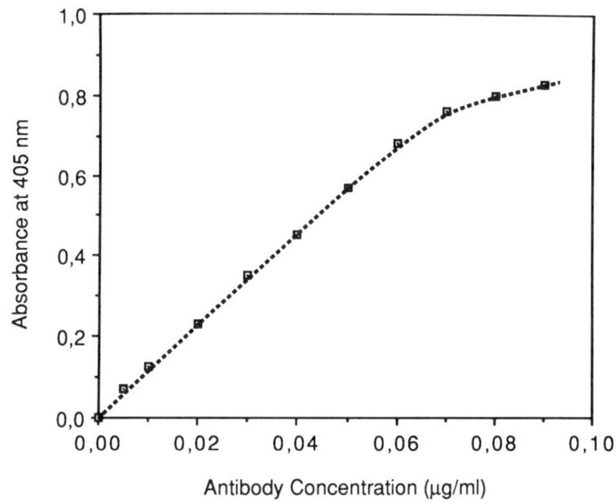

FIGURE 1. Calibration curve obtained with a monoclonal antibody directed against the β_2 subunit of *E. coli* tryptophan synthase. The coating was carried out with 0.1 ml of the antigen at 1 µg/ml. The absorbance obtained after the enzymatic reaction in the ELISA test is plotted as a function of the antibody concentration. The antibody concentration that has been chosen for the competition test is 0.05 µg/ml, i.e., at the top of the linear portion of the calibration curve.

III. RECOGNITION BY MONOCLONAL ANTIBODIES OF THE NATIVE AND A NON-NATIVE FORM OF THE ENZYME IN COMPETITIVE ELISA TESTS

A competitive ELISA assay was developed for both the native and a nonnative form of the enzyme to test whether some mAbs recognized antigenic determinants normally hidden in the native protein.[3] If our hypothesis was correct, it could be predicted that antibodies which react slowly in solution with native β_2 should react more rapidly with a nonnative form of the enzyme. The nonnative form was obtained after denaturation of the β-chain and reaction of its five thiols with *N*-ethymaleimide as described.[4] The resulting protein, β-NEM, is monomeric, soluble only at low protein concentration, and devoid of enzymatic activity. As evidenced by a red shift of its fluorescence emission spectrum, the tryptophan residue of the β-chain is considerably more accessible to the solvent in β-NEM than in native β_2. Consequently, β-NEM can be considered as a rather good nonnative polypeptide chain.

The competitive ELISA assay developed to test our hypothesis was based on the following rationale. (1) The mAbs and an excess of the enzyme (either native or nonnative) were first incubated for a short time (15 min) to allow only rapid binding to take place. The amount of antibody remaining free in the solution after a 15-min incubation was then determined using an indirect ELISA in which the enzyme was adsorbed onto a microtitration plate. Compared to the signal obtained with the antibody incubated in solution without antigen, the rapid binding of the antibody to the enzyme in solution (if it occurs) should result in a decrease in the amount of antibody which can be detected using the adsorbed enzyme. (2) The antibody concentration must first be determined in a previous ELISA calibration test. Indeed, in order to detect the binding of the antibody to the antigen in solution, i.e., to detect a decrease of absorbance measured in the ELISA test, one must be sure that the antibody is not used at a saturating concentration and that the absorbance measured is proportional to the concentration of the antibody. Figure 1 shows such a calibration plot.

Competitive ELISAs, using the optimal concentration for each mAb, were conducted with native β_2 and denatured NEM-treated β chains (β-NEM) as competiting antigens. The results

TABLE 1
Competition of Monoclonal Antibodies with the Native Enzyme β_2 and the Denatured form β-NEM

Antibodies[b]	Absorbance[a] No competitor	Absorbance[a] β_2	Absorbance[a] β-NEM	Recognition (in 15 min) of the enzyme coated β_2	Recognition (in 15 min) of the enzyme in solution β_2	Recognition (in 15 min) of the enzyme in solution β-NEM
19-1	0.90	0.10	0.40	+	+	±
31-2	1.30	0.10	0.70	+	+	±
68-1	1.47	0.06	0.65	+	+	±
93-6	0.94	0.10	0.55	+	+	±
30-2	1.00	1.10	1.00	+	–	–
69-4	0.63	0.65	0.23	+	–	±
169-3	1.28	1.29	0.72	+	–	±
172-3	0.97	1.07	0.50	+	–	±
B3-B5	0.34	0.48	0.05	+	–	±
D4-B6	0.80	0.64	0.03	+	–	+

[a] Values measured after 15 min of the β-galactosidase enzymatic reaction in the ELISA test.
[b] Each antibody has a distinct epitope specificity.

obtained (Table 1) confirm that mAbs which inactivate β_2 after short incubation time bind rapidly to the native β_2 in solution. Conversely, those mAbs which inactivate β_2 only after long incubation time fail to bind native β_2 after 15 min while they recognize (except 1) the denatured form of the antigen (β-NEM) even during the 15-min incubation.

IV. CONCLUSIONS

The results described provide convincing evidence that some of the mAbs selected as specific for the β_2 subunit and which recognize quite rapidly the enzyme adsorbed on a microtitration plate bind very slowly to the native enzyme in solution. Indeed these antibodies recognize the nonnative enzyme (β-NEM) better than the native β_2 and hence must be directed against antigenic determinants which become available only upon partial or total unfolding of the protein. If such antibodies had been considered as specific to the native enzyme, since native β_2 adsorbed on plastic was used in the hybridoma selection process, it could be erroneously concluded that some of these mAbs merely do not inhibit enzymatic activity.

Those mAbs which bind rapidly to the native protein, but which also recognize β-NEM, are dependent for their binding on the conformation state of the denatured enzyme, β-NEM. Indeed, various observation from our laboratory have lead to the conclusion that rather than being a fully denatured "random coil" polypeptide chain, β-NEM is a nonnative protein which may share some conformational features with the native β_2 subunit.

The results discussed in this chapter provide insight to the behavior of a protein during the immunization process. One might consider the possibility that denaturation occurs during emulsification with Freund adjuvant and/or that proteolysis occurs after injection in the mouse. Moreover, the antigen may be processed during the presentation to B and T cells, such that determinants not seen by the immune system on the native protein are now revealed. However, our finding neither suggests that all determinants on the native protein are altered nor that antigen processing by the immune systems modifies the conformation of the protein in the same manner as does adsorption of the protein on the plastic.

The hybridomas raised to β_2 produce a variety of antibodies recognizing different conformations of the antigen. Hence, the screening assay is a most crucial step which will often determine the specificity of the mAb used for subsequent studies. Antibodies recognizing the unfolded or degraded antigen could be selected if the screening procedure does not discriminate them from antibodies authentically specific to the native antigen. This would occur if the antigen undergoes partial or total denaturation during the antigen immobilization process which is part of the screening assay. Many results obtained with mAbs and reported to be unexpected or surprising when the tests were conducted by ELISA could be explained by our previously reported results,[3] those discussed above, as well as results reported by others[5-17] which have shown that proteins can undergo some denaturation during adsorption on microtitration plates. These observations also suggest the use of caution in the interpretation of results obtained with antibodies elicited by peptides from a given protein and which recognize the cognate native protein. One must ask how native is the protein which was used for characterizing specificity? In order to ascertain that antibodies recognize the native form of a protein, it is necessary to test their specificity in a liquid-phase assay that preserves the original conformation of the molecule.

The adsorption-induced denaturation of protein on plastic can be avoided via use of different agents, e.g., polyclonal capture antibodies. However, it must be established that the immobilizing reagent does not modify antigenicity. Conversely, if one wishes to select mAb to denatured protein a soluble denatured form of the protein can be used in solution or coated onto the plate. When a soluble denatured form of the protein cannot be prepared, one can try the procedure we used during the screening of monoclonal antibodies directed against the phage P_{22} tail spike endorhamnosidase.[22] Since proteins are not necessarily fully denatured on the plastic, we performed an ELISA in which the tail spike protein was first adsorbed on the microtitration plates, the plates sealed with adhesive tape, and then incubated for 5 min in a boiling waterbath. Several antibodies were selected which recognized the adsorbed, heat-denatured protein but failed to react with either the native antigen in solution or after it had been adsorbed on the plate. Since it is often easier to find a waterbath in a laboratory than a sophisticated, nonnative form of a given protein, such a simple procedure can provide valuable information about the specificity of even a large panel of antibodies prepared by monoclonal technology.

Once antibodies of known specificity have been obtained, caution must nevertheless be taken in interpreting quantitative assay based on solid-phase techniques such as those used to study the kinetics of antigen-antibody reactions[18,19] or to measure the equilibrium constant of antibody-antigen interaction.[20,21]

REFERENCES

1. **Friguet, B., Djavadi-Ohaniance, L., and Goldberg, M. E.,** A convenient enzyme linked immunosorbent assay for testing whether monoclonal antibodies recognize the same antigenic site, in *Immunoenzymatic Techniques, Proceedings of the Second International Symposium on Immunoenzymatic Techniques,* Avrameas, S. et al., Eds., Elsevier, Amsterdam, 1983, 171.
2. **Djavadi-Ohaniance, L., Friguet, B., and Goldberg, M. E.,** Structural and functional influence of enzyme-antibody interactions: effects of eight different monoclonal antibodies on the enzymatic activity of *E. coli* tryptophan synthase, *Biochemistry,* 23, 97, 1984.
3. **Friguet, B., Djavadi-Ohaniance, L., and Goldberg, M. E.,** Some monoclonal antibodies raised with a native protein bind preferentially to the denatured antigen, *Mol. Immunol.,* 21, 673, 1984.
4. **Chaffotte, A. F. and Goldberg, M. E.,** Immunochemical evidence for conformational flexibility and its modulation by specific ligands in the β_2 subunit of *Escherichia coli* tryptophan synthase, *Biochemistry,* 22, 2708, 1983.
5. **Mierendorf, R. C. and Dimond, R. L.,** Functional heterogeneity of monoclonal antibodies obtained using different screening assays, *Anal. Biochem.,* 135, 221, 1983.

6. **Finney, K. G., Messer, J. L., Dewitt, D. L., and Wilson, J. E.,** Monoclonal antibodies against rat brain hexokinase—effect on catalytic function and binding to the outer mitochondrial membrane, *J. Biol. Chem.,* 259, 8232, 1984.
7. **Vaidya, H. C., Dietzler, D. N., and Ladenson, J. H.,** Inadequacy of traditional ELISA for screening hybridoma supernatants for murine monoclonal antibodies, *Hybridoma,* 4, 271, 1985.
8. **Hollander, Z. and Katchalski-Katzir, E.,** Use of monoclonal antibodies to detect conformational alterations in lactate dehydrogenase isoenzyme 5 on heat denaturation and on adsorption to polystyrene plates, *Mol. Immunol.,* 23, 927, 1986.
9. **Smith, A. D. and Wilson, J. E.,** A modified ELISA that selectively detects monoclonal antibodies recognizing native antigen, *J. Immunol. Methods,* 94, 31, 1986.
10. **Solomon, B., Koppel, R., Benett, D., and Fleminger, G.,** Localization of a highly immunogenic region of carboxypeptidase A recognized by three different monoclonal antibodies and their use in detection of subtle conformational alterations in this enzyme region, *Biochemistry,* 28, 1235, 1988.
11. **Soderquist, M. E. and Walton, A. G.,** Structural changes in proteins adsorbed on polymer surfaces, *J. Colloid Interface Sci.,* 75, 386, 1980.
12. **Al Moudallal, Z., Altschuh, D., Briand, J. P., and Van Regenmortel, M. H. V.,** Comparative sensitivity of different ELISA procedures for detecting monoclonal antibodies, *J. Immunol. Methods,* 68, 35, 1984.
13. **Altschuh, D., Al Moudallal, Z., Briand, J. P., and Van Regenmortel, M. H. V.,** Immunochemical studies of tobacco mosaic virus. VI. Attempts to localize viral epitopes with monoclonal antibodies, *Mol. Immunol.,* 22, 329, 1985.
14. **McCullough, K. C., Crowther, J. R., and Butcher, R. N.,** Alteration in antibody reactivity with foot-and-mouth disease virus (FMDV) 146S antigen before and after binding to solid phase or complexing with specific antibody, *J. Immunol. Methods,* 82, 91, 1985.
15. **Dierks, S. E., Butler, J. E., and Richerson, H. B.,** Altered recognition of surface adsorbed compared to antigen bound antibodies in the ELISA, *Mol. Immunol.,* 23, 403, 1986.
16. **Suter, M. and Butler, J. E.,** The immunochemistry of sandwich ELISAs. II. A novel system prevents the denaturation of capture antibodies, *Immunol. Lett.,* 13, 313, 1986.
17. **Jemmerson, R.,** Antigenicity and native structure of globular proteins: low frequency of peptide reactive antibodies, *Proc. Natl. Acad. Sci. U.S.A.,* 84, 9180, 1987.
18. **Stenberg, M., Werthen, M., Theander, S., and Nygren, H.,** A diffusion limited reaction theory for a microtiter plate assay, *J. Immunol. Methods,* 112, 23, 1988.
19. **Stenberg, M. and Nygren, H.,** Kinetics of antigen-antibody reactions at solid-liquid interfaces, *J. Immunol. Methods,* 113, 3, 1988.
20. **Friguet, B., Chaffotte, A. F., Djavadi-Ohaniance, L., and Goldberg, M. E.,** Measurements of the true affinity constant in solution of antigen-antibody complexes by enzyme-linked immunosorbent assay, *J. Immunol. Methods,* 77, 305, 1985.
21. **Stevens, F. J., Iwo, J., Carperos, W., Köhler, H., and Schiffer, M.,** Relationship between liquid and solid-phase antibody association characteristics: implications for the use of competitive ELISA techniques to map the spacial location of idiotopes, *J. Immunol.,* 137, 1937, 1986.
22. **Friguet, B., Djavadi-Ohaniance, L., Haase-Pettingell, C. A., King, J., and Goldberg, M. E.,** Properties of monoclonal antibodies selected for probing the conformation of wild type and mutant forms of the P22 tailspike endorhamnosidase, *J. Biol. Chem.,* 265, 10347, 1990.

Chapter 11

STUDIES ON THE COVALENT ATTACHMENT OF ANTIBODIES TO CONTROLLED SURFACES

Patrick D. Mize, Ali Naqui, Catherine M. O'Connell, Jess N. Waller, Martha Fesler, Ronald G. Myatich, and William E. Keating

TABLE OF CONTENTS

INTRODUCTION

Increasing the sensitivity of a solid-phase immunoassay (e.g., ELISA) has been a goal of many research programs. Modifying the orientation of capture antibodies (CAbs) on surfaces to enhance their antigen capture capacity is a possible route to accomplish this goal. Several studies have described types of polymeric backbones, linking chains, and covalent functionality that perform best for immobilizing antibodies on surfaces.[1-7,9-11,14] (See Reference 1 for a general overview of covalent protein immobilization.) However, a variety of linking chains were used with different polymeric backbones and antibodies, making determination of direct effects difficult. Moreover, claims have been made to the effect that certain linkages on certain polymeric backbones yield antibodies with increased activity. These studies have failed to provide controls for the amount of surface area to which the antibody is bound.[5,8,14] Comparisons among existing studies are therefore difficult. There is no one study in which the same antibody and the same polymeric backbone were used to evaluate the role of different linking arms on antibody activity. Since the start of our study, one article has appeared that examines the effect of orientation and linking chain.[15]

The objective of our studies was to systematically investigate surfaces for use in solid-phase immunoassays. The goal was to develop surfaces that ultimately have two characteristics: (1) the ability to covalently bind antibodies in a more active orientation compared to the same antibody adsorptively bound to a surface and (2) the ability to prevent nonspecific binding of antibody-enzyme conjugates. Both of these characteristics are desirable in developing immunoassay systems. Increased antibody activity should provide an increase in the amount of antigen bound and thus increase the assay sensitivity. The reduction of nonspecific binding of antibody-enzyme conjugates allows higher concentrations of this reagent to be used in the assay to drive the immunological reaction, again improving sensitivity. The results of experiments designed to address these considerations, with emphasis on oriented covalent antibody capture, are presented below.

Our approach was to study a series of related polymers and determine which members of the series had low adsorptive binding of proteins, particularly antibodies. The surfaces with the lowest adsorptive characteristics would then be modified to covalently bind capture antibodies in preferred orientations. Any remaining reactive groups on the surface would be "capped" and the surface-bound antibody tested for both specific and nonspecific antigen binding characteristics ("signal" and "noise").

II. EXPERIMENTAL METHODS

PHPMA, PMMA, PMMA-co-MAA, PPMA, PPNPMA, and PODMA were purchased from Scientific Polymer Products. AEC, Dehibit 100, and HPMA were purchased from Polysciences. BHT was obtained from Aldrich Chemical Company. AEMA and APMA (HCl salts) were available from Kodak Chemical Company. Heterobifunctional cross-linking agents were obtained from Pierce Chemical Company. MilliQ water (18Mohm-cm) was used for preparation of buffers and for contact angle determinations. These and other reagents and solvents were used as received.

Activation of HPMA with *N*-hydroxysuccinimide chloroformate as well as the preparation of the chloroformate were accomplished using procedures similar to those of Miron and Wilchek.[3,17] BHT was employed to prevent spontaneous polymerization of NHS carbonate-modified HPMA. Activation of CAMA to yield the *N*-hydroxysuccinimide ester was accomplished by reacting the free acid with NHS and DCC.

Aldehyde-reactive polymers were prepared via dropwise addition of one equivalent of NHS carbonate polymer to a stirred, chilled (ice/MeOH bath) solution of hydrazine hydrate under inert (N_2) conditions. Purified product was obtained by recrystallization from 2-

methoxyethanol/ether. Complete conversion to hydrazide was ascertained by the disappearance of the succinimidyl carbonate singlet at 2.81 ppm.

Sulfhydryl-reactive polymer surfaces were prepared by overnight ambient incubation of polymer-coated glass rods in buffered heterobifunctional cross-linker solution. Specifically, glass rods coated with hydrazide-derivatized polymer or amine-containing polymer (PAEMA and PAPMA) were immersed in a 10% v/v acetonitrile/Na phosphate buffer (50 m*M*, pH 7.0) solution containing 4 m*M* NHS ester-maleimide cross linker. Maleimide-functionalized rods were then soaked in pH 8.0 Tris-HCl/EDTA/NaOAc buffer for 15 min, followed by a final rinse in MilliQ water.

An AECL ^{60}Co gamma cell was employed to initiate free radical homopolymerizations of AEMA, CAMA, and APMA, and copolymerizations of HPMA with HPMA-NHS carbonate derivative. Gamma-induced polymerizations were conducted under inert conditions (argon gas) at a dose rate of 0.1 Mr/h and a total dose of 0.25 to 1.0 Mr. Polymers were generally isolated via precipitation in suitable solvent (Et20, i-PrOH, etc.), followed by suction filtration and vacuum drying. Absence of monomer and molecular structure were verified by ^{1}H and ^{13}C NMR. Results obtained via elemental analysis for intermediates, modified monomers, and polymers were in agreement with theoretical values. All monomers were deinhibited prior to polymerization by passing 10 wt% monomer/solvent solutions through a column packed with Dehibit 100.

Uniform thin films of amine and aldehyde reactive (co)polymer were prepared by mechanical dipping of CH_4 plasma-treated glass rods in 3 wt% polymer solutions at 50 mm/min. Film thicknesses were approximately 3000 Å as determined by SEM. Glass rods were plasma treated using a capacitively coupled external electrode plasma system (IPC 2103B) at 13.56 MHz RF and 1000 mT CH_4 to improve adhesion of methacrylate/methacrylamide films to the glass substrate.

Hydrophobicity/hydrophilicity measurements were conducted using a Wettek Model 11100-03 Dynamic Contact Angle Balance. Samples were analyzed using a stage speed of 19.2 mm/min. Advancing and receding angles were determined for three cycles to ascertain film stability/interfacial rearrangement.

A. RADIOLABELING OF MAB

MAb (hCG 166; isotype = IgG1) was radiolabeled by using the "enzymobead" method (Biorad). In short, the protein to be labeled was incubated with $Na^{125}I$ (Amersham), "enzymobead" (glucose oxidase and lactoperoxidase immobilized on beads) and glucose, which was added last to initiate reaction. After 30 min of incubation at room temperature, the protein was separated by gel filtration using a PD 10 column (Pharmacia). Fractions were collected every minute and counted for radioactivity; 5 μl of each fraction was diluted into 0.5 ml of buffer and counted in a gamma counter (LKB 1274). Fractions were then subjected to TCA precipitation (final concentration of TCA = 8%, BSA = 17%) on ice. The supernatant and precipitate were separated by centrifugation and radioactive content of each determined. Only protein fractions that showed >90% TCA precipitability were used in further studies. Radiolabeled proteins were stored at 4°C and used within 2 weeks of labeling.

B. ANTIBODY BINDING STUDIES

Glass rods coated with specific polymers were dipped into a tube (Nunc "minisorb") containing capture antibody (hCG 166 mAb). Two sets of binding studies were performed in parallel: (l) binding of radiolabeled antibody to the surface to determine the total amount of Ab bound and (2) binding of unlabeled mAb which was used for the determination of radiolabeled antigen binding.

After overnight incubation at 4°C, the antibody-coated rods were washed (5 times) in PBS and the surface blocked (or capped, in the case of reactive polymers) in 2% BSA at 37°C for

1 h. The mAb/polymer-coated rods were again extensively washed (5 times) and incubated in radiolabeled antigen solution (50 ng/ml of antigen, 1 ml volume) (radiolabeled hCG from Chemicon) for 3 h at 37°C. The rods were washed and the amount of antigen bound was determined by measuring the radioactivity. On average, polystyrene bound about 10 ng of antigen under these experimental conditions. Control experiments were performed simultaneously with blocked rods without trapping antibody (to determine the nonspecific binding of antigen). All experiments were performed in duplicate and all data were normalized to the values obtained with polystyrene rods, i.e.,

$$\text{normalized value} = \frac{\text{value for a specific polymer*}}{\text{value for polystyrene*}}$$

MAb adsorption onto the polystyrene surface was usually 0.6 μg/cm^2.

C. OXIDATION AND REDUCTION OF ABS

The oxidation of hCG 166 mAb was performed according to the procedure described by O'Shannessy and Hoffman.[4,5] The oxidation of the antibody was verified by using a fluorescent-labeled hydrazide (Lucifer yellow hydrazide from Molecular Probes).

Reduction of the antibody was performed by using DTT as the reducing agent (20 m*M* final concentration, 1 h incubation at room temperature in dark). This was verified by using Ellman's reagent (Sigma).[16]

III. RESULTS AND DISCUSSION

The starting assumption was that the hydrophilicity of the surface plays a predominant role in protein adsorption and that, as hydrophilicity (wettability) of a polymer surface increases, protein adsorption decreases. This assumption was verified using a series of polymethacrylates with different ester side chains (see Scheme I).

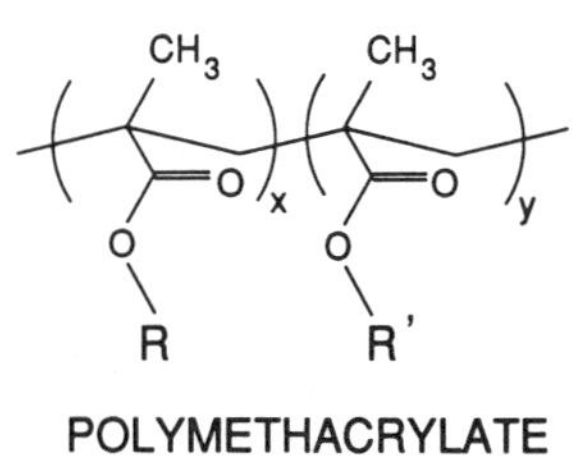

(I)

Use of this series allowed easy variation of the side chain arm to examine the properties of a variety of closely related polymers.

The series of polymethacrylates (Table 1) were coated onto glass rods. This format was used for several reasons: (1) a reproducible surface was readily obtained by the method, (2) the contact angle (wettability) of the surfaces could be easily measured using a Wilhelmy plate balance, (3) the surface area for different polymers could be held constant so that accurate measurements of the amount of antibody per unit surface area could be obtained, and (4) these surfaces could be easily washed and the amount of bound antibody or antigen measured (as radioactivity present).

The contact angle and the amount of protein absorbed by the different polymethacrylate surfaces was measured. The protein used for these studies was a monoclonal antibody for hCG (mAb-166). A positive relationship was observed relating contact angle of the polymer and

* Obtained in the same experiment.

TABLE 1
Contact Angles for Selected Polymethacrylates

Polymer	Contact Angle
Octadecyl Methacrylate	108
Isopropyl methacrylate	113
p-Nitrophenyl methacrylate	88
Methyl methacrylate	77
(Methyl methacrylate) (methacrylic acid) copolymer 75:25	65
Hydroxypropyl methacrylate	57

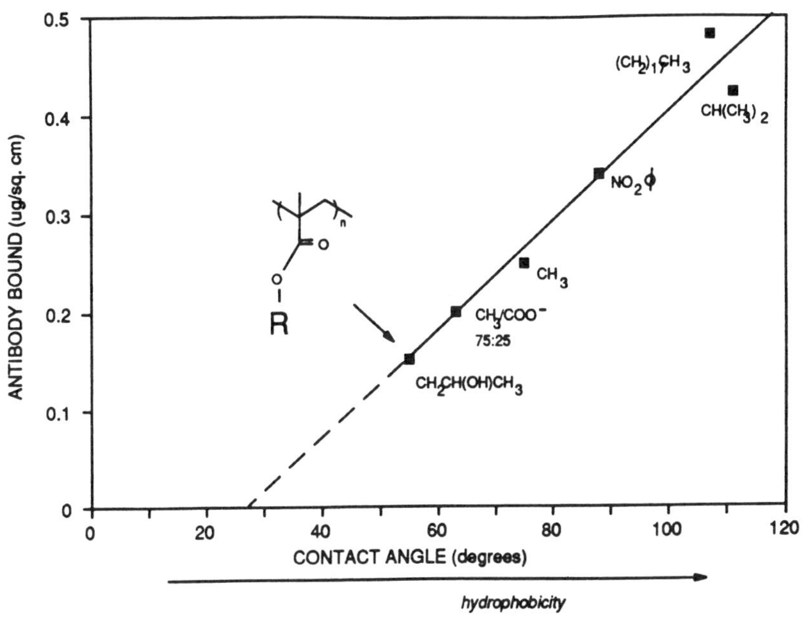

FIGURE 1. Antibody adsorption vs. contact angle.

the amount of antibody nonspecifically adsorbed (Figure 1). As the polymer surface became more hydrophilic (wettable), antibody adsorption decreased. The most hydrophilic and least adsorptive polymer tested in this series was PHPMA. Based on these properties, PHPMA was selected to be modified in order to produce a surface that would covalently bind protein in specified orientations.

PHPMA was modified to covalently bind protein through either amine, aldehyde, or sulfhydryl groups (Figure 2). A carbonate surface was obtained by reacting PHPMA with the chloroformate of *N*-hydroxysuccinimide.[17] This surface is designed to capture antibodies through the amine functions of lysines and arginines, which should give a random orientation. The carbonate polymer can be treated with hydrazine to form a hydrazide surface that reacts with aldehydes. The carbohydrate moiety on the F_c region of the antibody (Figure 1, Chapter 9) can be oxidized with periodate to yield aldehydes which react with the hydrazide surface to yield an oriented antibody. Antibodies can also be treated with DTT to break disulfide bonds, presumably in the hinge region. The free sulfhydryl groups are thus reactive with a maleimide surface to impart surface orientation.

The procedures described above thus yield three reactive surfaces that have the same

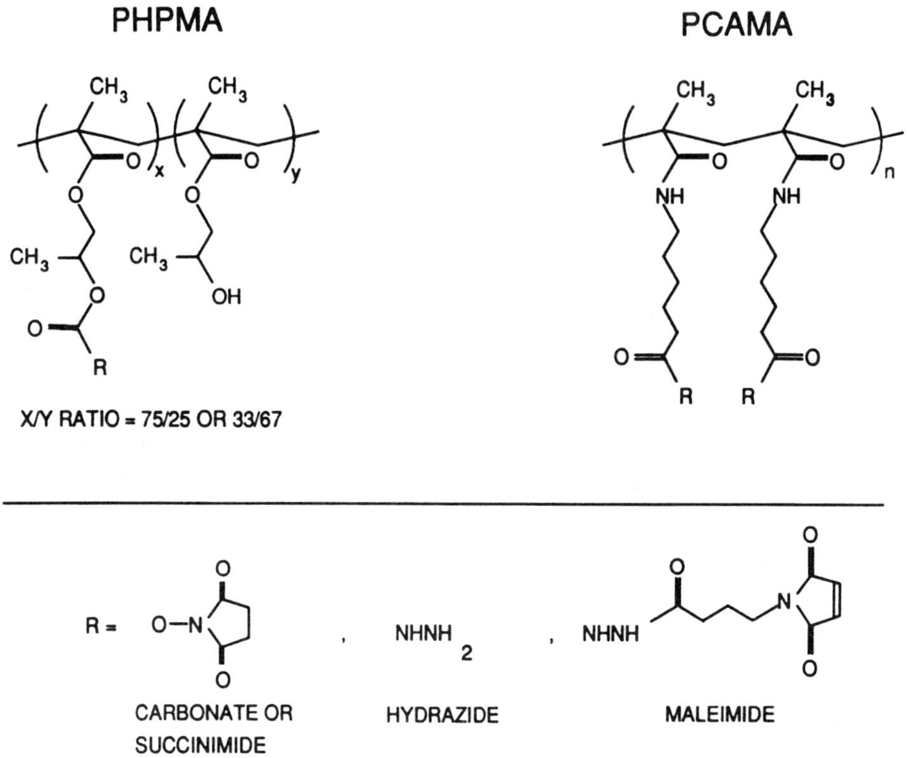

FIGURE 2. Polymethacrylates and polymethacrylamides used for covalent attachment of antibodies.

polymeric backbone (polymethacrylate) and similar sidechains (all based on the hydroxypropyl group; Figure 3). The PHPMA surfaces were prepared with 2 levels of coverage of the reactive groups (75 and 33%) to determine the effect of the density of reactive groups on the amount of antibody immobilized and its activity.

In addition to the PHPMA used for this study, another methacrylate polymer with a different side chain was also evaluated, (PCAMA; Figure 2). This polymer had 100% of the side-chain acid moiety activated with the three different functional groups described above for reaction with antibodies (Figure 3).

The efficacy of these modified surfaces for immunological reactions was determined using covalently immobilized hCG-166 antibody to capture radiolabeled hCG.

Three parameters that describe an immunological reaction were determined: (1) antibody bound = amount of mAb immobilized on the surface, (2) antigen specifically bound = antigen bound to the antibody surface — antigen bound without mAb present, and (3) antibody activity = nanograms of antigen bound/microgram of mAb bound. These values obtained for the experimental polymers were normalized to that obtained for polystyrene. Thus, if more antibody or antigen is immobilized on the experimental surface, values larger than 1.0 are observed. Since the surface area of all surfaces are held constant, results can be directly compared. The amount and activity of hCG-166 adsorbed on polystyrene (see Chapter 12) was used as the standard in these comparisons because of the wide uses of polystyrene as a surface for immunoassays and the ability to reproducibly prepare the polystyrene surface. The results of these studies are summarized in Table 2 and are expressed relative to polystyrene.

The PHPMA maleimide surfaces seem to be the most reactive as they bound the most hCG-166 antibody compared to the polystyrene surface. Specific signal for the PHPMA maleimide surfaces was also larger than that for the carbonate and hydrazide surface. The maleimide surfaces also provide the highest antibody activity. Since these three reactive surfaces all have the same polymer backbone and similar side chains, the difference in amount of antibody

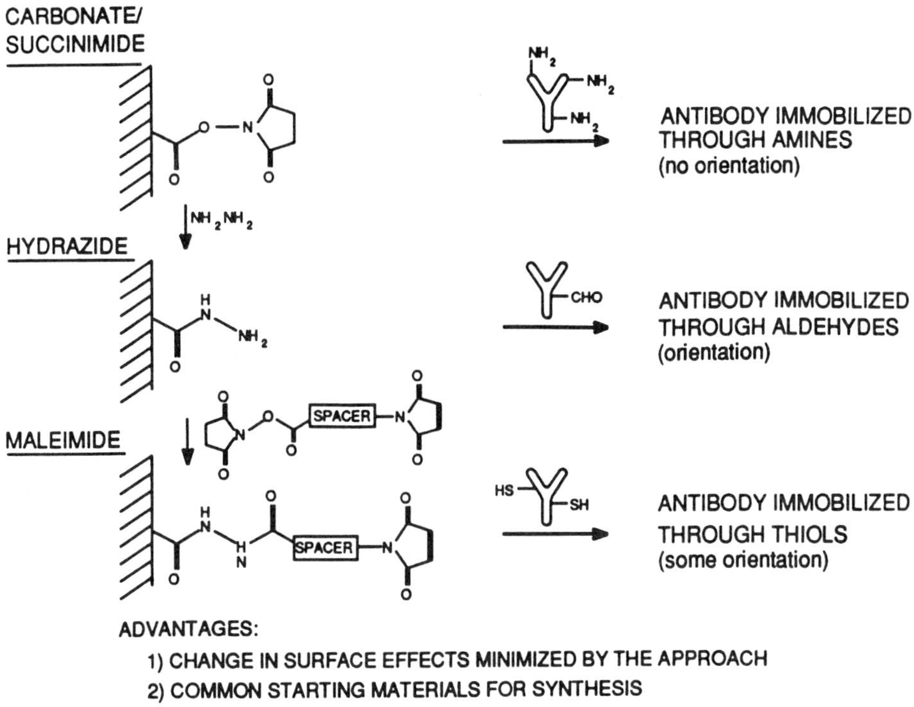

FIGURE 3. Synthesis of surfaces for oriented antibody immobilization.

TABLE 2
Results of mAb Immobilization and Activity on PHPMA and PCAMA

Polymer[a]	Reactive group: Carbonate/ succinimide	Hydrazide	Maleimide
PHPMA 75:25			
Antibody bound	0.38	0.42	0.63
Antigen bound	0.43	0.53	0.94
Antibody activity	1.10	1.25	1.49
PHPMA 33:67			
Antibody bound	0.18	0.25	0.46
Antigen bound	0.27	0.46	0.93
Antibody activity	1.45	1.83	2.02
PCAMA 100:0			
Antibody bound	0.36	0.39	0.34
Antigen bound	0.063	0.55	0.42
Antibody activity	0.10	1.40	1.24

[a] Results normalized to that seen for polystyrene.

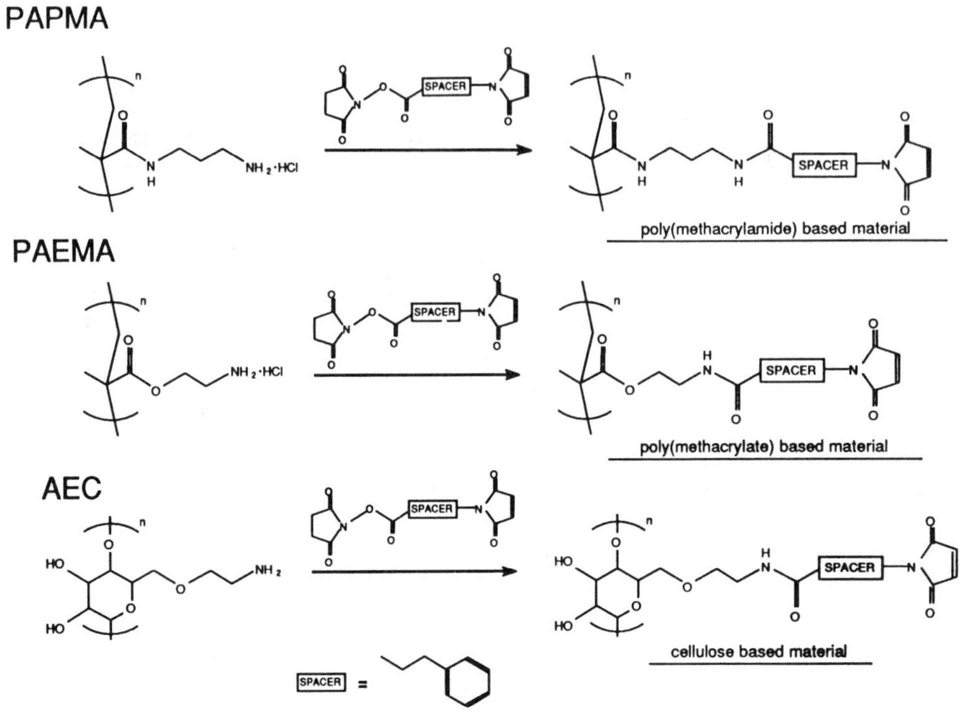

FIGURE 4. Synthesis of maleimide polymers with three different amine polymer backbones.

immobilized and its activity can be attributed to the preferred orientation of the reduced antibody on this surface.

In the case of PCAMA, very different results were observed. Similar amounts of antibody were covalently bound to the three different reactive surfaces. The succinimide ester surface, equivalent to the carbonate surface in the PHPMA series, bound antibody in a relatively inactive form. Slightly higher signal was seen for the PCAMA hydrazide surface than the PCAMA maleimide surface, but the signal was far less than that observed for the maleimide surface in the PHPMA series. The difference in the results for these two series cannot be explained by the authors, but highlights the fact that comparisons between different surfaces can only be made if the same experimental conditions and antibody are used.

What is also apparent from these studies is that less antibody is immobilized on reactive surfaces than on an adsorptive surface like polystyrene. This is probably due to the fact that adsorption to polystyrene is through many small nonspecific interactions while immobilization on these reactive surfaces are through a relatively few specific covalent bonds. Thus, no hydrophilic surface modified to provide covalent coupling of antibody provided higher amounts of antibody immobilized than the amount adsorbed to a hydrophobic surface like polystyrene. What these reactive hydrophilic surfaces do provide is a means of immobilizing antibodies on a surface with special properties or on a surface in special formats like microbeads, chromatographic supports, or membranes. Furthermore, directionalized covalent immobilization could overcome the adsorption-induced loss of activity experienced by mAbs (see Chapters 12 and 17).

Since maleimide-reactive surfaces appeared to hold the most promise for immobilization of antibodies, the effect of changes in the polymeric backbone and the maleimide cross linker on antibody immobilization was examined in two separate experiments.

First, the effect of three different polymeric backbones (PAPMA, PAEMA, and AEC) was examined (Figure 4). The reactive cross linker used in this experiment was SMPB. In this experiment the amount of native (unreduced) antibody that the reactive surfaces would

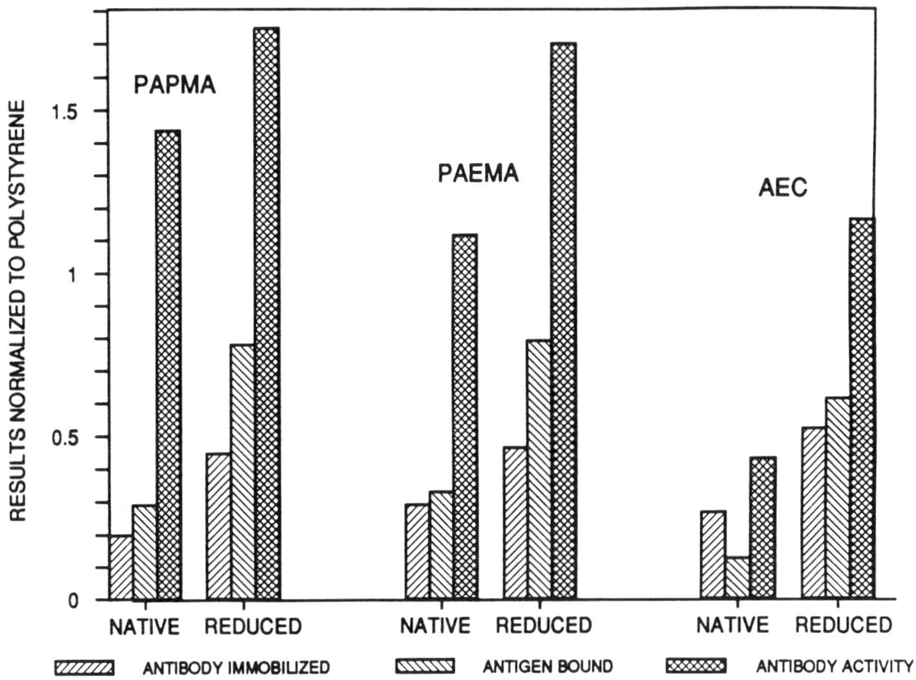

FIGURE 5. hCG antibody immobilization and activity on maleimide-modified amine surfaces.

nonspecifically adsorb was determined as a control. This gave an indication of how selective a reactive surface was for immobilization of a protein through covalent interactions.

All three maleimide surfaces demonstrated good selectivity, as specific immobilization of reduced antibody exceeded the nonspecific adsorption of native antibody (Figure 5). The native antibody also had lower activity on all three surfaces which suggests that it was adsorbed randomly, perhaps some of which is in nonactive orientations. The amount of native antibody adsorbed is fairly equivalent, even though the AEC surface is the most hydrophilic as measured by contact angle.

The amount of reduced antibody specifically immobilized on PAPMA, PAEMA, and AEC was essentially constant. The activity of reduced antibody was lower on AEC than the other two surfaces which resulted in lower antigen binding for the AEC surface. Again, more antibody was adsorbed to the polystyrene surface than was immobilized to the reactive surfaces and antigen binding (signal) was also largest for polystyrene.

In order to study the effects of the heterobifunctional cross-linker spacer group on the amount of antibody immobilized and its activity, PAEMA was modified with six different maleimide cross linkers (Figure 6) and used to prepare reactive surfaces.[12,13]

The three surfaces with linear alkyl chains in the cross linker (β-MPS, γ-MBS, and ε-MCS) all bound a very low amount of native antibody (Figure 7). In addition, the native antibody exhibited low activity on these surfaces which resulted in very low antigen binding. More native antibody was adsorbed nonspecifically to the surfaces with hydrophobic cross linkers like SMPB and SMCC.

The β-MPS and γ-MBS surfaces exhibited the highest amount of specific antibody immobilization, but all the maleimide surfaces showed similar amounts of antigen binding except the ε-MCS surface which was significantly lower. Once again, polystyrene showed the highest amount of antibody immobilization and antigen binding.

The data depicted in Table 2 and Figure 7 show an interesting similarity in that antibody activity is inversely proportional to the amount of antibody immobilized. Indeed, in the large variety of polymers studied thus far this seems to be a recurring theme. This may be due to

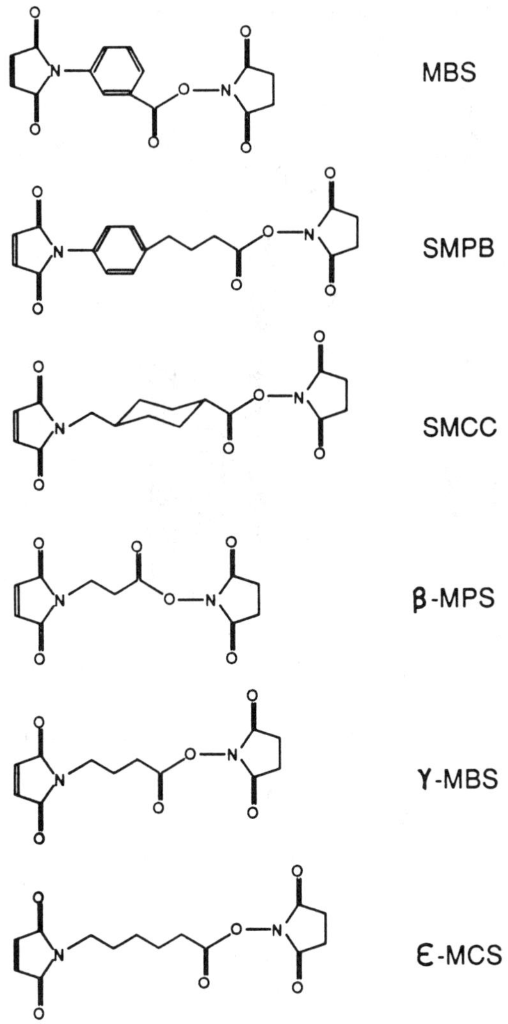

FIGURE 6. Heterobifunctional cross linkers used to modify PAEMA for the synthesis of maleimide surfaces.

a steric effect in which increasing amounts of immobilized antibody have progressively less room to adopt conformations that allow antigen binding.

IV. CONCLUSIONS

The objective of this study was to systematically investigate polymer surfaces for the oriented immobilization of capture antibodies. Results for the experimental surfaces were normalized to that obtained for polystyrene. Since the surface areas of the experimental polymers and polystyrene were identical, comparisons between the reactive hydrophilic surfaces and a hydrophobic surface could be made.

PHPMA and PCAMA were modified to yield three different reactive surfaces that would covalently immobilize antibody through amine, aldehyde, and sulfhydryl functionality on the protein. The maleimide surfaces, which react with sulfhydryl groups on reduced antibodies, yielded the most immobilized antibody with the highest activity.

The effect of varying the polymeric backbone on the ability of the maleimide surfaces to

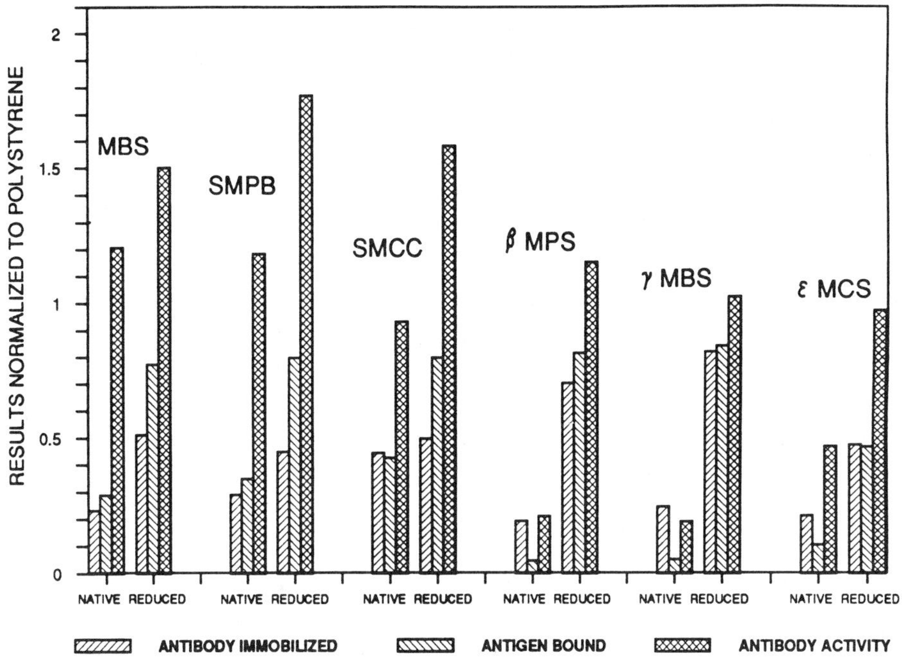

FIGURE 7. hCG antibody immobilization and activity on PAEMA modified with different maleimides.

serve as supports for immunological reactions was investigated using PAEMA, PAPMA, and AEC. The polymeric backbone had little effect on the amount of antibody immobilized, although the antibody was less active on the AEC maleimide surface.

The effect of the linking chain in the maleimide cross linker on antibody immobilization was investigated using PAEMA as the polymeric backbone. The maleimide cross linkers with short alkyl chains, like β-MPS and γ-MBS, demonstrated the highest amount of reduced antibody immobilization and specific antigen binding. In addition, the maleimide surfaces all bound low amounts of native antibody which exhibited low activity. These results suggest that the maleimide surfaces are both selective in immobilizing antibody through specific covalent interactions and that reduced antibody is immobilized in a more active orientation.

None of the reactive surfaces investigated bound more antibody than polystyrene, which was used as the standard in these experiments. This suggests that a hydrophobic surface like polystyrene, which binds proteins through many small nonspecific interactions, will always bind more protein than hydrophilic surfaces that have been modified to covalently immobilize antibody through a few specific interactions.

V. ABBREVIATIONS

CAb	capture antibody
mAb	monoclonal antibody
PHPMA	poly(hydroxypropyl methacrylate)
PAPMA	poly(aminopropyl methacrylamide)
PAEMA	poly(aminoethyl methacrylate)
PCAMA	poly(caproyl methacrylamide)
AEC	aminoethyl cellulose
β-MPS	*N*-(beta-maleimidopropionyloxy)succinimide
γ-MBS	*N*-(gamma-maleimidobutyryloxy)succinimide
ε-MCS	*N*-(epsilon-maleimidocapryloxy)succinimide

SMPB	succinimidyl (*p*-maleimidophenyl)butyrate
SMCC	succinimidyl (4-maleimidocyclohexyl)carboxylate
DTT	dithiothreitol
BSA	bovine serum albumin
TCA	trichloroacetic acid
PBS	phosphate buffered saline
SDS-PAGE	sodium dodecylsulfate-polyacrylamide gel electrophoresis
HPMA	2-hydroxypropyl methacrylate
PMMA	poly(methyl methacrylate)
PPNPMA	poly (*p*-nitrophenyl methacrylate)
PMMA-co-MAA	poly(methyl methacrylate-co-methacrylic acid (75% methylmethacrylate content)
PODMA	poly(octadecyl methacrylate)
PPMA	poly(iso-propyl methacrylate)
BHT	butylated hydroxytoluene (2,6-di-t-butyl-4-methylphenol)
NHS	*N*-hydroxysuccinimide
CAMA	caproyl methacrylamide
DCC	dicyclohexylcarbodiimide
APMA	aminopropyl methacrylamide
AEMA	aminoethyl methacrylate
SEM	scanning electron microscopy

REFERENCES

1. **Weetall, H. H.,** Insolubilized antigens and antibodies, in *The Chemistry of Biosurfaces,* Vol. 2, Hair, M. L., Ed., Marcel Dekker, New York, 1972, chap. 13.
2. **Rauch, P., Marek, M., Matas, V., and Vodrazka, Z.,** The use of antibody immobilized on supports coated with polyglycidyl methacrylate in saturation analysis, *Anal. Biochem.,* 152, 333, 1986.
3. **Wilchek, M. and Miron, T.,** Immobilization of enzymes and affinity ligands onto agarose via stable and uncharged carbamate linkages, *Biochem. Int.,* 4, 629, 1982.
4. **O'Shannessy, D. J. and Hoffman, W. L.,** Site-directed immobilization of glyoproteins on hydrazide-containing solid supports, *Biotech. Appl. Biochem.,* 9, 488, 1987.
5. **Hoffman, W. J. and O'Shannessy, D. J.,** Site-specific immobilization of antibodies by their oligosaccharide moieties to new hydrazide derivatized solid supports, *J. Immunol. Methods,* 112, 113, 1988.
6. **Kennel, S. J.,** Binding of monoclonal antibody to protein antigen in fluid phase or bound to solid supports, *J. Immunol. Methods,* 55, 1, 1982.
7. **Peterman, J. H., Tarcha, P. J., Chu, V. P., and Butler, J. E.,** The immunochemistry of sandwich-ELISAs. IV. The antigen capture capacity of antibody covalently attached to bromoacetyl surface-functionalized polystyrene, *J. Immunol. Methods,* 111, 271, 1988.
8. **Prisyazhnoy, V. S., Fusek, M., and Alakhov, Y. B.,** Synthesis of high-capacity immunoaffinity sorbents with oriented immobilized immunoglobulins or their F_{ab}' fragments for isolation of proteins, *J. Chromatogr.,* 424, 243, 1988.
9. **Usanov, S. A., Guzov,** V.M., **Chashchin, V. L.,** Synthesis, properties, and application of immunosorbent immunoassay, *F.E.C.S. 3rd Intl. Conf. Chem. Biotechnol. Biol. Act. Natl. Prod. Proc.,* 5, 327, 1985.
10. **Chu, V. P. and Tarcha, P. J.,** Protein-reactive, molded polystyrene surfaces having applications to immunoassay formats, *J. Appl. Polym. Sci.,* 34, 1917, 1987.
11. **Jimbo, Y. and Saito, M.,** Orientation-controlled immobilization of protein molecules on thin organic films deposited by the plasma technique, *J. Mol. Electron.,* 4, 111, 1988.
12. **Kitagawa, T., Shimozono, T., Aikawa, T., Yoshida, T., and Nishimura, H.,** Preparation and characterization of hetero-bifunctional cross-linking reagents for protein modifications, *Chem. Pharm. Bull.,* 29, 1130, 1981.

13. **Hashida, S., Imagawa, M., Inoue, S., Ruan, K.-H., and Ishikawa, E.,** More useful maleimide compounds for the conjugation of Fab′ to horseradish peroxidase through thiol groups in the hinge, *J. Appl. Biochem.*, 6, 56, 1984.
14. **Bhatia, S. K., Shriver-Lake, L. C., Prior, K. J., Georger, J. H., Calvert, J. M., Bredehorst, R., and Ligler, F. S.,** Use of thiol-terminal silanes and heterobifunctional crosslinkers for immobilization of antibodies on silica surfaces, *Anal. Biochem.*, 178, 408, 1989.
15. **Guzov, V. M., Rodionov, M. A., Golubovich, V. P., But'kova, L. S., Chashchin, V. L., and Usanov, S. A.,** Optimization of methods of immobilization of antiimmunoglobulins. Equilibrium parameters of the interaction of immobilized antibodies with antigen, *Biokhimiya*, 53, 238, 1988.
16. **Ellman, G. L.,** Tissue sulfhydryl groups, *Arch. Biochem. Biophys.*, 82, 70, 1959.
17. **Miron, T. and Wilchek, M.,** Immobilization of proteins and ligands using chlorocarbonates, in *Methods in Enzymology*, Vol. 135, Mosbach, K., Ed., Academic Press, New York, 1987, chap. 5.

Chapter 12

THE APPLICATION OF TRADITIONAL IMMUNOCHEMICAL METHODS TO EVALUATE THE PERFORMANCE OF CAPTURE ANTIBODIES IMMOBILIZED ON MICROTITER WELLS

John E. Butler, K. S. Joshi, and William R. Brown

TABLE OF CONTENTS

I. INTRODUCTION

Capture antibodies (CAbs) are a critical reagent in solid-phase sandwich immunoassays for the quantitation of multivalent antigens and in labeled-antigen, competitive immunoassays for measuring low molecular weight antigens. The performance of CAbs is usually indirectly inferred from the performance of the complete immunoassays. In the case of large, multivalent antigens like immunoglobulins, this procedure is generally valid because a direct relationship exists between the amount of bound antigen and the signal generated by the detection system provided steric problems are avoided.[1] Nevertheless, there has been no systematic study designed to evaluate the direct performance of CAbs by traditional immunochemical parameters.

Our laboratory has observed that adsorbed polyclonal CAbs are superior to monoclonal CAbs when evaluated in terms of antigen capture capacity (AgCC) and dynamic range;[1] this is at least in part due to the apparent propensity of monoclonal CAbs to be denatured during adsorption.[2] These observations suggest that immunochemical studies on solid-phase CAbs are needed to (1) gain insight into the events which take place when antibodies are immobilized, (2) better understand solid-phase ligand-receptor interactions, and (3) determine what properties of antibodies are most essential for good performance as CAbs. Information gained from such studies should be valuable in the selection of CAbs or in the design of CAbs produced by genetic engineering.

This chapter summarizes immunochemical studies on monoclonal and polyclonal CAbs immobilized by different methods to microtiter wells and includes performance data described by the same immunochemical criteria as used for solution-phase antibodies. In addition, an evaluation of the performance of polyclonal CAb prepared by various methods is summarized.

II. THE IMMOBILIZATION OF CAPTURE ANTIBODIES (CAB) IN MICROTITER WELLS

A. THE ADSORPTIVE CHARACTERISTICS OF CABS FOR POLYSTYRENE AND THEIR FUNCTIONAL ACTIVITY

Figure 1A illustrates the adsorptive properties of rabbit IgG for Immulon 2® (Dynatech, Alexandria, VA) polystyrene microtiter wells; polystyrene microtiter plates from other sources have similar properties (Figure 2A). As with other proteins[3] (Figure 2, Chapter 1) the amount of IgG adsorbed is linearly proportional to the amount added in a log-log plot until the concentration of 5 μg/ml of added IgG has been exceeded, i.e., 600 ng (4×10^{-3} n*M*) has become adsorbed. Increasing the concentration of added IgG results in a loss of log-log linearity in the binding plot and in the efficiency of adsorption (Figure 1A). The log-log linear binding of IgG over the low concentration range suggests homogeneous receptor-ligand interaction up to a point where the primary plastic binding sites are no longer in excess (see Chapter 8). Adsorption beyond this point could result from adsorption to heterogeneous, lower affinity binding sites or perhaps the formation of protein aggregates (see Chapter 1). Data shown here for rabbit IgG are characteristic of the IgGs of other species we have studied as well as for monoclonals IgGs.

Figure 1B demonstrates the functional activities of polyclonal antibodies (pAb) and monoclonals antibodies (mAb) when adsorbed as CAbs on Immulon 2® in relationship to the amount adsorbed. Beyond the addition of 5 μg/ml (600 ng of adsorbed IgG), the AgCC does not increase[4] and when expressed in terms of functional capture antibody concentration ($[CAb_t]$) per nanomolar of adsorbed IgG, performance efficiency actually decreases. Performance continues to decrease for polyclonal and monoclonal CAbs adsorbed at even higher concentrations (Figure 2B and C).

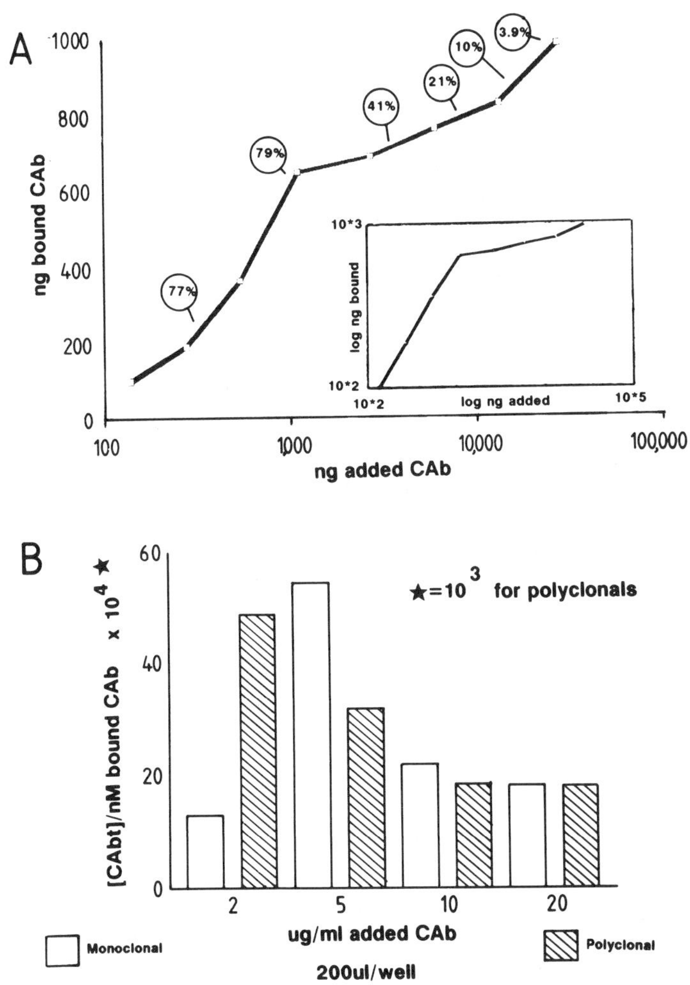

FIGURE 1. The relationship between the amount of adsorbed capture antibody (CAb) and its function. (A) The relationship between the amount of added CAb and the amount adsorbed on Immulon 2. The inset depicts this relationship in a log-log plot. Values encircled are the percentage of added CAb which becomes stably adsorbed at different points on the adsorption plot. (B) The functional activity of monoclonal and polyclonal anti-IgA when adsorbed at 2, 5, 10, and 20 μg/ml on Immulon 2.

B. DIFFERENCES IN THE FUNCTIONAL ACTIVITY OF CABS IMMOBILIZED BY VARIOUS METHODS

Rabbit IgG exhibits similar adsorptive characteristics for the four different plastic surfaces studied; Maxisorp® and polyvinyl showed a slightly higher capacity but only when incubated with 150 μg/ml of IgG (Figure 2A).* The Protein-Avidin-Biotin Capture (PABC) system[2] is a low capacity system and the use of higher concentrations of biotinylated CAbs would not have increased the amount of bound CAb.

The covalent immobilization of rabbit IgG CAb to Covalink-NH wells using disuccinimidyl suberate (DSS), appears less efficient at low protein concentration than adsorption

* Polysorp, Maxisorp, and Covalink-NH microtiter wells were kindly provided by A/S NUNC, Roskilde, Denmark.

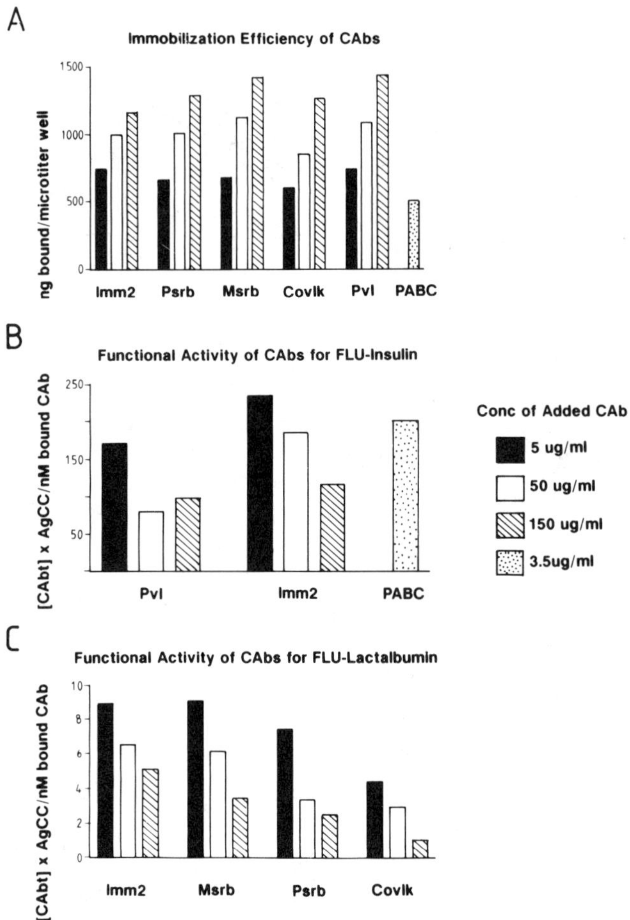

FIGURE 2. Functional differences among CAbs immobilized on different microtiter wells and using different methods. (A) The efficiency of immobilization of polyclonal anti-FLU on microtiter wells by different methods. The concentration in µg/ml of added CAb is indicated in the legend. Imm2 = Immulon 2 (Dynatech, Alexandria, VA); Psrb = Polysorp; Msrb = Maxisorp; Covlk = Covalink-NH (Psrb, Msrb and Covlk all from NUNC, Roskilde, DEN). PABC = Protein-avidin-biotin capture system on Imm2. PvL = polyvinyl. (B) The functional activity of polyclonal anti-FLU immobilized at different concentrations on Imm2, PvL, and by PABC in binding ^{125}-FLU insulin. Only a single, optimal concentration of CAb was immobilized by the PABC system. (C) Same as (B) but comparisons were made in the binding activity for FLU-lactalbumin among adsorbed polyclonal CAb (Imm2, Psrb, Msrb) or covalently bound CAbs (Covlk) at three different concentrations of added CAb. Legend is the same as in (A) and (B).

(Figure 2A). Furthermore, the binding isotherm is a continuously curved line thus providing no evidence for a homogeneous ligand-receptor interaction (data not shown). Immobilization to Covalink-NH wells at high input is equally efficient as adsorption and coupling at this concentration has been recommended.[5]

The concentration of antibody used for immobilization significantly effects the performance of polyclonal anti-fluorescein (anti-FLU) CAbs when expressed as $[CAb_t]$ X AgCC/nM of immobilized CAb (Figure 2B and 2C). Using either ^{125}I-FLU-insulin (MW 6 kDa) or ^{125}I-FLU-lactalbumin (MW 14 kDa), the best performance was seen when CAbs were adsorbed at a concentration of 5 µg/ml; this is consistent with data presented in Figure 1B for mAb and

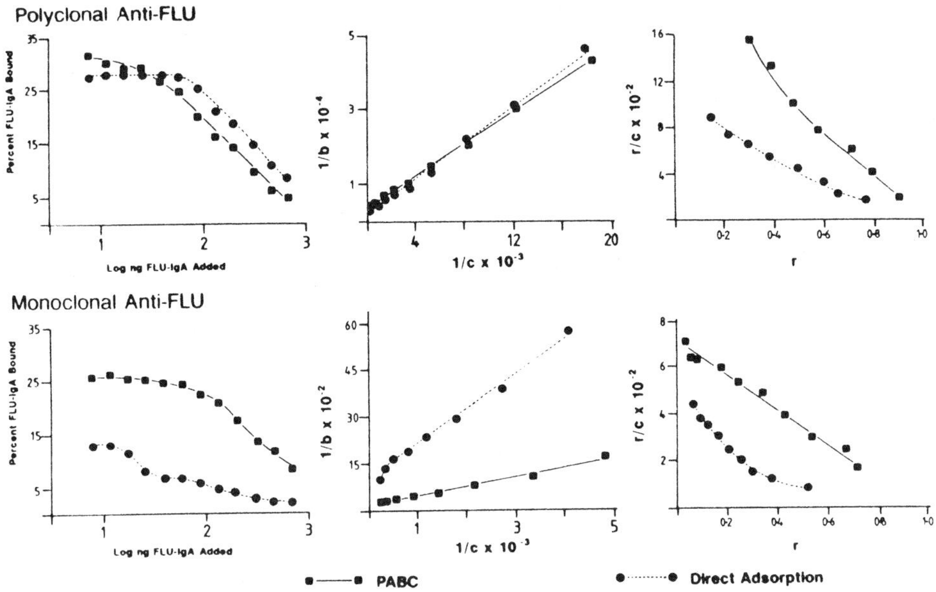

FIGURE 3. The comparative behavior of polyclonal and monoclonal anti-FLU of equal affinity when directly adsorbed on Immulon 2 (●---●) or immobilized on Immulon 2 using the PABC system (■—■). Left = percent binding plots, center = Steward-Petty plots, right = Scatchard analyses. ^{125}I-FLU-IgA (MW = 160 Kd) was used as the antigen in all cases.

pAbs specific for IgA. The performance of adsorbed polyclonal anti-FLU also differs depending on the type of plastic used. CAbs adsorbed at high concentration on polyvinyl behaved erratically. CAbs adsorbed on Polysorp® had poorer performance at all concentrations when compared to surface-treated polystyrene (Maxisorp and Immulon 2); this is probably due to its more hydrophobic character which increases the risk of protein unfolding.

The failure of pAbs to show improved performance when immobilized nonadsorptively and noncovalently using the PABC system (Figure 2B) is consistent with our earlier report and is in contrast to the effect observed when monoclonal CAbs are immobilized using the PABC system[2,6] (see Section III and Chapter 17).

The relatively poor performance of covalently bound CAbs is not unlike results that we and other have previously observed.[7] The finding suggests that although adsorption of proteins results in conformation changes which can effect their function (Chapters 1 and 10), adsorbed polyclonal CAbs generally function as well as those immobilized using the PABC system and have higher overall activity than those immobilized covalently using DSS.

III. ANALYSES OF MONOCLONAL AND POLYCLONAL CABS USING TRADITIONAL IMMUNO-CHEMICAL METHODS

A. THE BEHAVIOR OF CABS IN STEWARD-PETTY, SCATCHARD, AND PERCENT BINDING PLOTS

Monoclonal and polyclonal CAb of similar affinity (10^9 l/M) for fluorescein (FLU), when immobilized on microtiter wells using the nonadsorptive, noncovalent PABC system, and tested using FLU-IgA, *both* behave nearly homogeneously; some slight tendency toward heterogeneous behavior is observed for the polyclonal (Figure 3). The maximum percentage of antigen bound is similar for both CAbs, i.e., ca. 26%. When CAbs are immobilized by adsorption, polyclonal anti-FLU, when evaluated in percent binding plots, displays a more extensive and pronounced region over which the percent binding remains constant; this is typical of a homogeneous ligand-receptor interaction, i.e., a monoclonal antibody (Figure 3, top left). The percentage of antigen bound by adsorbed monoclonal anti-FLU is only half that seen with the polyclonal (ca. 12%; Figure 3, lower left) and there is a near absence of a region of constant percent binding. Evidence for the lower affinity of adsorbed monoclonal anti-FLU is also apparent from the Scatchard plot. Adsorbed monoclonal anti-FLU shows heterogeneity by Scatchard analysis when compared to mAb immobilized using the PABC system. Collectively, Figure 3 indicates that adsorption causes polyclonals to behave homogeneously and monoclonals to show lower affinity and perhaps heterogeneity, when compared to nonadsorptive immobilization. It is hypothesized that mAb adsorbed to plastic undergo a variety of conformational changes which correspondingly affect their affinity for FLU. The fact that the mAb behaves monoclonally at low antigen input but not at high antigen input (Figure 3), suggests that conformational distortion is progressive with a corresponding loss of affinity and increasing heterogeneity as higher concentrations of antigens are added. These fundamental differences between the behavior of adsorbed monoclonal vs. polyclonal CAbs (to large molecular weight antigens) have been observed in the majority of our studies.[1,6]

B. MONOCLONAL CABS PREFERENTIALLY UNDERGO CHANGES IN FUNCTIONAL [CAB_T], Keq AND VALENCE DURING ADSORPTION

The data presented in Figure 3 compare the activity of solid-phase mAbs and pAbs of similar affinity in solution, for the large multivalent antigen, FLU-IgA. The interpolated valence (n) should be regarded as relative because the method used to determine the functional [CAb_t] only estimates the molar binding site concentration at saturation. Presumably for steric reasons, an antigen the size of IgA complexes equimolarly with CAb so that the Y-intercept of the Steward-Petty plot equals 1/[CAb_t]. However, this cannot be proven from the data presented so that *relative differences* in valence between adsorbed and nonadsorbed CAbs become the important issue. The lower valence (Figure 3, lower right) AgCC (Figure 3, lower left) and lower Keq (Figure 3, lower center and right) of the adsorbed mAb, provide collective evidence that monoclonal anti-FLU undergoes adsorption-induced denaturation.

Figure 4 summarizes data on the functional [CAb_t] and relative valence of the pair of anti-FLU antibodies described in Figure 3 but tested with fluorescyl antigens of three different molecular sizes. Values for [Cab_t] were obtained from Steward-Petty plots and valence from Scatchard analyses. The nanomoles of bound antibody were determined in separate experiments using iodinated antibodies. The data indicate that adsorption lowers functional activity, especially that of mAbs. An 82% reduction is seen for the mAb vs. 40% for the pAb (Figure 4A).

The data in Figure 4B have been derived making certain assumptions, namely that 1 mol of CAb, regardless of the method of immobilization, can bind only 1 mol of FLU-IgA, but 2 mol of FLU-LA or FLU-insulin. Hence, [CAb_t] values interpolated from Steward-Petty plots have been adjusted accordingly. Independent of this assumption, it is clear that adsorption reduces the molar concentration of functional binding sites.

C. THE DISCREPANCIES IN THE FUNCTIONAL [AB_T] AND Keq OF SOLID-PHASE CABS DETERMINED BY METHODS USED FOR SOLUTION-PHASE LIGAND-RECEPTOR INTERACTIONS

The values from Steward-Petty plots for [CAb_t] are discrepant with the [CAb_t] calculated on the basis of the amount adsorbed; the former are >3 logs higher than possible (Table 1). Furthermore, the value for Keq are of a magnitude that would be unmeasureable in a solution phase system and would be regarded as nonspecific binding. A Keq value of $1.5 \times 10^{-4}\ m^{-1}$ for polyclonal anti-IgA is inconsistent with the ability of this antibody to capture 98% of the added

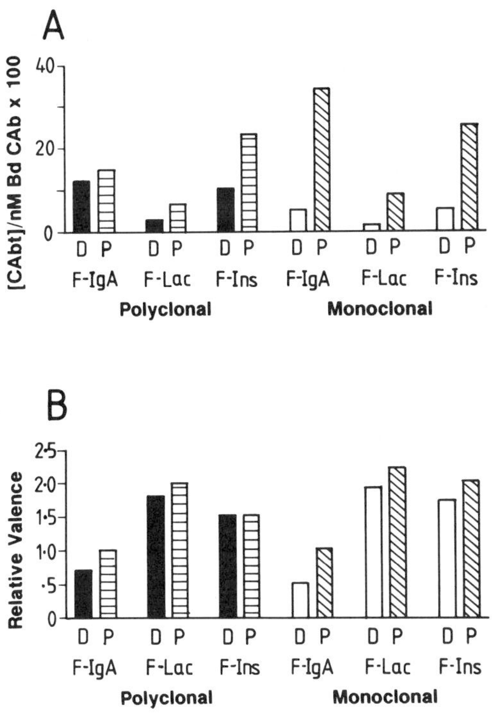

FIGURE 4. The influence of adsorption on relative activity and valence of polyclonal and monoclonal anti-FLU. The same CAbs described in Figure 3 were studied using three different fluorescyl-protein conjugates; FLU-IgA (F-IgA; 160 kDa MW), FLU-lactalbumin (F-Lac; 14 kDa MW) and FLU-Insulin (F-Ins; = 6 kDa MW). (A) Activity expressed as [CAb_t]/nM bound CAb. (B) Valence extrapolated from Scatchard plots using the assumption that for steric reasons, one CAb can bind only one F-IgA but could bind two molecules of F-Lac or F-Ins at saturation.

antigen! This discrepancy in calculated [CAb_t] could be a result of lack of knowledge about the correct reaction volume. The values in Table 1 were calculated using a reaction volume equal to the volume of liquid in the microtiter well, i.e., 200 μl. However, the actual reaction is presumably interfacial which mean it might represent only a thin film on the surface of the microtiter well; the remaining volume in the well would be irrelevant. Should this be the case, the value of "b" in the Steward-Petty plot would be much higher and the estimated [CAb_t] much lower. Assuming the true reaction volume is only 1 μl, the corrected value obtained for observed [CAb_t] would begin to approach the calculated value (Table 1). However, correcting the value for [CAb_t] for a much smaller reaction volume would not affect the estimated value of Keq unless the functional value of "c" (unbound ligand) is also in error. This would suggest that the free ligand concentration is not uniformly distributed within the 200-μl volume of the well.

The Keq (regardless of magnitude) is two to three times higher for pAb than mAb; this difference agrees in direction, but not always in magnitude, with the differences observed in AgCC (Table 1). In cases where the AgCC for polyclonal CAbs is much greater than expected on the basis of differences in Keq, the discrepancy could be due to a small population of polyclonal antibodies which are responsible for the high AgCC of the polyclonal at low input of antigen. If one estimates the Keq where only one tenth of the fractional binding sites are saturated, much better agreement is obtained between Keq and AgCC.[4]

TABLE 1
The Estimation of [CAb_t], Keq and Valence for Affinity-Purified Solid-Phase Capture Antibodies

Antibody tested	AgCC[c]	$[CAb_t]_{sp}$[a] (n*M*)	Keq[b] (l m^{-1})	Valence	$[CAb_t]_{calc}$[a] (n*M*)
Polyclonal swine anti-IgA	98	2.5×10^{-4}	1.5×10^{4}	1.0	2×10^{-8}
Monoclonal anti-IgA[d]	46	2.85×10^{-5}	8.5×10^{3}	1.0	2×10^{-8}
Polyclonal swine Anti-FLU-IgA	27	4.9×10^{-4}	5.8×10^{2}	0.7	2×10^{-8}
Monoclonal anti-FLU-IgA[d]	12.7	2.0×10^{-4}	2.0×10^{2}	0.5	2×10^{-8}

[a] Functional $[CAb_t]_{sp}$ was estimated from Steward-Petty plots whereas actual bound CAb $[CAb_t]_{Calc}$ was determined using iodinated antibodies.
[b] Estimated from Scatchard plots at one-half saturation of functional binding sites.
[c] AgCC = mean percentage of antibody bound in the linear binding region of the percent bound plot (see Figure 3).
[d] Monoclonals kindly provided by M. Scharf (Rockefeller University) and E. W. Voss, Jr. (University of Illinois, Chapaign), respectively.

IV. THE BEHAVIOR OF POLYCLONAL CABS PREPARED BY DIFFERENT METHODS

The superior performance of polyclonal CAbs, even when adsorbed, was considered sufficient to warrant investigation of their preparation. Suter et al. demonstrated little difference between affinity-purified CAb adsorbed on plastic, those immobilized using the PABC system, and those prepared merely by adsorbing the ammonium sulfate fraction of whole serum on plastic. Data illustrated in Figures 2B and 3 top, confirms and extends such observations.

Figure 5 illustrates the effectiveness of goat polyclonal antibodies when prepared by different methods. Expression as [CAb_t] × AgCC/ng IgG was selected because it considers the major ingredients which we have found most predictive of good performance in sandwich ELISAs. Furthermore, it is consistent with the Karush equation expressing "extent of reactivity", in which both affinity (AgCC in this case) and antibody concentration (CAb_t) are considered.[8] We have obtained similar data for swine antibodies to TEPC-15 mouse IgA and swine and rabbit antibodies to fluorescein.[4] Various concentrations of saturated ammonium sulfate ($AmSO_4$) from 25 to 50% were tested, and the antibody-enriched fraction precipitated at 30% saturation of $AmSO_4$ either performed best or as well, as any other $AmSO_4$ fraction when goat, rabbit, or swine sera were tested. This was generally the fraction containing the highest percentage of IgG. However, the data of Figure 5 indicate that there is no *simple* means to prepare polyclonal capture antibodies which is vastly superior to any other, although certain results may be suggestive for future studies. Affinity purification of polyclonals naturally results in the best CAb performance, but this method is generally considered impractical for large scale preparation of CAbs. High-pressure affinity chromatography on SpG and preparation using caprylic acid are rapid and showed the next best level of performance (Figure 5). Of these two methods, caprylic acid performed the best in all three species although by no means always yielding preparations with the highest proportion of IgG. It is possible that short chain fatty acid may stabilize IgG against adsorption induced loss of activity.

V. SUMMARY

The data presented in this chapter show that IgG CAb is initially adsorbed on plastic in a manner which suggests a homogeneous receptor-ligand interaction with plastic. At the point

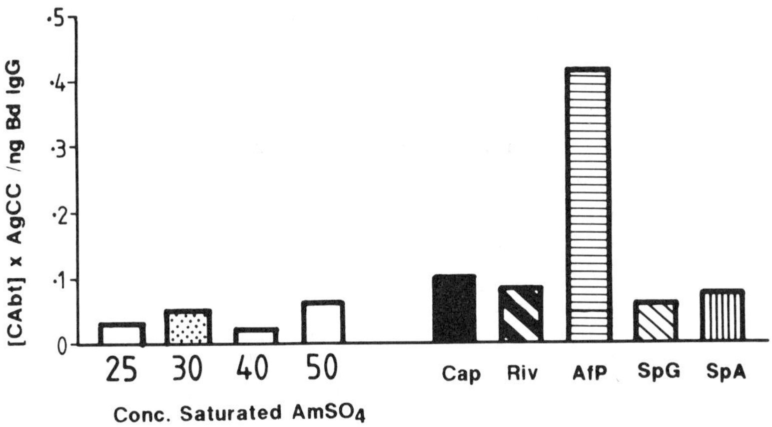

FIGURE 5. The functional activity of polyclonal swine anti-TEPC 15 IgA CAb prepared by various methods. The four bars on the left are the activity of CAb precipitated from serum using four different concentrations of saturated ammonium sulfate ($AmSO_4$) expressed as percent of saturation. The abbreviations under the bars on the right are as follows: Cap = caprylic acid; Riv = Rivanol, Afp = affinity-purified standard, SpG = high pressure protein G affinity chromatography, SpA = high pressure protein A affinity chromatography.

of inflection in the binding plot, 600 ng have been adsorbed although this would cover only one fifth of the plastic surface were it uniformly distributed. Adsorption above this concentration continues in a nonlinear fashion; it is noteworthy that functional activity fails to increase beyond the point where log-log linear adsorption ceases (Figures 1, 2B, and 2C). Therefore the procedure used by some individuals and firms, of adsorbing antigen and antibodies on plastic at high concentration, where <10% of the added material is adsorbed (Figure 1A) and then using the unbound supernatant to coat additional plates, could produce differences in the quality of the solid phase. This procedure makes the assumption that the quality, e.g., immunochemical activity, of the reactant adsorbed during the initial incubation is the same as the reactant which is adsorbed from supernatants transferred to subsequent plates. This is not supported by the data of Figures 1B, 2B, and 2C.

Three different immobilization methods are compared in this study. Simple adsorption on polystyrene yields the highest amount of bound CAb with some minor differences depending on the source of the microtiter plate used; this result is consistent with those of Mize et al. (Chapter 11). Random covalent immobilization using DSS to amino-derivatized microtiter plates does not yield amounts of bound IgG comparable to that obtained by adsorption except at very high input concentrations. This is consistent with the findings of the manufacturer, who finds it offers no advantage over adsorption for large molecular weight protein like polyclonal CAbs. The use of the PABC system, results in the lowest amount of immobilized CAb and also does not offer any advantage for pAb although it significantly prevents the adsorption-induced loss of function when monoclonal CAbs are used (Figures 3 and 4). Perhaps covalent coupling, especially if directionalized, could be advantageous for mAbs; evidence for this is presented in Chapter 11. There is also evidence that covalent immobilization may be more important in SpAbI for antigen and peptide immobilization than for immobilization of antibodies[5] (Butler, unpublished data).

The preferential susceptibility of mAb to surface-induced loss of activity for multivalent antigens could result from at least three, somewhat interrelated phenomena. While loss of function of pAb occurs during adsorption (Figure 3, top right and Figure 4), a small to moderate proportion of denaturation resistant antibodies of high affinity may survive whereas resistance or susceptibility, owing to protein homogeneity, is an "all-or-none" phenomenon for mAbs. The surviving subpopulations of pAbs appears to bind high molecular weight antigens homogeneously at low antigen input. Secondly, mAb may not all represent "healthy and complete" antibodies. Subtle differences in glycosylation may destabilize the mAb and permit it to be more susceptible to adsorption-induced conformational change and corresponding loss of activity. Third, CAbs may be adsorbed as clusters. The juxtaposition of a polyclonal population would allow antibodies recognizing different epitopes of the same multivalent antigen to act collectively and with high avidity. Clustered mAbs may not be able to form multiple interactions with the antigen due to the infrequency of identical determinants, thus lowering avidity and resulting in the "affinity effects" described in this report. Each of these hypotheses has its caveats and much work is still needed. For example, FLU-IgA is multivalent and expresses identical epitopes which should bind avidly to clustered pAbs or mAbs.

The "heterogeneous" behavior of adsorbed mAbs, combined with the loss of valence and functional $[CAb_t]$ (Figures 3 and 4), suggests a spectrum of conformationally distorted monoclonal CAbs. The extent to which adsorbed mAb lose activity can be quite variable[2] (Joshi and Butler, unpublished) but it is known "within the trade" that some hybridoma antibodies lose essentially *all* activity upon adsorption.

The analysis of CAbs by the traditional methods used for ligand-receptor interactions in solution results in values for $[CAb_t]$ which are much higher than possible and estimates of Keq which are unrealistic. Although the overestimation of $[CAb_t]$ could be explained by an overestimation of the true interfacial reaction volume, the low estimates of Keq cannot be readily explained unless the free ligand is nonrandomly distributed. Therefore, the application of traditional immunochemical methods to the study of solid-phase CAbs reveals logical differences in the nature of the binding plots although the magnitude of the absolute values for $[Ab_t]$ and Keq estimated by these procedures remain unreliable; a better understanding of interfacial ligand receptor interactions appears to be required.

The data summarized confirm the superior performance of adsorbed pAb, but fail to identify an entirely simple method of preparing polyclonal CAb which is significantly better than any other. The caprylic acid is rapid and does appear to yield CAbs with good AgCC and acceptable $[CAb_t]$. This has been confirmed in expanded studies[4] so that future studies on the effect of short-chain fatty acids in CAb stabilization seem worthwhile.

The data presented in this chapter offer insight into the behavior of monoclonal vs. polyclonal CAbs immobilized by various methods and offers several immunochemical criteria which are useful in evaluating their performance. On the other hand, studies summarized here also demonstrate the difficulties encountered when trying to apply the methods of solution-phase ligand-receptor interactions to a solid-phase system.

ACKNOWLEDGMENT

The authors wish to thank A/S NUNC of Roskilde, Denmark for supplying their plastic microtiter wells and ChromatoChem of Missoula, MT, for supplying the high pressure Protein-A and Protein-G affinity columns used in this research. The authors thank Drs. E. W. Voss, Jr., University of Illinois, Champaign, M. Scharf, Rockefeller University, and F. Klobasa, Institute of Animal Husbandry, Mariensee, Germany, for kindly supplying monoclonal anti-FLU, monoclonal anti-T15 IgA, and polyclonal swine anti-FLU, respectively. The authors also thank Ms. Marcia Reeve for preparing the typescript. This research was supported by a grant from Abbott Laboratories, North Chicago, IL.

REFERENCES

1. **Butler, J. E., Spradling, J. E., Dierks, S. E., Heyermann, H., Peterman, J. H., and Suter, M.,** The immunochemistry of sandwich ELISAs. I. The binding characteristics of immunoglobulins to monoclonal and polyclonal capture antibodies adsorbed on plastic and their detection by symmetrical and asymmetrical antibody-enzyme conjugates, *Mol. Immunol.,* 23, 971, 1986.
2. **Suter, M. and Butler, J. E.,** The immunochemistry of sandwich ELISAs. II. A novel system prevents denaturation of capture antibodies, *Immunol. Lett.,* 13, 313, 1976.
3. **Cantarero, L. A., Butler, J. E., and Osborne, J. W.,** The binding characteristics of various proteins to polystyrene and their significance in solid-phase immunoassay, *Anal. Biochem.,* 105, 375, 1980.
4. **Joshi, K. S. and Butler, J. E.,** The immunochemistry of sandwich ELISAs. V. The antibody performance of polyclonal antibody-enriched fractions prepared by various methods, *J. Immunol. Methods,* in press.
5. **Rasmussen, S. E.,** A/S NUNC, Roskilde, Denmark, personal communication, 1989.
6. **Suter, M., Butler, J. E., and Peterman, J. H.,** The immunochemistry of sandwich ELISAs. III. The stoichiometry and efficacy of the protein-avidin-biotin capture (PABC) system, *Mol. Immunol.,* 26, 221, 1989.
7. **Peterman, J. H., Tarcha, P. J., Chu, V. P., and Butler, J. E.,** The immunochemistry of sandwich ELISA. IV. A comparison of the antigen capture capacity of antibody adsorbed or covalently attached to polystyrene, *J. Immunol. Methods,* 111, 271, 1988.
8. **Karush, F.,** Affinity and the immune response, *Ann. N. Y. Acad. Sci.,* 169, 56, 1970.

Chapter 13

CONSIDERATIONS OF THE INTERPRETATION OF THE SPECIFICITY OF MONOCLONAL ANTIBODIES DETERMINED IN SOLID-PHASE IMMUNOASSAYS

Fred J. Stevens

TABLE OF CONTENTS

I. INTRODUCTION

Methods for evaluating interactions between macromolecules in which one component is immobilized onto a solid-phase matrix now play a major role in the analysis of the specificity of interactions between antibodies and antigens and between receptors and their specific ligands. The majority of immunochemical assay methods in current use involve measurement of the direct binding of antibody to adsorbed antigen (or vice versa) or of the ability of a solution-phase component to inhibit binding to the immobilized component. The contribution of the solid-phase matrix as a ternary participant in these interactions has received relatively little consideration in such studies. In this report, some aspects of the physical chemistry of interaction between protein and the solid-phase support are explored. It is suggested that in some cases the perturbation in overall reaction that results from the presence of the solid phase may be sufficient to generate significantly misleading results if these methods alone are used to evaluate binding specificity.

II. PHYSICAL CHEMISTRY IN PROTEIN: SOLID-PHASE INTERACTION

Solid-phase immunoassays (SPIAs) are based on the ability of polymers such as polypeptides to adsorb onto the surfaces of glass, plastics, or various synthetic membranes. Particularly in the case of adsorption to plastic, the energy of binding is nominally provided by what are thought to be hydrophobic interactions.[1] In effect, the energy of interaction arises from the ability of the solvent to occlude the surface of the plastic with protein and thereby exclude water from the interacting surfaces. Thus, the physical basis for protein adsorption to the solid phase originates from the same forces that are responsible for the stabilization of protein structure and for the interaction between protein subunits.[2-4] To a first approximation, the energy resulting from these interactions may be correlated with the amount of excluded surface.

The interaction between protein and solid phase is intrinsically reversible and dynamic; both lateral and rotational movement of protein at the interface may be anticipated.[5-7] If the excluded surface area may be increased by compression or unfolding of protein, the adsorption of a polypeptide is consequently associated with denaturation of the native conformation.[8-12] In addition, the solid-phase matrix may itself sterically block a significant portion of epitopes.

The upper panel of Figure 1 represents a schematic for an idealized SPIA. Antigen is shown to be dispersed uniformly across the surface and is fixed to the plate without conformational distortion. Epitopes are randomly oriented. The surface of the solid-phase that does not contain antigen is blocked by protein such as albumin. All interaction between solution-phase antibody and the solid phase is accomplished via discrete specific interaction between antibody paratope and antigen epitope.

The lower panel of Figure 1 depicts an approximation of a murkier reality. The solid-phase surface is unlikely to be smooth at the macromolecular scale and thus might better be represented as an irregular porous environment. The distribution of antigen may be uniform, but only if the macromolecule exhibits no tendency to aggregate, which would lead to formation of clusters on the solid-phase surface by provision of an additional mechanism for the exclusion of water from surfaces. In many cases, denaturation of protein may be expected; the presence of disulfide bonds may be found to stabilize local epitopes but not preserve other epitopes expressed at more distal locations. Although multiple modes of association of protein with solid phase are likely to exist, unless the protein comprises a spheroid of uniform surface chemistry, randomized orientation of epitopes is unlikely. The distribution of the blocking reagent, such as albumin, is also unlikely to be uniform, complete, or immobile.

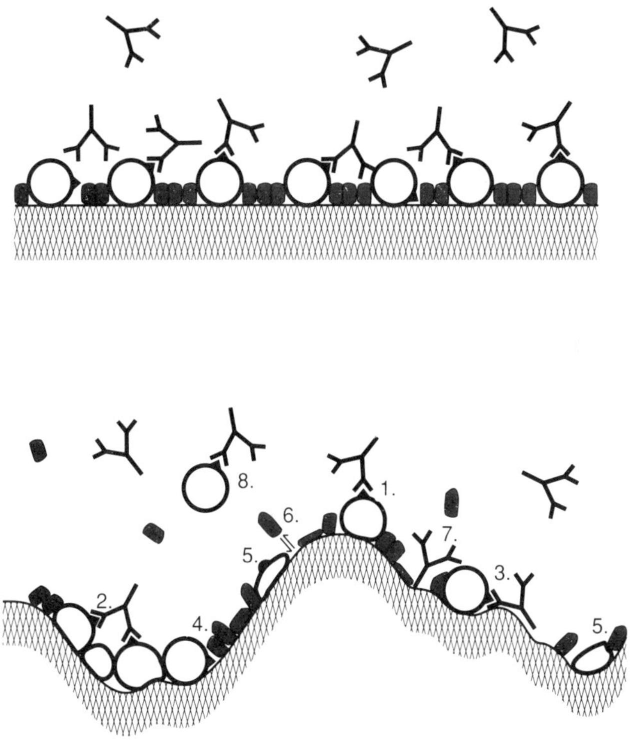

FIGURE 1. (Upper) Traditional idealized representation of interface interactions in the solid-phase assay configuration. Antigen particles are uniformly distributed along the surface with random distribution of epitopes. No denaturation of antigen is evident although as a consequence of the "randomized" orientation of epitopes not all epitopes are accessible to solution-phase antibody. Portions of the solid-phase surface that are free of antigen are uniformly and completely blocked by non-antigenic protein to prevent non-specific adsorption of antibody. All interactions between antigen and antibody are discrete paratope:epitope interactions. (Lower) Illustration of some complicating factors that may be involved in detailed interpretation of the interaction of antibody with solid-phase adsorbed antigen. These factors include the convoluted shape of the solid-phase surface, presence of both monomeric (1) and bivalent (2) interactions between antibody and antigen, interactions between antigen and antibody enhanced by ternary (3) antibody-solid phase interactions, inaccessibility (4) or denaturation of epitopes (5), disadsorption of blocking reagent (6) resulting in exchange with and nonspecific adsorption of antibody (7), and disadsorption of antigen leading to competing solution-phase interactions (8).

III. CONTRIBUTION OF INTERFACIAL DYNAMICS TO APPARENT IDIOTOPE SPECIFICITY

The interfacial dynamics are complex even if the solid-phase matrix is viewed as an inert participant in the interaction between the immobilized and solution-phase component, and the adsorbed constituent is statically fixed without conformational change. However, the immobility of one component leads to intrinsic alterations in the kinetics of interaction relative to that which may be anticipated on the basis of observations obtained when both components are freely dispersed in solution.[13-19] At the interface between solid- and solution-phase components, the dynamics of interaction between the immobilized component and the solid-phase matrix, and that between immobilized component and solution-phase component, may also be significant contributors to the overall reaction.

The dynamics of interaction between the immobilized component and solid-phase matrix were incorporated in a recent reanalysis[20] of observations previously derived in a solid-phase study of the specificities of a panel of antiidiotopic monoclonal antibodies (mAbs) that recognized the murine phosphorylcholine-binding IgA myeloma protein (T15) produced by the TEPC-15 plasmacytoma.[20] The reanalysis was prompted by size-exclusion chromatographic[22,23] examination of mixtures of T15 with antiidiotopic antibodies F6 and 4C11 in which strong interaction was observed for the former but no interaction was seen for the latter.[20] In the solid-phase experiments, phosphorylcholine had completely inhibited binding of solution phase antiidiotope 4C11 to immobilized T15. This inhibition suggested that 4C11 recognized determinants that comprised all or part of the hapten binding site. In contrast, although the binding of antibody F6 and several others were inhibited by solution-phase phosphorylcholine, even at antibody binding-site concentrations 10^6 to 10^7-fold below that of hapten, inhibition levels only on the order of 20% were observed. These results were interpreted to indicate that these antibodies recognized idiotopic determinants near the hapten binding site.

In the size-exclusion chromatography method, interaction between macromolecules is inferred from differences in the chromatography profile of a mixture relative to a computer-generated summation of the elution profiles of the two components chromatographed separately.[24] Since formation of a complex results in an aggregrate with shorter elution time, this approach potentially represents a direct method for observation of homogeneous solution-phase interaction that avoids complications introduced by solid-phase immobilization. Moreover, in many applications it is not necessary to use radio- or enzyme-labeled reagents to monitor the interaction.

Figure 2a illustrates interaction during size-exclusion chromatography in a series of simulated chromatograms[22] generated as a function of affinity. The near elution profile is that of a mixture of intact antibody and (antiidiotopic) Fab for which the association constant is set at $K_a = 0\ M^{-1}$. In this contour representation of the simulated data, the affinity of interaction increases logarithmically from front to rear. Interaction is first noticable as apparent forward shifts of the Fab and antibody peaks. Subsequently, a continuum of elution profile changes is observed in which the elution at the Fab position is depleted and an early eluting peak corresponding to an Fab:Ig:Fab complex is obtained. The forward shift of eluting material is clearly seen from the viewpoint presented in the figure inset.

A convenient method for depicting changes in elution behavior as a result of interaction is represented by delta profiles constructed by the difference between the elution profile of a mixture relative to that expected in the absence of interaction. As seen in Figure 2b, delta profiles sensitively reveal the forward elution shifts resulting from interaction and the corresponding depletions at the original positions of antibody and Fab. Experimental correlates of these simulated interactions have been demonstrated.[22,23,25] It should be noted that for the short run times used in high-performance liquid chromatography, the assumption of rapid equilibration used in the above simulation may not be appropriate for high-affinity interactions typically characterized by slow dissociation rate constants. The influence of interaction kinetics has been addressed in a nonequilibrium simulation of protein-protein interaction during size-exclusion chromatography.[26]

From the perspective of the chromatographic observations, then, the fundamental thermodynamics of the interaction between T15 and the antiidiotopes F6 and 4C11 were quite different, although both antiidiotopes reacted well in solid-phase assays. The interaction between T15 and F6 was clearly high affinity, as shown by the absence of free reactants in the initial mixture and by little change in the elution profile of the diluted mixture after a 24-h incubation. In contrast, the elution profile of a mixture of T15 and 4C11 was not distinguishable from the reference "zero-interaction" profile generated by the summation of the individually chromatographed samples. This suggested that relatively insignificant interaction between the two antibodies was present.

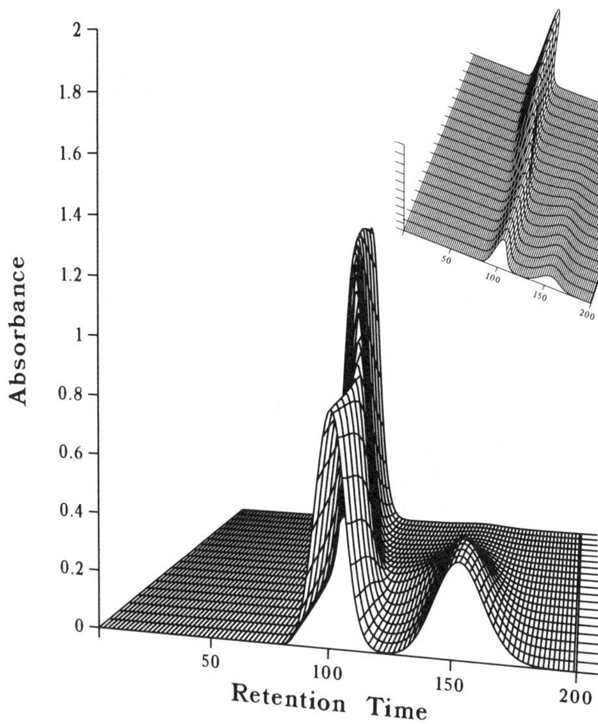

A

FIGURE 2. (A) Simulated size-exclusion chromatograms depicting affinity dependence of elution profiles of interacting macromolecules. The frontmost elution profile is that obtained for a hypothetical antibody, eluting at position 100 (arbitrary time units) and antiidiotopic Fab eluting at position 150. Other contours from front to rear are those generated as affinity of interaction is systematically increased from $5 \times 10^3\ M^{-1}$ to $1 \times 10^8\ M^{-1}$ at equimolar initial reactant concentrations of $4 \times 10^{-5}\ M$. (B) Delta profile representation of elution profiles depicted in panel A. Each contour is generated by subtraction of the "zero-interaction" reference profile from the respective affinity-dependent profile. This representation highlights low-affinity interactions.

Rationalization of the differences between the chromatographic observations and those previously obtained by the solid-phase methods was aided by the concentration-dependence of the hapten-competition data presented in the careful study by Wittner et al.[21] As noted above, T15 was immobilized on the solid phase; the ability of six antiidiotopes at varying concentration in the presence of fixed concentration of excess hapten to bind to T15 was examined. With two of the antiidiotopes, including 4C11, dilution of antibody in the presence of phosphorylcholine resulted in complete inhibition of interaction. In the case of the remaining four antiidiotopes, including F6, inhibition curves plateaued at 15 to 20%. The extent of hapten inhibition was essentially independent of antiidiotope concentration in a range of one to two orders of magnitude.

A simple computer simulation of the competition between solution-phase hapten and antiidiotopic antibody during interaction with immobilized idiotope was constructed. For the purposes of this simulation, binding of hapten or antiidiotope to immobilized component was evaluated on the basis of "binding potentials" represented as a product of the individual intrinsic affinity constant and the effective concentration of available idiotope. Appropriately chosen values of binding potentials were shown to qualitatively reproduce the plateau inhibi-

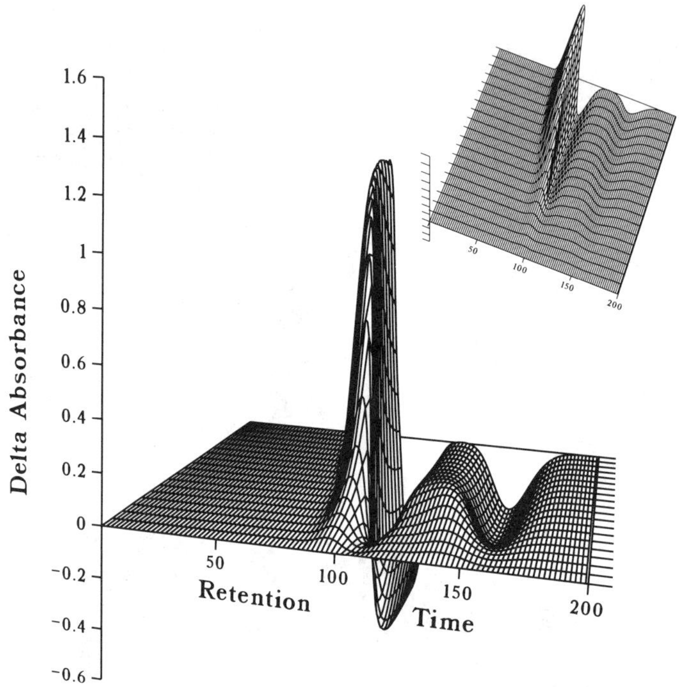

FIGURE 2B.

tion profiles observed by Wittner et al.[21] Because both idiotope (IgA T15) and the IgG antiidiotopes were bivalent molecules, the formation of a monogomous bivalent complex would be expected to lead to an effectively stable complex. Therefore, the effective binding potential was postulated to be a function of the ratio τ_1/τ_2, in which τ_1 represented the lifetime of the idiotope:antiidiotope bond and τ_2 represented a "rotational" period for exposure of an idiotope required for formation of the stabilizing second bond. This formulation suggested that the dispersity of observed solid-phase antiidiotope:hapten competition results could be explained on the basis of differences of instrinsic affinity, revealed by chromatography, as well as by the originally suggested site and near-site specificity properties. Kohler et al.[27] have recently incorporated considerations of antiidiotope affinity as an important factor in understanding the operation of an idiotype regulatory network.

It is not intended here to imply that one method of analyzing idiotope:antiidiotope interaction is correct and another is wrong. Although it can be argued that the observation of any inhibition of antiidiotope binding by hapten represents site specificity (or conformational change) rather than near-site specificity, the observation of inhibition plateaus by Wittner et al.[21] is a very interesting result. Regulation of the immune response by a network of idiotype and antiidiotypic antibodies as originally proposed by Jerne[28] involves interaction between solution-phase antibodies and receptor antibodies immobilized on lymphocyte surfaces. In this context, binding of high-affinity antiidiotopes to lymphocytes in a manner insensitive to antigen concentration would be expected to demonstrate functionally distinct regulatory characteristics compared to lower affinity antiidiotopes whose binding is dependent upon the concentration of antigen. Moreover, in a physiological system, regulation would be effected by a polyclonal mixture of antiidiotypic antibodies. As discussed elsewhere,[29] the characteristics of the polyclonal mixture is determined by the relative contributions of the constituent monoclonal antiidiotopes each having idiosyncratic specificity, affinity, kinetic, and stoichiometric properties. A combination of solid- and solution-phase assay methods may be necessary for comprehensive analysis of these molecules.

IV. CONTRIBUTION OF INTERFACIAL COOPERATIVITY TO POSSIBLE FALLACIES OF APPARENT PEPTIDE-BINDING SPECIFICITY

Synthetic peptides comprising amino acid sequences of known or potential epitopes have become important tools for the analysis of antigenic determinants and as reagents for the development of synthetic vaccines. In many of the assays for evaluation of the specificity of interaction between peptide and antibody, the peptide is constrained to the surface of a solid-phase matrix. In some cases this is effected by adsorption; in other situations, the peptide remains covalently attached to the solid-phase matrix upon which it was synthesized. In either event, the close proximity of the matrix may significantly alter the nature of the association event.

In general, a short peptide segment does not exhibit a single well-defined conformation in solution but rather exists as an equilibrated mixture of isomerizing conformers,[30,31] few of which necessarily resemble the conformation of the cognate segment as expressed in the context of the original intact antigen. When adsorbed to the solid-phase matrix, the conformation, or spectrum of conformations, attained by the peptide is determined by free energy minimization processes that include contributions from the interaction of peptide and matrix. Moreover, because of the small size of the typical peptide it is highly probable that at least some amino acid side chains, those in the intact antigen interacting with the antibody binding site, are now unavailable due to contact with the solid phase. Thus, the correlation between binding of antibody to adsorbed peptide and binding of antibody to the corresponding epitope in the native protein may be fortuitous.

Nevertheless, numerous reports have documented the use of synthetic peptides as immunogens in which the antibodies evoked apparently bind the parent, native antigen as well as the peptide when immobilized on a solid phase. Jemmerson[32] examined the ability of a large number of mAbs generated in a splenic focus assay for their ability to bind intact and fragmented cytochrome c bound to a solid phase. The ability of each mAb to bind native protein was tested by addition of free solution-phase cytochrome c to the reaction mixture. Inhibition by cytochrome c of the binding of antibodies to adsorbed intact protein was taken to indicate the ability of the mAb to recognize native antigen.

He found that most of the antibodies for which binding to the solid phase was inhibitable by soluble, presumably native, cytochrome c were not able to bind to the peptide fragments of cytochrome c. In contrast, antibodies for which binding to the solid phase was not inhibited by soluble cytochrome c and therefore presumably bound a denatured form of the adsorbed protein, typically were able to bind the peptide fragments. These studies with peptides suggest the possibility that polyclonal antiserum or mAbs which bind adsorbed native protein form complexes reflecting the interaction of antibody with denatured portions of the polypeptide rather than with epitopes relevant to the antigenic determinants of intact solution-phase proteins.

Another physical chemical aspect of antibody interaction with peptide adsorbed to a solid-phase matrix should also be considered. In order to bind to adsorbed peptide, the antibody Fab must itself come into close proximity to the solid-phase surface which may in turn lead to exclusion of water and thus contribute additional free energy to the interaction.

Upon binding to a solid-phase adsorbed peptide, an Fab surface area of 500 to 1000 Å^2, involving paratope and adjacent segments, may be brought into contact with the solid phase. Exclusion of water from this contact region may contribute 25 cal/mol Å^2 additional decrease in free energy.[3] The accumulated supplementary free energy change is, thus, comparable to that of many interactions between oligomeric protein subunits. As a result, a weak intrinsic interaction between peptide and paratope may be converted into a strong interaction in a ternary complex of peptide, Fab, and solid-phase matrix. Further stabilization of interaction may be realized through the ability of bivalent antibody to form a second bond or through additional contact between other antibody surfaces and that of the solid-phase matrix. Thus, physiologically inconsequential affinities may mimic significant binding capability to peptide.

V. IMPLICATIONS OF INTERFACIAL PHYSICAL CHEMISTRY FOR MONOCLONAL ANTIBODY TECHNOLOGY

Fundamental differences in solid- and solution-phase reaction performance may result not only from alterations evoked directly in the presentation of the immobilized component, but in the heterogeneity of the solution-phase reactant. Consider, for instance, immobilized IgG randomly adsorbed to a solid phase matrix in a manner that maximizes contact. To a first approximation, access to roughly 50% of the epitopes on the bilaterally symmetrical molecule might be resticted by steric influences and other epitopes may be lost due to denaturation. A polyclonal antisera may contain antibodies that compete with each other for binding, either by recognizing overlapping epitopes or by steric interference resulting from recognition of neighboring epitopes. The significance of steric interference might be enhanced when the epitope presenting antigen is immobilized, both because of restricted access and because of the smaller number of available epitopes. Therefore, when a polyclonal antisera is used to quantify a fixed number of IgG molecules in free solution or adsorbed to a solid phase, a significant concentration-dependent attenuation of the reaction, in the latter case, may be anticipated. These factors may have played a role in the observations of Dierks et al.[33] in which, in some cases, more than 95% of binding observed in a solution-phase immunoprecipitation assay was lost when the IgG was immobilized directly to a solid phase. In contrast, a homogeneous mAb reagent is intrinsically less vulnerable to effects that originate from competition between antibodies recognizing overlapping epitopes that may be present in a polyclonal antiserum. Moreover, mAbs have typically been prescreened for their ability to recognize solid-phase-presented epitopes.

The prevalent contemporary method for selecting hybridoma-produced mAbs that recognize protein or peptide antigens involves testing hybridoma cell culture supernatants for the presence of antibody that binds to solid-phase adsorbed antigen. Implicit in the results of Jemmerson,[32] and as explicitly discussed by Vaidya et al.,[34] denaturation of protein at the interface can be expected to lead to selection of antibodies that are not suitable for analysis of antigen in free solution (see Chapter 11). This complication might be expected to interfere with selection of optimal mAbs for use in immunodiagnostic assays or in immunotherapy applications. Modified selection/analysis methods have been described.[35,36]

In some cases the use of the term "denaturation" may be misleading as it tends to imply obliteration of the global structure. Apparent denaturation might arise from more subtle origins. For instance, antigenic proteins may be present in solution in multiple isomeric conformers. If these conformers are associated with significantly different free energies of association with the matrix, then it would be expected that adsorption to a solid-phase matrix preferentially selects one of these conformers. Functional denaturation in this case reflects the appearance of an alternative native structure; mAbs will then be selected for which interaction with antigen alters the isomeric distribution of antigen in solution. If the antigen is an enzyme that exhibits allosteric conformers that are catalytically active or inactive, then interaction with appropriate mAbs will appear to either activate or inactivate the enzyme.

VI. CONCLUSIONS

SPIAs are a powerful tool of immunochemistry. Although some caution may be advised in the application of these methods, the considerations presented above are also intended to suggest that much may be learned from further detailed theoretical analysis of the complex multiple interactions involved in these assays. As recently noted by Nygren and Stenberg,[19] many physiologically relevant processes involving antibodies are effectively solid-phase interactions. These processes include binding to epitopes expressed on viral particles and bacterial cell surfaces or interactions with lymphocytes. Therefore, solid-phase immunochemical assays may represent model systems that can be studied to provide insight into the macromolecular physics involved in the function and regulation of the immune system.

The physical chemistry associated with the interaction between an adsorbed ligand and a solid-phase matrix generates mechanisms that can obscure detailed interpretation of the specificity of interaction with antibody or receptor. In some cases, these mechanisms can lead to substantial enhancement of intrinsically weak or nonrelevant interactions. Because the proportional contribution of the solid-phase component is inversely related to the size of the antigen under study, it may be suggested that work with antigenic peptides, such as that involving the use of these synthetic reagents to develop vaccines for protection against toxins and viruses, may be particularly vulnerable to this influence. Appropriate confirmatory assays involving inhibition of binding by solution-phase antigen or direct demonstration of homogeneous solution-phase binding by chromatographic methods may minimize the consequences of artificially enhanced solid-phase binding.

ACKNOWLEDGMENT

This work was supported by the U.S. Department of Energy, Office of Health and Environmental Research under Contract No., W-31-109-ENG-38.

REFERENCES

1. **Tanford, C.,** Interfacial free energy and the hydrophobic effect, *Proc. Natl. Acad. Sci. U.S.A.*, 76, 4175, 1979.
2. **Miller, S., Lesk, A. M., Janin, J., and Chothia, C.,** The accessible surface area and stability of oligomeric proteins, *Nature*, 328, 834, 1987.
3. **Janin, J., Miller, S., and Chothia, C.,** Surface, subunit interfaces and interior of oliogmeric proteins, *J. Mol. Biol.*, 204, 155, 1988.
4. **Kellis, J. T., Nyberg, K., Sali, D., and Fersht, A. R.,** Contribution of hydrophobic interactions to protein stability, *Nature*, 333, 784, 1988.
5. **Michaeli, I., Absolom, D. R., and Van Oss, C. J.,** Diffusion of adsorbed protein within the plane of adsorption, *J. Colloid Interface Sci.*, 77, 586, 1980.
6. **Andrade, J. D.,** Principles of protein adsorption, in *Surface and Interfacial Aspects of Biomedical Polymers*, Vol. 2, Andrade, J. D., Ed., Plenum Press, New York, 1985, 1.
7. **Ansari, A. A., Hattikudur, N. S., Joshi, S. R., and Medeira, M. A.,** ELISA solid phase: stability and binding characteristics, *J. Immunol. Methods*, 834, 117, 1985.
8. **Soderquist, M. E. and Walton, A. G.,** Structural changes in proteins adsorbed on polymer surfaces, *J. Colloid Interface Sci.*, 75, 386, 1980.
9. **Kennel, S. J.,** Binding of monoclonal antibody to protein antigen in fluid phase or bound to solid supports, *J. Immunol. Methods*, 55, 1, 1982.
10. **Friguet, B., Djavadi-Ohaniance, L., and Goldberg, M. E.,** Some monoclonal antibodies raised with a native protein bind preferentially to the denatured antigen, *Mol. Immunol.*, 21, 673, 1984.
11. **Hollander, Z. and Katchalski-Katzir, E.,** Use of monoclonal antibodies to detect conformational alterations in lactate dehydrogenase isoenzyme 5 on heat denaturation and on adsorption to polystyrene plates, *Mol. Immunol.*, 23, 927, 1986.
12. **Darst, S. A., Robertson, C. R., and Berzofsky, J. A.,** Adsorption of the protein antigen myoglobin affects the binding of conformation-specific monoclonal antibodies, *Biophys. J.*, 53, 533, 1988.
13. **DeLisi, C.,** The effect of cell size and receptor density on ligand-receptor reaction rate constants, *Mol. Immunol.*, 18, 507, 1981.
14. **Mason, D. W. and Williams, A. F.,** The kinetics of antibody binding to membrane antigens in solution and at the cell surface, *Biochem. J.*, 187, 1, 1980.

15. **Wiegel, F. W. and DeLisi, C.,** Evaluation of reaction rate enhancement by reduction in dimensionality, *Am. J. Physiol.,* 243, R475, 1982.
16. **Mason, D. W. and Williams, A. F.,** Kinetics of antibody reactions and the analysis of cell surface antigens, in *Handbook of Experimental Immunology,* Vol. 1, Weir, D. W., Herzenberg, L. A., Blackwell, C., and Herzenberg, L. A., Eds., Blackwell Scientific, Oxford, 1986, 38.1.
17. **Stenberg, M., Werthen, M., Theander, S., and Nygren, H.,** A diffusion limited reaction theory for a microtiter plate assay, *J. Immunol. Methods,* 112, 23, 1988.
18. **Stenberg, M. and Nygren, H.,** Kinetics of antigen-antibody reactions at solid-liquid interfaces, *J. Immunol. Methods,* 113, 3, 1988.
19. **Nygren, H. and Stenberg, M.,** Immunochemistry at interfaces, *Immunology,* 66, 321, 1989.
20. **Stevens, F. J., Jwo, J., Carperos, W., Kohler, H., and Schiffer, M.,** Relationships between liquid- and solid-phase antibody association characteristics: implications for the use of competitive ELISA techniques to map the spatial location of idiotopes, *J. Immunol.,* 137, 1937, 1986.
21. **Wittner, M. K., Bach, M. A., and Kohler, H.,** Immune response to phosphorylcholine. IX. Characterization of hybridoma anti-TEPC15 antibodies, *J. Immunol.,* 128, 595, 1982.
22. **Stevens, F. J.,** Analysis of protein-protein interaction by simulation of small-zone size-exclusion chromatography: application to an antibody-antigen association, *Biochemistry,* 25, 981, 1986.
23. **Stevens, F. J., LeBuis, D. A., Eisler, W. J., and Ainsworth, C. F.,** Macromolecular interactions: application of microcomputer-controlled high speed size-exclusion chromatography, *LCGC,* 4, 340, 1986.
24. **Stevens, F. J.,** Applications of size-exclusion HPLC in the analysis of protein and peptide epitopes, *Methods Enzymol.,* 178, 107, 1989.
25. **Stevens, F. J., Carperos, W. E., Monafo, W. J., and Greenspan, N. S.,** Size-exclusion HPLC analysis of epitopes, *J. Immunol. Methods,* 108, 271, 1988.
26. **Stevens, F. J.,** Analysis of protein-protein interaction by simulation of small-zone size-exclusion chromatography: stochastic formulation of kinetic rate contributions to observed HPLC elution characteristics, *Biophys. J.,* 155, 1155, 1989.
27. **Kohler, H., Kaveri, S., Kieber-Emmons, T., Morrow, W. J. W., Muller, S., and Raychaudhuri, S.,** Idiotypic networks and nature of molecular mimicry: an overview, *Methods Enzymol.,* 178, 3, 1989.
28. **Jerne, N.,** Towards a network theory of the immune system, *Ann. Immunol. Inst. Pasteur,* 125C, 373, 1974.
29. **Stevens, F. J. and Schiffer, M.,** Immunoglobulin structural diversities and idiotope expression, *Int. Rev. Immunol.,* 2, 357, 1987.
30. **Sachs, D. H., Schecter, A. N., Eastlake, A., and Anfinsen, C. B.,** An immunologic approach to the conformational equilibria of polypeptides, *Proc. Natl. Acad. Sci. U.S.A.,* 69, 3790, 1972.
31. **Crumpton, M. H.,** The importance of conformation and of equilibria in the interaction of globular proteins and their fragments with antibodies, in *Synthetic Peptides as Antigens,* Porter, R. and Whelan, J., Eds., John Wiley & Sons, Chichester, England, 1986, 93.
32. **Jemmerson, R.,** Antigenicity and native structure of globular proteins: low frequency of peptide reactive antibodies, *Proc. Natl. Acad. Sci. U.S.A.,* 84, 9180, 1987.
33. **Dierks, S. E., Butler, J.E., and Richerson, H. B.,** Altered recognition of surface-adsorbed compared to antigen-bound antibodies in the ELISA, *Mol. Immunol.,* 23, 403, 1986.
34. **Vaidya, H. C., Dietzler, D. N., and Ladenson, J. H.,** Inadequacy of traditional ELISA for screening hybridoma supernatants for murine monoclonal antibodies, *Hybridoma,* 4, 271, 1985.
35. **Smith, A. D. and Wilson, J. E.,** A modified ELISA that selectively detects monoclonal antibodies recognizing native antigen, *J. Immunol. Methods,* 94, 31, 1986.
36. **Kruger, U., Wichert, L., and Wagener, C.,** Determination of epitope specificities and affinities of monoclonal antibodies in solution phase using biotin-labeled carcinoembryonic antigen and avidin as precipitating agent, *J. Immunol. Methods,* 117, 25, 1989.

Chapter 14

MONOSIZED POLYMER PARTICLES IN IMMUNOASSAYS: APPLICATIONS AND IMMUNOCHEMISTRY

Kjell Nustad, Elisabeth Paus, and Ole Børmer

TABLE OF CONTENTS

I. INTRODUCTION

This chapter summarizes the application of a special type of polymer particles in immunoassays for the measurement of soluble and particle-bound antigens. The advantages and limitations of both magnetizable and nonmagnetic particles are stressed. Before proceeding further, one might ask "Why use particles as the solid phase in immunoassays?" While the complete answer is rather complex, it suffices here to say that particles permit solid-phase assays to be conducted more rapidly. A comparison of particle-bound antibodies vs. an equal amount of antibody on the wall of microtiter wells demonstrated that the particles captured all the antigen in 5 to 10 min, whereas a microtiter well required 24 h.[1] Other advantages of using particles in immunoassays will be discussed later.

II. PREPARATION OF MONOSIZED POLYMER PARTICLES

A detailed description of the preparation and properties of these particles is outside the scope of this chapter; a short summary is provided. Uniform latex particles were a serendipitous product of paint manufacture at Dow Chemical Co. in 1947.[2] Since then, uniform latex particles with diameters ranging from 0.05 to 1 μm have been regularly produced. Their principal application in immunology was in latex agglutination assays, a topic which will not be covered in this chapter.[3]

Monosized nonmagnetic polymer particles with diameter larger than 1 μm can now be produced by the activated swelling method developed by Ugelstad and co-workers.[4] These particles have the following properties:

1. Extremely uniform size; the coefficient of variation (CV) of their diameter is 1 to 3% as measured by flow cytometry.
2. Surface chemistry can be prepared according to requirements because production processes permit a wide variety of polymeric materials to be used.
3. Surface topography can either be smooth or macroporous so that surface area and particle density can be controlled.
4. High mechanical strength and resistance to organic solvents can be achieved through extensive cross linking.

New possibilities were opened for immunoassays when Ugelstad and associates found a method to produce magnetizable particles with the same features as those listed for the nonmagnetic particles.[5] Iron is precipitated as fine grains throughout the interior of the monosized porous particles. Thus, truly superparamagnetic particles are made which show no remnant magnetism, i.e., they can be readily dispersed after exposure to a magnetic field.[5]

III. APPLICATION OF MONOSIZED POLYMER PARTICLES

A. MONOSIZED PARTICLES IN FLOW CYTOMETRY FOR SIMULTANEOUS MEASUREMENT OF MULTIPLE ANALYTES

When a series of particles of different sizes, each size coated with an antibody of unique specificity, are permitted to interact with a complex antigen mixture, a multianalyte immunoassay is configured. Each particle-bound antibody is allowed to capture its specific antigen from the mixture, while a fluorescent-labeled antibody is used for detection. A flow cytometer is used to measure particle size and amount of fluorescence on each particle. The particle size serves to identify the analyte while the fluorescence intensity measures the amount of the analyte bound to the particle.[6-12] This assay demands that the particles used are monodisperse. It is the only assay format which allows one to simultaneously measure a number of different analytes. Alternatively, multiple forms of a single analyte can be measured.

The assay range of each analyte can be enhanced using antibodies of different affinity to the same epitope on different particles. The high-affinity antibody will give good sensitivity in the lower range of the standard curve, and the low-affinity antibody on a larger particle size will extend the assay range at high antigen concentrations.[13] Such assays can be fully automated, they need no washing step, and will have a throughput of at least one sample per minute measuring as much as ten parameters simultaneously. A very important advantage is that we can include a measurement of interfering substances present in individual serum samples (see Section IV.B). The automated multianalyte system described represents a goal for large-scale assays of soluble analytes.

B. MAGNETIZABLE PARTICLES AS THE SOLID PHASE FOR IMMUNOASSAYS

1. Application to Particulate Antigens

Magnetizable particles used in conjunction with monoclonal antibodies offer a unique tool for rapid isolation of antigens present on the surface of cells, bacteria, viruses, and membranes. Special properties of the isolate can then be analyzed.

The advantages of immunomagnetic isolation of the cells are (1) pure cells are isolated in 10 to 15 min compared with hours using centrifugation techniques, (2) the isolated cells will be mostly living cells due to the rapid isolation and the ability to choose antigens present on living cells only, and (3) the process is a batch procedure which can easily be scaled up if large quantities of cells are needed. This is in contrast to cell sorting by flow cytometry which is expensive and can handle only a limited amount of cells.

The overall impact of immunomagnetic cell separation is increased speed, precision, and quality in cell analysis. The prime example of this kind of immunomagnetic cell analysis is HLA tissue typing.[14] The analysis of the ratio between T4- and T8-lymphocytes using immunomagnetic isolation is also a fine example of a specific assay which can substitute for assays using fluorescent-activated flow cytometry.[15]

Bacteria have been similarly isolated using magnetizable particles coated with the appropriate antibodies.[16] The method permits the rapid isolation and concentration of a specific strain or type of bacteria from complex microbial cultures or from media components which might inhibit bacterial growth. The isolated species can then be further cultivated and, for example, their antibiotic resistance determined. Furthermore, as defined monoclonal antibodies for bacterial surface antigens become more readily available, these reagents should permit direct immunological identification of the isolated bacteria on the magnetizable particles.

Viruses and cellular membranes might also be isolated by immunomagnetic procedures. It might be important to use a special free-flow washing device which will not damage these fragile membranes.[17]

2. Application to Soluble Antigens

Although magnetizable particles are ideal for the isolation of particulate antigens such as cells and bacteria, modifications are necessary to successfully apply them in immunoassays of soluble antigens. Size, density, and surface modifications were necessary to obtain a useful particle for this type of immunoassay.[18,19] Interestingly, these modified particles were also very well suited for use with cells and DNA probes.[20,21]

At present, a convenient, automated device for the washing of magnetizable particles is not on the market. Hence, manual magnetic racks devised for certain commercial kits have been adopted. These work well only when relatively small-scale assays are performed. We are presently experimenting with the use of magnetizable particles in microtiter wells, including the modification of microtiter washers, so that magnetizable particles can also be used in existing systems.

IV. IMMUNOCHEMICAL PHENOMENA ON POLYMER PARTICLE SURFACES

Immunochemical observations made in our laboratory regarding polymer particles are most likely also valid for other solid phase systems such as plastic test tubes and microtiter wells.

A. IMMOBILIZATION OF ANTIBODIES ON PARTICLES

The binding of antibodies to the solid phase is an important step in the development of an assay. The goal is to firmly immobilize the antibody while at the same time, preserving its antigen-binding properties.

1. Physical Adsorption of Antibodies

The physical adsorption of antibodies to polymer particles results in solid-phase antibodies with excellent antigen-binding properties. The explanation for this phenomenon is possibly that the process favors the hydrophobic interaction between the Fc region of IgG antibodies and the polymer surface. This phenomenon is even more pronounced if the antibody has been partially denatured before adsorption.[22-26] Such treatment is believed to preferentially expose hydrophobic areas in the Fc region of the antibody.[24]

2. Chemical Immobilization of Antibodies

Hydrophobic binding can be combined with the establishment of a chemical linkage using sulfonyl chloride-activated particles.[27-29] This seemingly unlikely combination occurs because the activated particle is hydrophobic and the hydrophobic interaction between antibody and particle is much faster than the establishment of the chemical bonds. Activation of the particles with carbonyldiimidazole results in a more rapid antibody immobilization, yielding significantly lower antigen-binding properties.[27] Increased antigen-binding capacity was found for partially denatured antibodies immobilized on glutaraldehyde-activated silica. The phenomenon was explained by higher density and better orientation of the antibodies on the solid phase.[26]

B. ASSAY INTERFERENCE AT THE POLYMER SURFACE

1. Cross-Linking Antibodies

It is well known that patient sera may contain antibodies to animal proteins.[30] These antibodies often cross react with mouse immunoglobulins. The net effect is a linkage of the labeled mouse antibody with the solid-phase mouse antibody in the absence of antigen, i.e., a short circuitry of the immunoassay. The incidence of this phenomenon may be as high as 40% in a healthy population.[31] The addition of normal mouse, rat, and bovine immunoglobulins will in most instances neutralize this effect. However, it is certainly a problem to determine the correct amounts and mixtures of immunoglobulins that are necessary to block these unwanted reactions. Most commonly this will be done by performing the actual assay on a panel of patient's sera in the presence and absence of blocking immunoglobulins. However, the results become ambiguous as the actual level of true antigen in each serum sample varies. An assay which utilizes the same antibody as capture and labeled antibody, i.e., symmetrical sandwich, provides a solution to this problem when the antigen contains only one epitope for that antibody.[31,32]

2. Complement Binding and Activation on the Solid Phase

In the process of developing an assay for CEA, increased CEA concentrations were found when sera was reanalyzed after storage.[33] The explanation was that complement in fresh serum interfered with the measurement of CEA by binding to the solid-phase capture antibody and thereby reducing its ability to bind CEA. The effect was dependent on the subclass of mouse IgG used as capture antibody. Mouse IgG2a and IgG2b can activate complement by the classical pathway, whereas IgG1 antibodies functions weakly via the alternative pathway. The activation of complement by mouse IgG2a/b can be so marked that assay performance suffers. This was observed on particles as well as on antibody-coated microtiter wells.

Surprisingly, this phenomenon has been hardly recognized as an analytical problem by others. Several investigators have utilized the phenomenon to measure complement factors.[34,35] Interference was noted in particle counting immunoassays using polyclonal antibodies as well as interference in assays for immune complexes.[36-38] A recent report noted assay interference even when mouse IgG1 antibody was used in an immunometric thyroid-stimulating hormone (TSH) assay.[39]

3. Other Interfering Reactions

Interference in solid-phase-based immunoassays by other substances which can bind to polymer surfaces have not yet been documented, but the potential for such phenomena exists. Possibilities include (1) activation of Factor XII by the negatively charged surfaces and (2) activation of plasminogen by immobilized IgG.[40]

Autoantibodies to the analyte are a recognized problem in the measurement of thyroglobulin.[41] Similar interference may occur in several assay systems. The net effect is that the analyte may be altered so its immunoreactivity differs from the standard. The same phenomenon occurs when the analyte complexes with inhibitors in serum as in the case of prostate specific antigen and kallikrein.[42,43] Interestingly, the latter can be overcome by utilizing a two-stage capture-antibody system.[43]

Under certain circumstances desorption of solid-phase antibodies can occur through the action of other substances present in the sample.[44]

The examples provided serve as a reminder for the need to be mindful of the potential for interference in solid-phase based immunoassays and provide an incentive to prepare polymer surface and assay designs which will be least affected by interfering factors.[45,46]

V. WHY USE POLYMER PARTICLES AS THE SOLID PHASE?

Polymer particles as the solid phase for immunoassays offer numerous advantages.

1. The speed of the assay compared with microtiter based assay increases considerably as mentioned in the introduction.
2. Batchwise preparation of solid phase antibody allows easy monitoring and quality control.
3. Freedom to modify surface characteristics in order to reduce interference and allow antibody to be immobilized with a variety of methods.
4. No limitation in the amount of solid-phase antibody which can be used in an assay and no antibody variation between samples.
5. Magnetizable particles are applicable to rapid isolation of particulate antigens.
6. The combined use of defined particle-size mixtures, each coated with its own uniquely specific capture antibody and flow cytometry allows development of multianalyte assays. These also allow for the inclusion of an interference assay in the package.
7. No limitation to sample size.

Monosized particles also have major drawbacks: particles are not readily adapted to automated washing procedures. Furthermore, they tend to sediment during the incubation period, a property which necessitates continuous shaking during the incubation period.

REFERENCES

1. **Millán, J. L., Nustad, K., and Nørgaard-Pedersen, B.,** Highly sensitive solid-phase immunoenzymometric assay for placental and placental-like alkaline phosphatases with a monoclonal antibody and monodisperse polymer particles, *Clin. Chem.,* 31, 54, 1985.
2. **Bangs, L. B.,** *Uniform Latex Particles,* Seragen Diagnostics, Indianapolis, IN, 1985, 3.
3. **Singer, J. M.,** The use polystyrene latex in medicine, in *Future Directions in Polymer Colloids,* El-Asser, M.S. and Fitch, R.M., Eds., NATO ASI Series E, Applied Sciences No. 138, Martinius Nijhoff, Amsterdam, 1987, 371.
4. **Ugelstad, J., Mørk, P. C., Kaggerud, K. H., Ellingsen, T., and Berge, A.,** Swelling of oligomer-polymer particles. New methods of preparation of emulsions and polymer dispersions, *Adv. Colloid Interface Sci.,* 13, 101, 1980.
5. **Ugelstad, J., Berge, A., Ellingsen, T., Aune, O., Kilaas, L., Nilsen, T.N., Schmid, R., Stenstad, P., Funderud, S., Kvalheim, G., Nustad, K., Lea, T., Vartdal, F., and Danielsen, H.,** Monosized magnetic particles and their use in selective cell separation, *Makromol. Chem. Macromol. Symp.,* 17, 177, 1988.
6. **Lisi, P. J., Huang, C. W., Hoffmann, R. A., and Teipel, J. W.,** A fluorescence immunoassay for soluble antigens employing flow cytometric detection, *Clin. Chim. Acta,* 120, 171, 1982.
7. **Saunders, G. C., Jett, J. H., and Martin, J. C.,** Amplified flow-cytometric separation-free fluorescence immunoassays, *Clin. Chem.,* 31, 2020, 1985.
8. **McHugh, T. M., Stites, D .P., Casavant, C. H., and Fulwyler, M. J.,** Flow cytometric detection and quantitation of immune complexes using human Clq-coated microspheres, *J. Immunol. Methods,* 95, 57, 1986.
9. **McHugh, T. M., Miner, R. C., Logan, L. H., and Stites, D. P.,** Simultaneous detection of antibodies to cytomegalovirus and herpes simplex virus by using flow cytometry and a microsphere-based fluorescence immuno assay, *J. Clin. Microbiol.,* 26, 1957, 1988.
10. **Wilson, M.R. and Wotherspoon, J. S.,** A new microsphere-based immunofluorescence assay using flow cytometry, *J. Immunol. Methods,* 107, 225, 1988.
11. **Wilson, M. R., Mulligan, S. P., and Raison, R. L.,** A new microsphere-based immunofluorescence assay for antibodies to membrane-associated antigens, *J. Immunol. Methods,* 107, 231, 1988.
12. **McHugh, T. M., Wang, Y. J., Chong, H. O., Blackwood, L. L., and Stites, D. P.,** Development of a microsphere-based fluorescent immunoassay and its comparison to an enzyme immunassay for the detection of antibodies to three antigen preparations from *Candida albicans, J. Immunol. Methods,* 116, 213, 1989.
13. **Lindmo, T., Børmer, O., Ugelstad, J., and Nustad, K.,** Immunometric assay by flow cytometry using mixtures of two particle types of different affinity, *J. Immunol. Methods,* 126, 183, 1990.
14. **Vartdal, F., Gaudernack, G., Funderud, S., Bratlie, A., Lea, T., Ugelstad, J., and Thorsby, E.,** HLA Class I and II typing using cells positively selected from blood by immunomagnetic isolation — a fast and reliable technique, *Tissue Antigens,* 28, 301, 1986.
15. **Brinchmann, J. E., Vartdal, F., Gaudernach, G., Markussen, G., Funderud, S., Ugelstad, J., and Thorsby, E.,** Direct immunomagnetic quantification of lymphocyte subsets in blood, *Clin. Exp. Immunol.,* 71, 182, 1988.
16. **Lund, A. Hellmann, A. L., and Vartdal, F.,** Rapid isolation of K88$^+$ *Escherichia coli* by using immunomagnetic particles, *J. Clin. Microbiol.,* 26, 2572, 1988.
17. **Howell, K. E., Ansorge, W., and Gruenberg, J.,** Magnetic free-flow immunoisolation system designed for subcellular fractionation, in *Microspheres: Medical and Biological Applications,* Rembaum, A. and Tökés, Z. A., Eds., CRC Press, Boca Raton, FL, 1988, 33.
18. **Börmer, O. P. and Nustad, K.,** Selection of monoclonal antibodies for use in an immunometric assay for carcinoembryonic antigen, *J. Immunol. Methods,* 127, 171, 1990.
19. **Paus, E. and Nustad, K.,** Immunoradiometric assay for $\alpha\gamma$- and $\gamma\gamma$-enolase (neron-specific enolase), with use of monoclonal antibodies and magnetizable polymer particles, *Clin. Chem.,* 35, 2034, 1989.
20. **Kvalheim, G., Fjeld, J. G., Pihl, A., Funderud, S., Ugelstad, J., Fodstad, Ø., and Nustad, K.,** Immunomagnetic removal of B-lymphoma cells using a novel mono-sized magnetizable polymer bead, M-280, in conjunction with primary IgM and IgG antibodies, *Bone Marrow Transplantation,* 4, 567, 1989.
21. **Uhlen, M.,** Magnetic separation of DNA, *Nature,* 340, 733, 1989.
22. **Ishikawa, E., Hamaguschi, Y., Imagawa, M., Inada, M., Imura, H., Nakasawa, N., and Ogawa, H.,** An improved preparation of antibody-coated polystyrene beads for sandwich enzyme immunoassay, *J. Immunoassay,* 1, 385, 1980.
23. **Vandenbranden, M., Coen, J. L. De, Jeener, R., Kanarek, L., and Ruyschaert, J. M.,** Interaction of γ-immunoglobulins with lipid mono- or bilayers and liposomes. Existence of two conformations of γ-immunoglobulins of different hydrophobicities, *Mol. Immunol.,* 18, 621, 1981.
24. **Vanderbranden, M., Kayser, G., Banerjee, S., and Ruysschaert, J. M.,** Immunoglobulin-lipid interaction; a model membrane study, *Biochim. Biophys. Acta,* 685, 177, 1982.
25. **Conradie, J. D., Govender, M., and Visser, L.,** ELISA solid phase: partial denaturation of coating antibody yields a more efficient solid phase, *J. Immunol. Methods,* 59, 289, 1983.

26. **Lin, J. N., Andrade, J. D., and Chang, I.-N.,** The influence of adsorption of native and modified antibodies on their activity, *J. Immunol. Methods,* 125, 67, 1989.
27. **Nustad, K., Ugelstad, J., Berge, A., Ellingsen, T., Schmid, R., Johansen, L., Børmer, O.,** Use of monosized polymer particles in immunoassays, in *Radioimmunoassay and Related Procedures in Medicine* 1982, International Atomic Energy Agency, Vienna, 1982, 45.
28. **Nustad, K., Johansen, L., Ugelstad, J., Ellingsen,T., and Berge, A.,** Hydrophilic monodisperse particles as solid-phase material in immunoassays: comparison of shell-and-core particles with compact particles, *Eur. Surg. Res.,* 16(Suppl. 2), 80, 1984.
29. **Nustad, K., Danielsen, H., Reith, A., Funderud, S., Lea, T., Vartdal, F., and Ugelstad, J.,** Monodisperse polymer particles in immunoassay and cell separation, in *Microspheres: Medical and Biological Applications,* Rembaum, A. and Tökés, Z. A., Eds., CRC Press, Boca Raton, FL, 1988, 53.
30. **Foucard, T., Bennich, H., Johansson, S. G. O., and Lundkvist, U.,** Human antibodies to bovine γ-globulin. Occurrence in immunological disorders and influence on allergy radio-immunoassays, *Int. Arch. Allergy Appl. Immunol.,* 48, 812, 1975.
31. **Boscato, L. M. and Stuart, M. C.,** Incidence and specificity of interference in two-site immunoassays, *Clin. Chem.,* 32, 1491, 1986.
32. **Butler, J. E., Spradling, J. E., Dierks, S. E., Heyermann, H., Peterman, J. H., and Suter, M.,** The immunochemistry of sandwich ELISA. I. The binding characteristics of immunoglobulins to monoclonal and polyclonal capture antibodies adsorbed on plastic and their detection by symmetrical and asymmetrical antibody-enzyme conjugates, *Mol. Immunol.,* 23, 971, 1986.
33. **Börmer, O. P.,** Interference of complement with the binding of carcinoembryonic antigen to solid-phase monoclonal antibodies, *J. Immunol. Methods,* 121, 85, 1989.
34. **Zwirner, J., Felber, E., Reiter, C., Riethmüller, G., and Feucht, H. E.,** Deposition of complement activation products on plastic-adsorbed immunoglobulins. A simple ELISA technique for the detection of defined complement deficiencies, *J. Immunol. Methods,* 124, 121, 1989.
35. **Larsson, A. and Sjöquist, J.,** Binding of complement components Clq, C3, C4 and C5 to a model immune complex in ELISA, *J. Immunol. Methods,* 119, 103, 1989.
36. **Masson, P. L., Cambiaso, C. L., Collet-Cassart, D., Magnusson, C. G. M., Richards, C. B., and Sindic, C. J. M.,** Particle counting immunoassay (PACIA), *Methods Enzymol.,* 74, 106, 1981.
37. **Schifferli, J. A., Woo, P., and Peters, D. K.,** Complement-mediated inhibition of immune precipitation. I. Role of the classical and alternative pathways, *Clin. Exp. Immunol.,* 47, 555, 1982.
38. **Baatrup, G., Svehag, S.-E., and Jensenius, J. C.,** The attachment of serum- and plasma-derived C3 to solid-phase immune aggregates and its relation to complement-mediated solubilization of immune complexes, *Scand. J. Immunol.,* 23, 397, 1986.
39. **Käpyaho, K., Tanner, P., and Weber, T.,** Effect of complement binding on a solid-phase immunometric TSH assay, *Scand. J. Clin. Lab. Invest.,* 49, 211, 1989.
40. **Harpel, P. C., Sullivan, R., and Chang, T.-S.,** Binding and activation of plasminogen on immobilized immunoglobulin G. Identification of the plasmin-derived Fab as the plasminogen-binding fragment, *J. Biol. Chem.,* 264, 616, 1989.
41. **Feldt-Rasmussen, U. and Schlumberger, M.,** European interlaboratory comparison of serum thyroglobulin measurement, *J. Endocrinol. Invest.,* 11, 175, 1988.
42. **Lilja, H.,** A kallikrein-like serine protease in prostatic fluid cleaves the predominant seminal vesicle protein, *J. Clin. Invest.,* 76, 1899, 1985.
43. **Johansen, L., Nustad, K., Berg, T., and Pierce, J. V.,** Excess antibody immunoassay for rat glandular kallikrein. Measurement of kallikrein complexed with inhibitors and in plasma, *J. Immunol. Methods,* 69, 253, 1984.
44. **Hanvanich, M., Viscidi, R., Laughon, B. E., Bartlett, J. G., and Yolken, R. H.,** Stoll desorbing activity: a possible cause of false-positive reactions in competitive enzyme immunoassay, *J. Clin. Microbiol.,* 21, 184, 1985.
45. **Graves, H. C. B.,** Noise control in solid-phase immunoassays by use of a matrix coat, *J. Immunol. Methods,* 111, 167, 1988.
46. **Graves, H. C. B.,** The effect of surface charge on non-specific binding of immunoglobulin G in solid-phase immunoassays, *J. Immunol. Methods,* 111, 157, 1988.

Chapter 15

ENZYME IMMUNOFILTRATION ASSAY FOR HUMAN CREATINE KINASE ISOENZYME MB

Fritz Poulsen and Ole J. Bjerrum

TABLE OF CONTENTS

I. INTRODUCTION

Enzyme immunofiltration assays (EIFA) utilize microporous membranes as the receptor-bearing solid phase and employ filtration as a means to hasten contact with the soluble sample ligand and the signal-generating reagents.

The diffusion limitation of the reaction rate seen for conventional solid-phase immunoassays[1] is minimized in EIFA.[2] This is due to the flow of reactants through the receptor-bearing membrane solid phase and the high ratio of microporous membrane surface area-to-liquid volume. Thus, EIFA permits rapid tests to be developed which reach completion in minutes.[2,3] Furthermore, the sensitivity of EIFA can easily be improved by simply filtering a larger sample volume through the membrane.[2] The antigen-antibody reactions in EIFA are visualized directly by immunostaining, in which the signal-generating conjugate yields colored spots at the reaction sites on the membrane.[3,4] The color intensity of these spots can be quantitated by reflectance photometry.[2,4]

In previous reports, various EIFA methods have been described for the detection of antigens by means of direct binding of sample to the membrane[4] or by employing two antibodies in a sandwich.[2,5] Detection of antibodies by permutations of this method has also been described.[3] EIFAs differ in the use of flow during the reactions. In the sandwich assay described by Valkirs and Barton, rapid flow followed by a short incubation period was used to give a total assay time of 5 min.[2] Quantitative assays based on EIFA have reproducibility and sensitivity comparable to that of other enzyme-linked immunosorbent assay (ELISA) techniques.[4,5]

The EIFA system can be incorporated in a unit which, besides the antibody-bearing solid phase, includes an absorbing material for drawing liquid through the membrane and a waste reservoir. Because of their convenience, simplicity, and speed EIFA devices can be used in technically unsophisticated patient environments, i.e., as near-patient tests. Commercially available EIFAs of this type include those from Hybritech, Inc. (Icon®), V-Tech, Inc. (Target™), and Abbott (Test Pack™).

This chapter describes a thoroughly optimized EIFA for the determination of human creatine kinase isoenzyme MB (CK-MB) in serum. The assay is based on the sandwich principle using two monoclonal antibodies. The capture antibody (CAb) immobilized on the microporous membrane binds to the B-subunit of the enzyme, while the signal-generating antibody-enzyme conjugate reacts with the M-subunit. The assay employs filtration in all assay steps.

II. MATERIALS AND METHODS

The solid phase, on which the CAb was immobilized, was a nylon-based, microporous membrane preactivated for covalent coupling of proteins (Immunodyne immunoaffinity membrane, Pall BioSupport Division, Portsmouth, England). This type of membrane was found to be superior to nitrocellulose for quantitative assays.[6] The pore size of the membrane used in the present study was 1.2 μm.

The membrane preparation and the assay was performed in a microfiltration apparatus, a vacuum manifold containing 96 filtration wells (Bio-Dot, Bio-Rad), connected to an adjustable vacuum source. The well diameter at the bottom of the wells was 3 mm.

The CAb-containing membrane was prepared by applying 3 μl of 3 mg/ml (in PBS) monoclonal anti-human CK-BB antibody (Novo Nordisk) to the membrane in a single well. After 5 min at room temperature, 200 μl of 0.2% (w/v) casein (Hammarsten, BDH Chemicals, Ltd.) in PBS pH 7.2 was added and the membrane incubated for 20 min at 37°C. Finally, 100 μl of PBS pH 7.2 containing 0.05% (v/v) Tween 20 was added under vacuum (–0.22 bar).

Standard assays were performed by applying vacuum during all steps; –0.13 bar was used for sample and reagent steps and –0.22 bar for washing steps. All steps were carried out at room temperature. Test samples were added to each well under vacuum followed by horseradish peroxidase conjugated monoclonal (Fab') anti-human CK-MM (Novo Nordisk) without an intermediate wash step. After an interval, the membranes in all wells were washed with 3 × 300 µl PBS pH 7.2 containing 0.05% (v/v) Tween 20 and 0.25 *M* $(NH_4)_2SO_4$. Substrate [0.006% (w/v) hydrogen peroxide and 65 mg/l tetramethylbenzidine (TMB)] was then applied. After 3 min the membrane was washed with 3 × 300 µl distilled water containing 0.01% (w/v) sodium azide per well.

Reflectance photometry was employed to quantitate the colored spots developed on the membrane. The principles of this technique have been described in detail by Werner and Rittersdorf.[7] The actual measurements were performed using a spectrophotometer (Shimadzu MPS-2000) equipped with a reflectance unit (Shimadzu Integrating Sphere Assembly IRS-240). This unit has been provided with specially designed holders and apertures rendering it suitable for the measuring of 3- to 4-mm circular spots on membranes. The reflectance was read at 652 nm against a white reflectance standard ($BaSO_4$, Wako Pure Chemical Industries Ltd.).

The recorded reflectance values were transformed according to the Kubelka-Munk relationship:

$$\frac{(1 - R_\infty)^2}{2R_\infty} = \frac{K}{S} = \frac{\varepsilon C}{S} \tag{1}$$

where $R\infty$ = the reflectance, K = the absorption coefficient, S = the scattering coefficient, ε = the extinction coefficient, and C = the concentration of the absorbing material.[7] If certain conditions are fulfilled, there is a linear relationship between the transformed reflectance and the concentration of the absorbing material.[7]

III. RESULTS AND DISCUSSION

A. CALIBRATION CURVE

Human CK-MB in bovine serum was used as model sample in our EIFA. A duplicate analysis of samples with increasing CK-MB concentration gave rise to spots on the membrane with correspondingly increasing color intensity and no background staining (Figure 1). The spots, having a diameter of 3 mm, appeared uniform and well defined.

Figure 2 shows a calibration curve, with K/S plotted against the CK-MB concentration. Each point on the curve represents the mean of four determinations. The curve covering the range of clinical interest[8] shows a linear relationship between K/S and concentration in the concentration range 0 to 50 ng/ml and a slight arch in response to CK-MB at 50 to 100 ng/ml. However, the assay is applicable for detection of CK-MB in serum at concentrations up to 100 ng/ml. The results also show the detection limit to be less than 10 ng/ml, i.e., less than 0.5 ng CK-MB.

B. CAPTURE ANTIBODY

The quantity of CAb immobilized on the membrane is important for the performance of the assay. The effect of added CAb concentration on assay performance is shown in Figure 3. When 25 ng/ml of CK-MB is used, the K/S increases nearly linearly up to 2 mg/ml of added CAb after which a plateau develops. Similar results were obtained using 10, 50, and 100 ng CK-MB per milliliter, while the background response was unaffected (data not shown). On the assumption that all applied antibody is bound, the CAb density at 3 mg/ml is about 45 µg/cm^2 (the area coated was larger than the area exposed at the bottom of the wells).

The susceptibility of the assay to variations in CAb concentration points to the importance of uniform coupling capacity of the membrane in order to avoid assay variations.

FIGURE 1. EIFA of human CK-MB calibrators in bovine serum. Fifty µl of CK-MB sample was added per well followed by 50 µl of conjugate (containing 5 µg conjugate + 14 µg unlabeled conjugate antibody/ml). At 1 min after addition of conjugate the membrane was washed; 50 µl of color development solution (H_2O_2 + TMB) was then added to each well, and 3 min later the reaction was stopped by washing. The calibrators (run in duplicate) were (from left to right) 0, 10, 25, 50, and 100 ng CK-MB per milliliter.

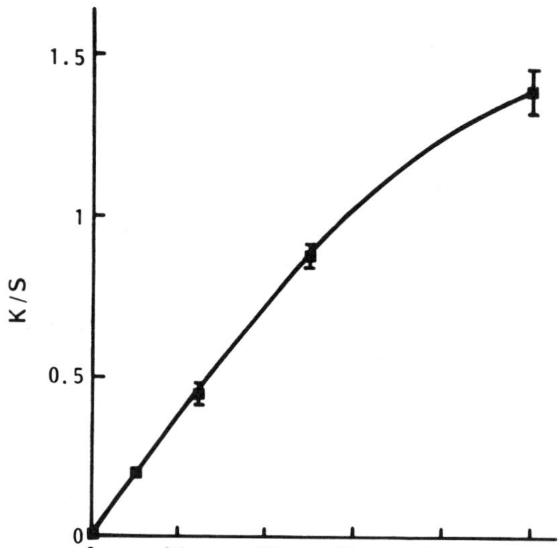

FIGURE 2. Calibration curve generated using human CK-MB in bovine serum. The reflectance of the spots was read at 652 nm and transformed to K/S. Each point represents the mean of four determinations. Vertical lines indicate SD; the SD at 0 and 10 ng/ml was too small to be indicated.

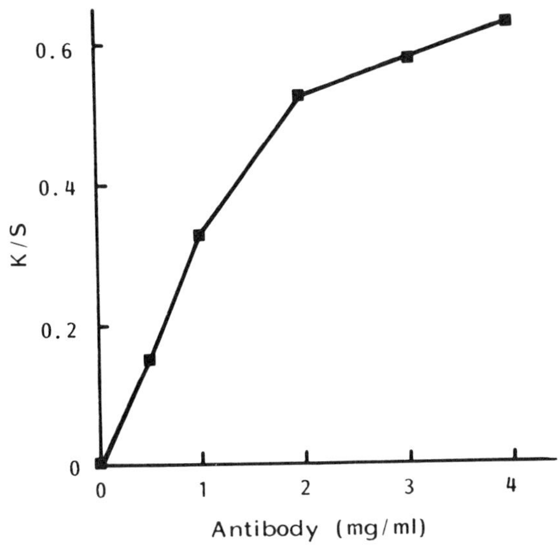

FIGURE 3. The effect of added CAb on the performance of the EIFA. A constant concentration of CK-MB (25 ng/ml) was used. Each point was the mean of duplicate determinations.

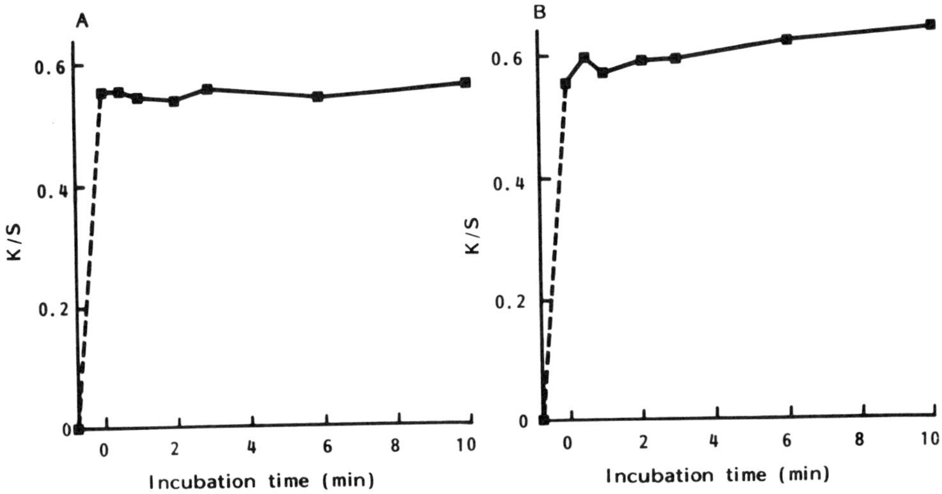

FIGURE 4. Time course of the antigen-antibody reactions. (A) The antigen was incubated for 0 to 10 min after filtration, before addition of the conjugate. (B) The conjugate was incubated 0 to 10 min after filtration before washing. The flow-through time of 45 s is indicated by connecting the points with a dotted line. The incubation time is measured from the point at which the liquid had passed the membrane (time 0) until addition of the next solution. The results were based on duplicate experiments at an antigen concentration of 25 ng/ml.

C. ANTIGEN-ANTIBODY REACTIONS

1. Kinetics

The time course of the binding of CK-MB to immobilized CAb and of the antibody-peroxidase conjugate to the captured CK-MB was studied (Figures 4A and B). The results show that CK-MB was bound during the 45-s passage through the membrane. Furthermore, no change in K/S was observed after a longer incubation time. For the conjugate step, the response increased slightly during the 10-min incubation period. However, the response

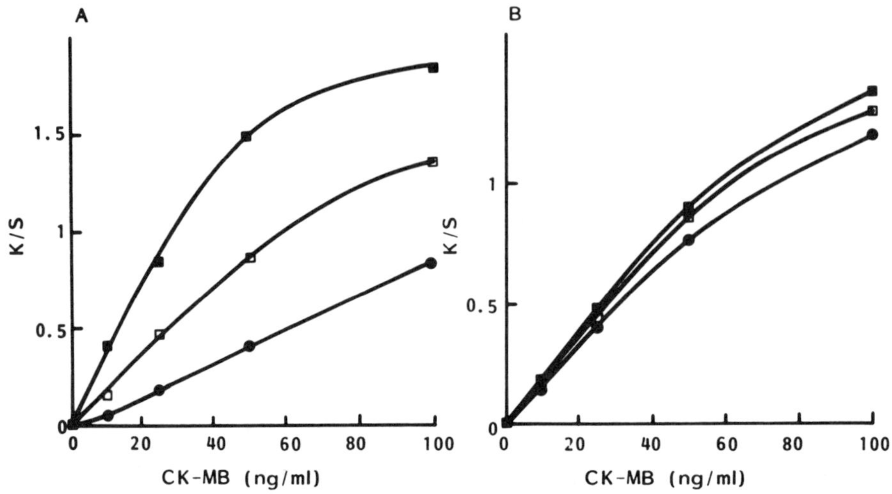

FIGURE 5. Effect of sample volume (A) and conjugate volume (B) on the EIFA for CK-MB. The volumes were 25 μl (●), 50 μl (□), and 100 μl (■). The results are the mean of two determinations.

obtained during the flow-through is 86% of the response obtained after incubation for an additional 10 min.

Thus, in practice, the extent of reaction in both antigen-antibody binding steps does not increase with incubation time subsequent to the filtration step. The data presented are based on experiments using 25 ng/ml CK-MB, but application of 10, 50, and 100 ng/ml CK-MB did not alter the conclusion.

2. Sample and Conjugate Volume

The binding of CK-MB to the CAb and the binding of conjugate to the captured CK-MB were quite different with respect to volume dependence (Figure 5).

Figure 5A shows that K/S increases by increasing the sample volume. The assay responded equally to changes in CK-MB concentration and sample volume, e.g., the same K/S was obtained with 25 μl of 100 ng/ml, 50 μl of 50 ng/ml, and 100 μl of 25 ng/ml. This is in accordance with the results of Valkirs and Barton.[2] Thus, the assay response seems to be dependent on the total amount of antigen rather than on the concentration. These results mean that the sensitivity of the assay can be increased by increasing the sample volume. On the other hand, it also makes precise sample volume crucial for quantitative assays.

In contrast to sample volume the assay is relatively insensitive to differences in the conjugate volume (Figure 5B). This result is possibly due to the high content of unlabeled antibody in the conjugate preparation.

D. GENERATION OF COLOR

1. Kinetics

The effect of varying the substrate reaction time on the response in EIFA is shown in Figure 6. It can be seen that color is generated just as well during flow of substrate through the membrane as during the following 10-min incubation period.

The various curves appear to be linear over different ranges of time. Nonlinearity appears earlier and is more pronounced at high concentration of CK-MB, i.e., when more conjugate is bound. This is possibly caused by substrate exhaustion during the reaction period.

Since the response of a zero sample does not markedly increase during a 10-min incubation period (data not shown), extension of the color development time will increase the sensitivity of the assay. Thus, variation of the substrate reaction time is a factor for controlling assay sensitivity.

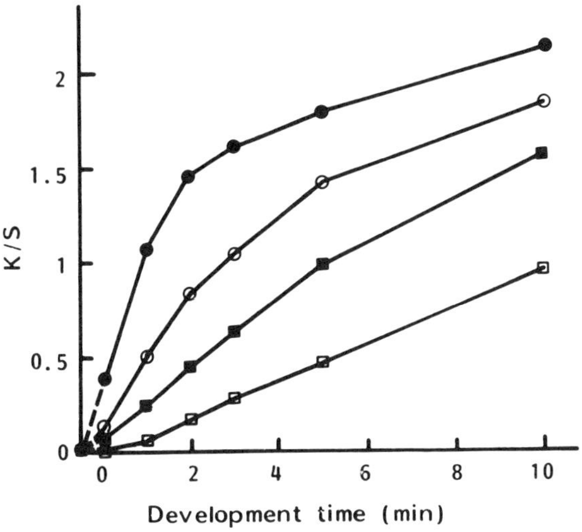

FIGURE 6. Time course of the substrate reaction. The CK-MB concentrations analyzed were 10 ng/ml (□), 25 ng/ml (■), 50 ng/ml (○), and 100 ng/ml (●). The results are based on duplicate determinations. Otherwise, conditions and designations as for Figure 4.

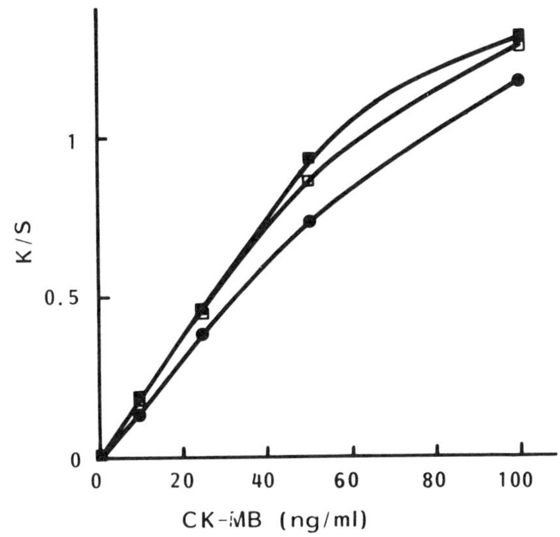

FIGURE 7. The effect of substrate volume on the detection signal. The following substrate volumes were used: 25 μl (●), 50 μl (□), and 100 μl (■). The results were based on duplicate determinations.

The fact that color development continues with time means that it is essential for the reproducibility of the response that the reaction is stopped after a fixed time.

2. Substrate Volume

Investigations of the effect of substrate volume on the assay response indicate that variation in volume does not markedly effect the K/S over the range of CK-MB concentrations tested (Figure 7). Thus, the substrate volume is not an optimization parameter.

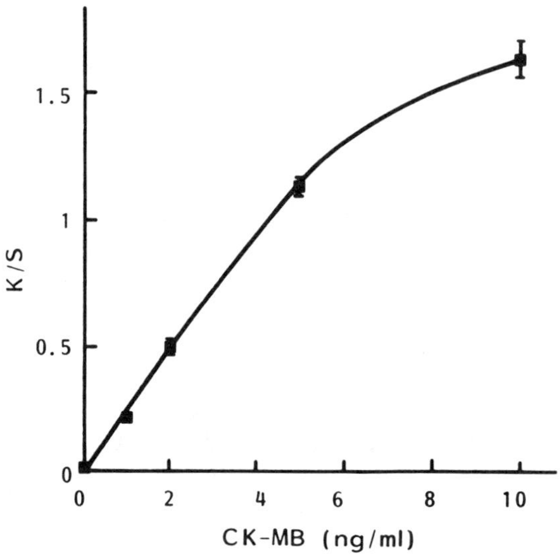

FIGURE 8. Calibration curve for a sensitive CK-MB EIFA. The volume of the CK-MB calibrators was 100 μl, the conjugate concentration 10 μg/ml, and the color development time 5 min. The calibrators used were 0, 1, 2, 5, and 10 ng/ml of CK-MB. Each point represents the mean of four determinations.

TABLE 1
Precision Data for EIFA of Human CK-MB in Bovine Serum

Sample (ng/ml)	Measured concentration mean ng/ml	Within assay SD ng/ml	Within assay CV %	Between assay SD ng/ml	Between assay CV %	Total Sd ng/ml	Total CV %
1 (20)	20	1.3	6.5	0.85	4.4	1.5	7.9
2 (40)	39	1.3	3.2	1.4	3.5	1.9	4.8

Note: The samples were assayed in duplicate in ten consecutive assays.

E. PRECISION

The reproducibility of the assay was evaluated by measuring the CK-MB concentration of two samples using the standard procedure (see Figure 2). The CK-MB concentration was calculated from a calibration curve established for each assay. The samples and calibrators were assayed in duplicate in ten consecutive assays. The precision data based on variance analysis are shown in Table 1. The precision of the assay was good, the total CVs for samples containing 20 and 40 ng CK-MB/ml being 7.9 and 4.8%, respectively. The precision obtained in this study is comparable to that obtained with conventional ELISA techniques.

F. SENSITIVITY

The EIFA described here has been designed to provide an assay covering the CK-MB concentration range of clinical interest[8] (Figure 2). However, the sensitivity of the assay can be controlled by manipulation of three variables: the sample volume, the composition of the conjugate preparation, and the color development time. Optimization of these variables yields an assay covering a concentration range 10 times lower than the standard version of the assay (Figure 8). The shapes of the two calibration curves are very similar. This also applies to the precision profiles.

Thus, the EIFA is potentially a very sensitive assay with a detection limit for CK-MB of less than one ng/ml, i.e., less than 50 pg.

IV. CONCLUDING REMARKS

The aim of this chapter has been to demonstrate the properties of the EIFA technique. Within the context of a membrane-based sandwich assay for CK-MB employing filtration in all assay steps, we have demonstrated that EIFA is very rapid. The antigen-antibody binding steps essentially attain their final level of reactivity, and the color development is already detectable during the rapid flow at reactants through the membrane. Separation of bound and unbound material is made easy using filtration, and only 6 min are required for the standard version of the EIFA described here for CK-MB.

The rapidity of the assay is not achieved at the expense of reproducibility and sensitivity. EIFA is a quantitative assay with which one can reproducibly determine the concentration of CK-MB in serum samples with sensitivity comparable to that obtainable with other immunoassays.

ACKNOWLEDGMENT

The skillful technical assistance of Mrs. Berit Svendsen is highly appreciated.

REFERENCES

1. **Nayak, P. N.,** The kinetics of solid-phase immunoassays, *Ligand Q.*, 4, 34, 1981.
2. **Valkirs, G. E. and Barton, R.,** Immunoconcentration™ — a new format for solid-phase immunoassays, *Clin. Chem.*, 31, 1427, 1985.
3. **Ijsselmuiden, O. E., Meinardi, M. M. H. M., Van der Sluis, J. J., Menke, H. E., Stolz, E., and Van Eijk, R. V. W.,** Enzyme-linked immunofiltration assay for rapid serodiagnosis of syphilis, *Eur. J. Clin. Microbiol.*, 6, 281, 1987.
4. **Ogata, F.,** Quantitative dot-blot enzyme immunoassay for serum amyloid A protein, *J. Immunol. Methods*, 116, 131, 1989.
5. **Anderson, R. R., Lee, T. T., Saewert, D. C., Sowden, K. M., and Valkirs, G. E.,** Internally referenced immunoconcentration assays, *Clin. Chem.*, 32, 1692, 1986.
6. **Marlow, S. J. and Handa, A. K.,** Immuno slot-blot assay using a membrane which covalently binds protein, *J. Immunol. Methods*, 101, 133, 1987.

Chapter 16

THE USE OF SYNTHETIC PEPTIDES AS SOLID-PHASE ANTIGENS

Martial Lacroix, Gervais Dionne, Maan Zrein, Robert J. Dwyer, and Robert J. Chalifour

TABLE OF CONTENTS

I. INTRODUCTION

This chapter illustrates the use of synthetic peptide epitopes in the development of a highly sensitive and specific enzyme immunoassay (EIA) for the detection of antibodies.

Antibodies specifically recognize only small regions (epitopes) present on the surface of antigenic macromolecules (see Chapter 9). For diagnostic test development, it is of major interest to locate and define these epitopes, to synthesize them, and to use them as target antigens in the specific detection of their corresponding antibodies.

The advantages of using synthetic peptides as solid-phase antigens are numerous and include increased specificity and sensitivity. Because only those synthetic epitopes specific for a particular antigen are employed, specificity is much improved over the use of cell lysates or even biosynthetic protein antigens which contain structures and sequences irrelevant to the epitopes of importance. For example, a region of homology has been found between the HIV-1 and HIV-2 transmembrane proteins and the N-terminus of the F1 glycoprotein of some paramyxoviruses[1] which might be responsible for some of the false-positive results observed with diagnostic kits using biosynthetic proteins as antigens. These problematic regions can be eliminated when using synthetic peptide-based EIAs.

II. METHODOLOGY

A. PEPTIDE SEQUENCE SELECTION

HIV-1 and HIV-2 protein sequences were analyzed using a combination of algorithms allowing for the prediction of epitopes potentially useful in the development of a diagnostic test. Based on secondary structure predictions,[2] hydropathicity profiles,[3,4] and surface accessibility calculation[5,6] amino acid sequences were selected, synthesized,[7] and purified to homogeneity by high-performance liquid chromatography (HPLC)[6] before being evaluated by EIA. As more sequences for various HIV isolates became available, the identification of conserved regions became an extremely important selection criterion.[8]

Synthetic peptides were initially screened using a small number (around 50) of seronegative and seropositive samples and reactive synthetic peptides were then further submitted to a second screening process using a panel of 200 HIV-1 and 16 HIV-2 positive serum samples. Specificity was evaluated using a panel of over 300 seronegative samples including 50 which had been found falsely positive with viral lysate-based EIAs. Using this procedure, a number of promising peptides and analogues were synthesized and tested. Finally, three synthetic peptides were selected as representing immunodominant epitopes of HIV-1 and HIV-2. One peptide represents the C-terminal 22-amino acids of gp120 of HIV-1 while the other two peptides include two neighboring cysteine residues of the transmembrane proteins of HIV-1 and of HIV-2. These peptides were cyclized according to the procedure described by Felix et al.[9]

B. PEPTIDE SOLUBILIZATION AND COATING

Although it was previously thought necessary to couple each peptide to a carrier protein in order to ensure its binding to the wells of a microtiter plate, it was found that most synthetic peptides which have at least 15 amino acid residues do bind to plastic plates in sufficient quantity to allow for their evaluation by EIA.

Synthetic peptides were solubilized at a final concentration of 10 μg/ml in PBS (pH 7.2) or in 0.05 *M* carbonate buffer (pH 9.6). In some cases it was necessary to use a minimal volume of glacial acetic acid to bring a poorly soluble peptide into solution before diluting it to the final working concentration. Acidic conditions (acetic acid 1%) were used during the coating step when a peptide containing two cysteine residues had to be kept in its linear form. The plates containing 0.1 ml peptide solution per well were left overnight at 4°C after which time they were emptied and filled with 300 μl saturation buffer. After 1 h at 4°C the wells were emptied and washed 5 times with PBS containing 0.05% Tween-20. The plates were dried and stored individually in a plastic bag containing a desiccant. Prepared under these conditions, peptide-coated microtiter plates are stable for more than 1 year at room temperature. The choice of the microtiter plate is important. Certain peptides bind poorly to some plates. Attempts to covalently link peptides to "specially activated" microtiter plates obtained from two different suppliers have not been successful.

C. ENZYME IMMUNOASSAYS

Conditions employed in the peptide-based EIA were similar to those used in other EIAs. The high purity obtained when using synthetic peptide antigens allows for the testing of less-diluted specimens, without encountering serious nonspecific binding problems. For example, the EIA described here uses samples that are diluted by 50 when most viral lysate-based tests use dilutions by 100 and 400. Incubation with diluted samples last 30 min at room temperature. After five other wash cycles, a peroxidase-labeled goat anti-human IgG is added. Following a 30 min incubation at room temperature and 5 other wash cycles, a substrate solution containing the tetramethylbenzidine chromogen and H_2O_2 is added. After 30 min at room temperature, the solution is acidified and the absorbance at 450 nm is measured.

III. RESULTS

A. EFFECT OF PEPTIDE COATING DENSITY ON ANTIBODY BINDING-SITE OCCUPANCY

Because synthetic peptide antigens ideally represent only the minimal size necessary to mimic a given antigenic determinant, the density of epitope which can be coated on the plastic surface is maximized. One advantage of this high epitopic density is that it leads to bivalent attachment of antibodies; this condition can result in a 1000-fold increase in functional affinity (avidity) relative to monovalent antibody attachment.[10] Bivalent capture of antibodies results in an EIA which is very sensitive even for the detection of low-affinity antibodies.

To demonstrate bivalent attachment under the conditions of our EIA, we have measured the signal obtained in a direct and an indirect EIA with increasing peptide density (Figure 1). The direct EIA exhibits a maximum signal at 5 ng/cm^2 of plastic surface. At higher coating densities the signal was progressively reduced presumably due to the loss of available antigen binding sites (required to capture the peroxidase-labeled peptide) resulting from bivalent binding of the human IgG on the plate. We have calculated that the average distance between two peptide molecules is within the reach of the two binding sites of an IgG (87 Å[11]) when the coating density is above 9 ng/cm^2, a value in agreement with our experimental findings. These free antigen binding sites are required for attachment of the peroxidase-labeled peptide. The response of the indirect EIA to the coating density was quite different because the amount of peroxidase-labeled goat anti-human IgG retained in each well is only dependent on the quantity of human IgG present. As would be expected the indirect EIA signal increases with the amount of bound peptide, reaching a plateau corresponding to the saturation of the plastic with peptide.

IAF BioChem's DETECT-HIV™ is an indirect EIA kit using microtiter wells coated with greater than 1000 ng/cm^2 of peptide, a coating density which insures bivalent antibody binding.

B. COMPARATIVE PERFORMANCE OF CYCLIC VS. LINEAR SYNTHETIC PEPTIDES IN THE DETECTION OF HIV-SPECIFIC ANTIBODIES

Serially diluted HIV-1 and HIV-2 seropositive samples were tested on wells coated with linear and cyclic peptides. Sensitivity of detection of HIV-1 specific antibodies was 1.4 to 6.0

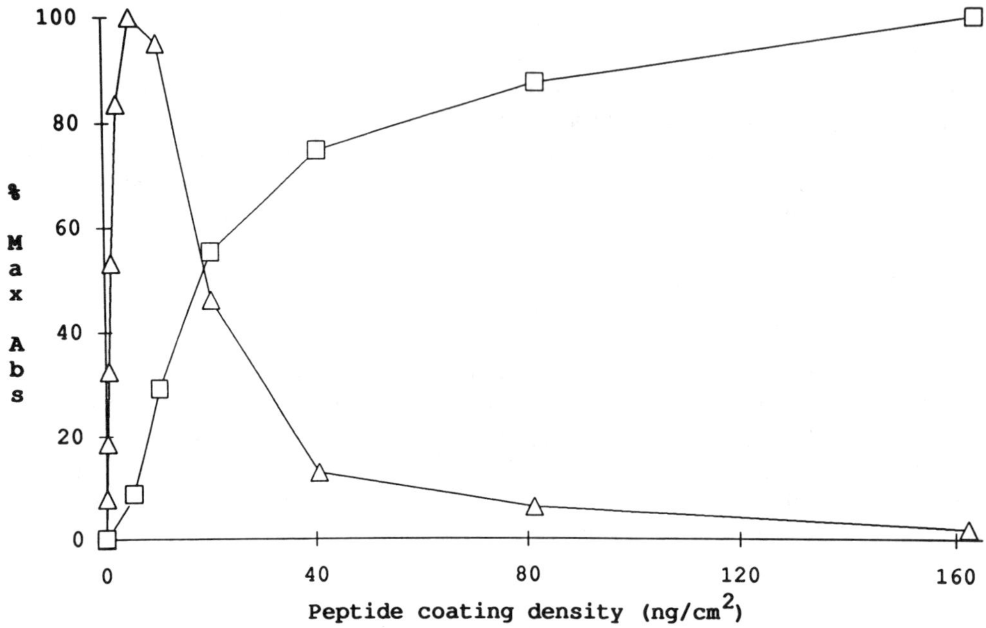

FIGURE 1. Polystyrene microtiter wells were coated with the indicated amounts of HIV-1 cyclic peptide (BCH-87c). An HIV-1 positive serum sample was used at a dilution of 1/50 for the direct EIA (Δ) and at a dilution of 1/15,000 for the indirect EIA (□). The direct EIA used a BCH-87c-peroxidase conjugate as a probe. The indirect EIA used the previosuly described peroxidase-labeled goat anti-human IgG.

times higher with the cyclic synthetic peptide than with its linear counterpart (Table 1). A greater difference in sensitivity was observed when HIV-2 positive serum samples were tested on linear and cyclic synthetic peptides (Table 1). In this case, sensitivity of detection of HIV-2-specific antibodies was 1.14 to 49.5 times higher with the cyclic synthetic peptide. The amounts of linear and cyclic peptides bound in each case were found not to be significantly different.[12]

Similar results were obtained when samples from a seroconversion panel were tested (Figure 2). The fourth sample of this HIV-1 seroconverter was detected as the first being positive (Abs > 0.200) using the cyclic peptide-based test. In comparison, only the fifth sample was detected as the first positive when using the linear peptide.

C. COMPARATIVE PERFORMANCE OF A SYNTHETIC PEPTIDE-BASED EIA WITH COMMERCIALLY AVAILABLE EIAS USING VIRAL LYSATES AND BIOSYNTHETIC ANTIGENS

Aliquots of serum samples tested in different centers using three different viral lysate-based EIAs (A, B, and C in Table 2) were retested using the IAF BioChem synthetic peptide-based EIA (DETECT-HIV™). This panel of 2223 plasma samples included many that were found to give falsely positive or falsely negative results on test A, B, C, or by Western Blot (W.B.). For example, Table 2 shows that using the "A" EIA only 979 out of 981 confirmed HIV-positive were found to be positive. This particular EIA thus missed two HIV-positive samples. Fortunately, because these patients presented clinical symptoms, their blood samples were retested by one of the three confirmatory tests described and correctly found to be positive. These 981 samples are included within the 1378 that were correctly identified as positive with the DETECT-HIV™ EIA. As can be seen in Table 2 the specificity (843 out of 845 HIV-negative samples were found negative; 99.8%) and sensitivity (1378 out of 1378 HIV-positive samples were found positive; 100%) of DETECT-HIV™ were both excellent despite the difficult composition of this panel of samples.

TABLE 1
The Effect of Peptide Cyclization on the Titer of HIV-Seropositive Samples

HIV-1 seropositive samples				HIV-2 seropositive samples	
Serum no.	R_1	Serum no.	R_1	Serum no.	R_2
M-0 (negative)	0.82	M-25	1.54	M-0 (negative)	1.00
M-1	2.19	M-26	1.40	F-63	49.50
M-2	2.54	M-27	5.20	F-70	13.17
M-8	4.20	M-28	6.02	F-71	35.19
M-10	1.89	M-32	1.51	B-M-2	8.21
M-11	2.32	M-33	2.21	B-0-1	12.16
M-12	2.51	M-34	1.97	B-2-65	9.96
M-14	1.69	M-35	1.80	B-2-69	1.14
M-15	2.35	M-36	1.78	B-2-72	2.60
M-17	2.88	M-38	1.52	B-2-81	7.87
M-19	1.84	M-39	3.18	B-2-103	6.61
M-20	1.40	M-41	1.69	B-2-114	7.08
M-21	1.92	M=45	2.10	B-2-115	2.02
M-22	1.67	M-46	2.29	B-2-116	2.43
M-23	2.10	M-47	1.99	B-2-117	5.93
M-24	1.98	M-49	1.59		

Note: R_1 — is the ratio of the titers of serially diluted HIV-1 seropositive samples measured with the HIV-1 peptides (BCH-87 cyclic and BCH-87 linear). R_2 — is the ratio of the titers measured with the HIV-2 peptides (BCH-201 cyclic and BCH-201 linear).

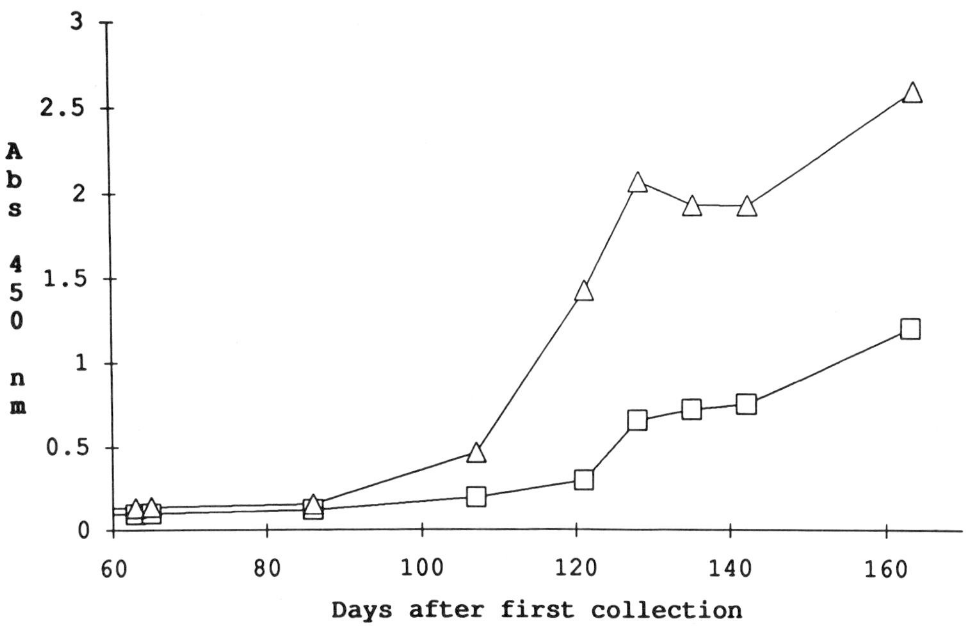

FIGURE 2. Differential reactivity of successive plasma samples taken from a patient undergoing seroconversion (panel A; Boston Biomedica Inc.) measured on a linear (□) and a cyclic (Δ) synthetic peptide each coated at 10 μg/ml. (From Lacroix, M., Chalifour, R. J., Stern, M. D., and Dionne, G., Comparative Performance of Cyclic Versus Linear Peptides in an ELISA for HIV-I and HIV-2 Specific Antibodies, presented at 5th Int. Conf. AIDS, Montreal, June 4 to 9, 1989. With permission.)

TABLE 2
Comparative Performance of a Synthetic Peptide-Based EIA (DETECT-HIV™) with Three Viral Lysate-Based Kits and Three Confirmatory Tests

Number of samples tested that were:	EIA DETECT-HIV™	EIA A	EIA B	EIA C	WB	IFA	RIPA
Confirmed positive	1378	981	193	177	166	801	481
Found positive	1378	979	192	174	158	800	481
Negative	845	93	29	27	51	111	106
Found negative	843	39	4	24	46	111	106
Found indeterminate	0	44	5	0	1	24	0

Note: A total of 2223 samples tested by 1 or 2 of the EIA tests were retested by IAF BioChem DETECT-HIV™ EIA. All those samples that were positive or indeterminate and about 250 negative samples of this panel were confirmed by at least 1 of the following tests: Western Blot (WB), immunofluorescence assay (IFA), and radioimmunoprecipitation assay (RIPA).

Table 3 summarizes the data obtained with two panels of longitudinally collected samples from two plasmapheresis donors. IAF BioChem's EIA for the detection of antibodies to HIV-1 and to HIV-2 gave higher signal-to-cutoff ratios than the other commercially available viral lysate-based kits dedicated only to the detection of antibodies to HIV-1. This demonstrates the improved sensitivity obtained with this synthetic peptide-based EIA.

More recently, Fenouillet et al.[14] have further illustrated the capacity of DETECT-HIV™ for detecting seroconversions earlier than a viral lysate-based EIA and earlier also than W.B.

IV. CONCLUSION

Carefully chosen synthetic peptide antigens mimicking only immunodominant and specific epitopes can form the basis of extremely sensitive and specific antibody detection tests. The use of highly purified peptides yields a low cutoff value contributing to a clearer distinction between negative and positive samples.

The configuration of the synthetic peptide epitopes can be modeled (e.g., by cyclization) to fit as closely as possible the native epitopes present on the virus and thus mimic epitopes which are lost or hidden when whole viral lysates or biosynthetic antigens are used.

In the specific examples described here, synthetic peptide sequences taken from the immunodominant region located on the transmembrane proteins of HIV-1 and of HIV-2 are more antigenic when their two cysteine residues are bridged. This possibly reflects the fact that on the virus this region may indeed be cyclic and antibodies raised against it consequently prefer cyclic peptide antigens. In another system however (HTLV-I) we observed that the antigenicity of one peptide was much lower in the cyclic than in the linear form. Cyclization is beneficial only when it reduces the number of possible conformations a synthetic peptide can adopt and does not exclude those that are well recognized by the antibodies.

The use of synthetic peptides as solid-phase antigens in the development of an EIA provides a way to routinely insure that the required epitope density needed for bivalent antibody binding can be reached. When large proteins are employed as capture antigen, it is uncertain that bivalent attachment is possible. Previous studies have demonstrated the relationship between high epitope density and increased sensitivity of EIA for low-affinity antibodies.[11,13] This point is clearly of importance in the screening of HIV antibodies, since low-affinity antibodies may predominate in the initial phase of the immune response.

TABLE 3
Detection of Antibodies to HIV in Longitudinally Collected Samples from Two Plasmapheresis Donors

Anti-HIV ELISA Results
(Signal/Cutoff)[1]

Sample I.D.	Collection date	Abbott	Cellular products	DuPont	ENI	Genetic Systems	Ortho	Organon Teknika	IAF BioChem
1A	05-04-81	0.32	0.30	0.25	0.34	0.39	0.30	0.56	0.3
2A	07-08-81	0.54	0.35	0.25	0.34	0.32	0.36	0.49	0.6
3A	07-29-81	0.30	0.34	0.42	0.35	0.73	0.38	1.06	1.7
4A	08-19-81	2.88	1.49	1.35	1.50	3.97	2.61	2.73	>12
5A	09-02-81	4.56	2.00	2.56	2.90	4.22	3.82	2.81	>12
6A	09-09-81	6.39	2.19	3.34	3.73	5.41	4.45	3.12	>12
7A	09-16-81	8.04	2.78	3.54	4.51	6.96	5.71	3.15	>12
8A	09-23-81	8.97	2.39	3.21	5.43	4.44	6.18	2.96	>12
9A	10-14-81	10.81	2.79	3.75	6.53	5.74	7.14	3.10	>12
1C	07-16-85	0.39	0.05	0.03	0.16	0.29	0.01	0.42	0.2
2C	07-23-85	0.32	0.04	0.04	0.17	0.14	0.03	0.47	0.2
3C	07-25-85	0.33	0.07	0.07	0.20	0.20	0.01	0.65	0.4
4C	07-30-85	0.47	0.33	0.24	0.28	0.35	0.08	1.02	3.5
5C	08-01-85	0.51	0.28	0.26	0.35	0.50	0.09	1.17	5.9
6C	08-06-85	0.85	0.82	0.75	0.64	0.82	0.36	1.60	8.
7C	08-08-85	0.91	1.04	0.89	0.84	0.83	0.54	2.04	7.
8C	08-13-85	1.04	1.53	1.34	1.13	1.13	1.40	2.18	9.
9C	08-15-85	1.81	2.65	1.83	1.60	1.98	2.13	2.25	>12
10C	08-20-85	4.65	3.57	2.77	3.14	2.63	3.83	2.65	>12
11C	08-27-85	6.25	5.09	3.50	3.78	3.28	5.05	3.02	>12
12C	08-29-85	5.36	5.25	3.38	3.71	3.29	4.33	2.70	>12
13C	09-10-85	7.08	4.79	3.40	4.47	3.30	4.41	2.94	>12
14C	09-12-85	6.67	5.44	3.39	4.18	3.66	5.10	2.90	>12
15C	09-17-85	5.99	4.17	3.35	3.62	3.01	4.32	2.86	>12
16C	09-19-85	7.46	4.87	3.50	4.25	2.81	4.66	2.99	>12
17C	09-24-85	7.36	4.97	3.50	4.52	4.13	4.90	3.03	>12
18C	09-26-85	8.76	5.28	3.60	4.64	3.67	4.76	2.91	>12

Note: Ratios >1.00 are considered positive for anti-HIV. Samples and results (with the exception of those of IAF BioChem) were obtained from Boston Biomedica Inc.

REFERENCES

1. **Gallaher, W. R.,** Detection of a fusion peptide sequence in the transmembrane protein of HIV, *Cell,* 50, 327, 1987.
2. **Chou, P. Y. and Fasman, G. P.,** Prediction of protein conformation, *Biochemistry,* 13, 22, 1974.
3. **Hopp, T. P. and Woods, K. R.,** Prediction of protein antigenic determinants from amino acid sequences, *Proc. Natl. Acad. Sci. U.S.A.,* 78, 3824, 1981.

4. **Kyte, J. and Doolittle, R. F.,** A simple method for displaying the hydropathic character of a protein, *J. Mol. Biol.,* 157, 105, 1982.
5. **Emini, E. A., Hughes, J. V., Perlow, D.S., and Boger, J.,** Induction of hepatitis A virus-neutralizing antibody by a virus-specific synthetic peptide, *J. Virol.,* 55, 836, 1985.
6. **Talbot, P., Dionne, G., and Lacroix, M.,** Vaccination against lethal coronavirus-induced encephalitis with a synthetic decapeptide homologous to a domain in the predicted peplomer stalk, *J. Virol.,* 62, 3022, 1988.
7. **Merrifield, R. B.,** Solid phase peptide synthesis, *Science,* 232, 341, 1986.
8. **Myer, G., Rabson, S. F., Josephs, T. F., Smith, T. F., and Wong-Staal, F.,** Human retroviruses and AIDS, 1989 Data base, in *Theoretical Biology and Biophysics,* Los Alamos National Laboratory, Los Alamos, New Mexico, 1989.
9. **Felix, A. M., Jimenez, M. H., Wang, C. T., and Meienhofer, J.,** Synthesis of somatostatin and (D-TRP)-somatostatin, *Int. J. Peptide Protein Res.,* 15, 342, 1980.
10. **Devey, M. E. and Steward, M. W.,** The role of antibody affinity in the performance of solid phase assays, in *ELISA and Other Solid Phase Immunoassays,* Kemeny, D. M. and Challacombe, S. J., Eds., John Wiley & Sons, New York, 1988, chap. 6.
11. **DeLisi, C.,** Hemolytic plaque inhibition: the physical chemical limits on its use as an affinity assay, *J. Immunol.,* 117, 2249, 1976.
12. **Lacroix, M., Chalifour, R. J., Stern, M. D., and Dionne, G.,** Comparative performance of cyclic versus linear peptides in an ELISA for HIV-1 and HIV-2 specific antibodies, presented at the 5th Int. Conf. AIDS, Montreal, June 4 to 9, 1989.
13. **Lew, A. M.,** The effect of epitope density and antibody affinity on the ELISA as analysed by monoclonal antibodies, *J. Immunol. Methods,* 72, 171, 1984.
14. **Fenouillet, E., Sorensen, A.-M., Lacroix, M., Coutellier, A., Herson, S., Fretz-Foucault, C., and Gluckman, J.-C.,** Early and specific diagnosis of seropositivity to HIVs by an enzyme-linked immunosorbent assay using env-derived synthetic peptides, *AIDS,* 4, 1137, 1990.

Chapter 17

STREPTAVIDIN: PRODUCTION, PURIFICATION, AND USE IN ANTIBODY IMMOBILIZATION

Mark Suter

TABLE OF CONTENTS

I. INTRODUCTION

The development of monoclonal antibodies (mAb) by Köhler and Milstein[1] lead to the expanded use of antibodies (Ab) for solid-phase immunoassays, affinity chromatography, immunohistology, and immunofluorescence including fluorescence-activated cell sorting. The same Ab may be used in all of these applications but the method of coupling to the various solid phases or different signal generators (see Chapter 1) may require extensive chemical modification of the Ab. The coupling of Ab with different signal generators may be costly and time comsuming and may impair the function of the Ab. Hence, systems have been developed in which various markers can be linked to the Ab using an indirect method. The high-affinity (10^{-15} l/*M*) interaction of biotin with streptavidin (SA) provides the basis for one such system.

In most common applications, biotinylated Ab[2-4] are first permitted to react with the antigen. Subsequently, the biotin moieties on the antigen-antibody (Ag-Ab) complex are exposed to SA that has previously been covalently coupled to the signal generator (e.g., enzymes, fluorescent dyes) of the particular system. Thus, the signal source coupled to SA now becomes a marker for the biotinylated Ab used in the assay.

This review will focus on the production, purification, and characterization of SA and concludes with selected examples of the use of SA in various solid-phase assays.

II. THE STREPTAVIDIN MOLECULE

A. HISTORICAL BACKGROUND

SA was first described by Stapley et al.[5] who isolated the molecule from culture supernates of the actinomycete *Streptomyces avidinii.* The ability of eumycetes to produce a biotin binding molecule appears to be restricted to *S. avidinii* since 11 other actinomycete isolates (representing 2 genera) and 134 eumycete isolates (representing 65 genera) did not produce a biotin-binding molecule.[6] SA is a 70-kDa tetrameric nonglycosylated protein. Each monomer can bind one molecule of biotin.[7] Recently, the gene coding for this protein has been cloned[8] and the molecular weight obtained by analyzing the SA protein has been confirmed.[9]

B. PRODUCTION, ISOLATION, AND CHARACTERIZATION OF STREPTAVIDIN

SA is synthesized by *S. avidinii* and is secreted into liquid media.[5] The maximum yield using nonsynthetic media is approximately 10 to 50 mg SA per liter of supernate.[10]

The purification of SA from culture supernates has been greatly facilitated by the development of a biotin analogue (iminobiotin) that reversibly binds SA.[11] However, commercial preparations of SA produced by iminobiotin chromatography exhibit considerable variations in size and in their binding characteristics for biotin.[8,10] Postsecretory modifications may be responsible for these variations.[12]

The requirement for large quantities of SA of defined quality for certain applications[13] encouraged the search for alternative procedures for the production and purification of SA.[6,14] Therefore, *S. avidinii* was adapted to grow in synthetic culture media and the constituents of the media were adjusted to achieve enhanced production of SA (100 to 120 mg/l) as well as reproducible yields. The use of chemically defined media also allowed the isolation of SA by a simple salt precipitation procedure followed by centrifugation.[14] When harvested by this method during the first 8 d of culture, SA was more than 90% pure. Within the detection limits of an inhibition ELISA,[14] no monomeric form(s) of the SA molecule was detected.

A biotin-binding molecule of approximately 150 kDa was present in cultures over 6 to 8 d old.[9,14] This molecule showed a binding avidity for biotin similar to that of the tetravalent, 70-kDa SA. Cell-free culture supernates from *S. avidinii* can be used without purification as a very inexpensive source of SA in assays that exploit the cross-linking ability of SA.[13,15] This is because of the absence of monomeric forms of biotin binding molecules in culture supernatants during the production of SA and the apparent similar biotin binding avidity of the 70 kDa and the low amounts of the 150-kDa form of SA which may be present. Moreover, the contents of SA in the culture supernates could easily be determined by an ELISA[14] facilitating the use of SA in the supernates at optimal dilutions.

Homogeneous SA molecules of 70 kDa are required for quantitative studies of SA,[15] for assays in which SA is limiting,[16] and for determination[9] of the stoichiometry of the coupling of SA to different markers. Isolation procedures to obtain homogeneous SA of 70 kDa have been described.[15]

C. PRACTICAL CONSIDERATIONS FOR HANDLING STREPTAVIDIN

Lyophilized SA can be stored at room temperature in a desiccator or kept at 4°C dissolved in distilled water in the presence of inhibitors such as 1 to 10 m*M* NaN_3[6,10] and 10 m*M* benzamidine[15] to prevent bacterial growth and proteolysis.[10] Proteolytically altered SA may have different binding properties for biotin.[12] If inhibitors are not desired, purified SA or supernatants containing SA can be kept in a 1:1 (v/v) glycerol/distilled water or in a synthetic medium solution at –20°C for at least 6 months without appreciable loss of activity.

III. APPLICATION OF STREPTAVIDIN IN SOLID-PHASE ASSAYS

A. GENERAL FEATURES

SA has found wide usage in solid-phase assays. It can be coupled to various enzymes, fluorochromes, and radioactive markers; both the activity of the marker as well the biotin binding activity of SA is minimally affected under a wide variety of conditions.[4,15] Because SA has a multivalent form, it can also cross link two biotinylated molecules. The cross-linking capability has been used to avoid the denaturation of proteins that occurs when they are adsorbed hydrophobically to solid-phase surfaces.[13,15] In one such system, a biotinylated monoclonal Ab (mAb) is cross linked by SA to a biotinylated carrier protein adsorbed to a solid phase (the PABC system[15]). Some mAbs used in this system had an antigen capture capacity (AgCC) up to 400 times greater than when the same mAbs were adsorbed directly to a solid phase (Figure 1).[13] Studies demonstrated that the AgCC of a particular mAb adsorbed directly to a solid phase could not be predicted from its isotype, light chain, or affinity as measured by dissociation rate analysis (Table 1);[13] rather, small differences in V-region hydrophobicity are thought to influence the conformation of the mAb after adsorption and hence influence its AgCC. In general, the AgCC of mAb was more sensitive to loss during solid-phase adsorption than that of polyclonal antibodies (pAb).[15,17] Stoichiometric analysis of the PABC system suggests that biotin substitution of mAb alters AgCC. Specifically, increasing the biotin substitution of a pAb from 2 to 15 mol biotin per mole Ab caused a parallel increase in the amount of pAb bound in the PABC system but led to a substantial decrease in maximal AgCC. In all cases, however, the AgCC of optimally biotinylated mAb used in the PABC system was higher when compared to the same amount of mAb adsorbed directly to the solid phase.[13,15]

B. PRACTICAL CONSIDERATIONS IN THE USE OF STREPTAVIDIN IN SOLID-PHASE ASSAYS

SA bound directly to plastic surfaces is impaired in both its maximal ability to bind to biotinylated mAb and in its functional affinity when compared to SA used in the PABC system.[24] Hence, SA may undergo adsorption induced alterations accompanied by functional loss as noted with some mAbs. This observation could also explain the apparent functional

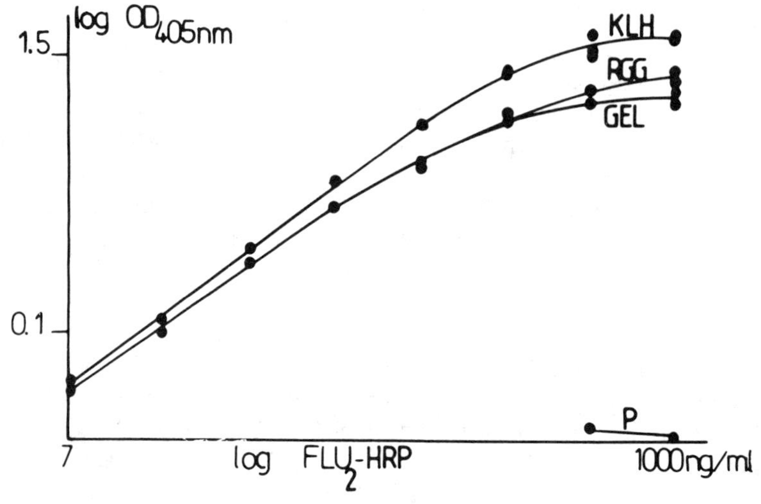

FIGURE 1. The binding of various amounts of FLU_2-HRP by mAb 10-25 specific for fluorescein adsorbed directly to plastic (P) or immobilized using the PABC system is shown. The resulting OD obtained using three different biotinylated carriers (KLH, RGG, and Gel) in the PABC system are presented. All date are corrected for negative controls.

TABLE 1
The Comparative Activity of Monoclonal Antibodies to Fluorescein when Adsorbed on Plastic or Immobilized by the PABC System

Monoclonal antibody tested		K_a (l/*M*)	OD 405/50 ng mAb	
			Plastic	PABC
9—40	(γ l/k)	3.0×10^7	0.0151	0.1836
10—25	(γ l/k)	5.4×10^7	0.0015	0.5409
4—4—20	γ 2a/k)	2.0×10^{10}	0.0259	0.7916

Note: Monoclonal antibody to fluorescein of known heavy and light chains and affinities were used adsorbed to plastic or bound to the PABC system.[13] The substrate conversion of fluorescein-labeled peroxidase captured by mAb is indicated

diminution of SA when blotted to nitrocellulose solid supports.[12] Therefore, it is not recommended to use SA hydrophobically adsorbed to solid supports to immobilize biotinylated molecules.

Desorption of hydrophobically adsorbed proteins may impair the sensitivity of ELISA system.[18] Hence, solid supports have been developed which offer the possibility to biotinylate covalently bound reactive groups in microtiter plates (i.e., Covalink, Nunc, Denmark). The biotinylated support could be used to employ SA as a stable cross-linking agent to immobilize biotinylated molecules in which direct contact with the solid phase is considered to be a disadvantage. This approach was shown to yield capture Ab with highest functional activity in a solid-phase system.[19]

In most experiments, SA was used to cross link two biotinylated molecules. Egg white avidin (Vector Laboratories, Burlington, VT) can also be used in the PABC system as an alternative to SA with virtually the same results.[24] Recently, SA has been chemically modified which result in a nearly eightfold increase in its ability to immobilize biotinylated enzyme used in an ELISA when compared to nonmodified SA.[12] Hence, chemically modified SA could be beneficial in immobilizing increased amounts of biotinylated molecules such as mAbs in the PABC system.

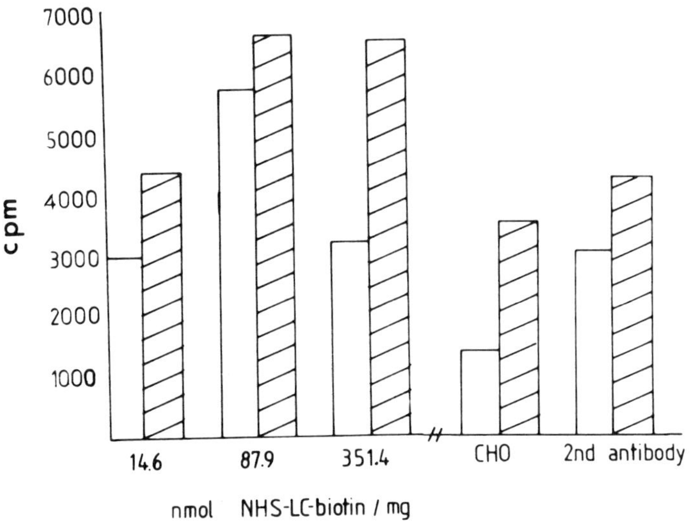

FIGURE 2. Relationship between the degree of biotinylation and the amount of antigen recovered. Two mAb (,) were biotinylated with 14.6, 87.9, or 351.4 nmol NHS-LC-biotin or 10 m*M* biotin-hydrazide (CHO) per milligram protein. Alternatively, biotinylated second antibody was employed (2nd antibody). Both biotinylated and unbiotinylated mAb (5 μ g of each) was reacted with [^{35}S]methionine-labeled HCMV antigen. The immune complexes were recovered using SA-agarose and the recoveries of [^{35}S]methionine-labeled antigen were determined by liquid scintillation counting.

C. AFFINITY PURIFICATION OF ANTIGEN-ANTIBODY COMPLEXES USING STREPTAVIDIN

The exquisite specificity of pAbs or mAbs has been extensively used for affinity purification of molecules from a complex mixture. The usual procedure has been to immobilize Ab directly to a solid-phase matrix.[20] Recently, systems have been developed to retrieve Ag-Ab complexes that were formed in fluid phase.[22] Biotinylated Ab is incubated first with the antigen (Ag) in the fluid phase, and subsequently the Ag:Ab complex is retrieved using SA coupled to a solid-phase matrix such as Agarose (SA-agarose). However, biotin substitution of amino groups has been shown to decrease the AgCC of some Ab.[3,22] Loss of AgCC can be minimized by lowering the biotin substitution of the Ab.[3] Biotin substitution as low as 8 *M* biotin per molar Ab resulted in approximately 100% binding of Ab to SA-agarose and hence all Ag-Ab complexes could be retrieved by SA-agarose. The use of directly biotinylated mouse mAb in combination with SA-agarose was compared to (1) the same mAb used in combination with a biotinylated rabbit anti mouse mAb, (2) mAb biotinylated at the sugar residues,[22] (3) mAb used in combination with Staphylococcal protein A coupled to agarose, or (4) mAb coupled directly to agarose (Figure 2).[3] The results indicated that directly biotinylated mAb retrieved four to five times more Ag when compared to SA-agarose systems using biotinylated rabbit anti mouse mAb, mAb substituted with biotin at the sugar residues, or compared to mAb used with a Staphylococcal protein A system. Directly biotinylated mAb retrieved 20 to 50 times more Ag when compared to mAb coupled directly to agarose.[3] Thus, any possible damage to directly biotinylated mAb used in combination with SA-agarose is clearly offset by the increase in recovery of the Ag.[3]

D. BIOTINYLATION OF PROTEINS

mAbs used in the PABC system or used for affinity purification require a defined degree of biotin substitution. We have standardized biotinylation[3,15,20] using NHS-LC-biotin [sulfosuccinimidyl 6-(biotin amido)-hexanoate, Pierce Chemicals]. All proteins are adjusted to 1 mg/ml in 0.1 *M* $NaHCO_3$, pH 8. The protein solution is freed of aggregates by centrifugation at 15,000 × g for 15 min at 4°C. Immediately before use, 1.63 mg of NHS-LC-biotin is dissolved in 1 ml dimethyl sulfoxide and 5 μ l (14.6 nmol), 50 μ l (146.4 nmol), or 120 μ l (351 nmol) of this mixture is used to biotinylate 1 mg of protein. After 1 h at room temperature, the reaction mixture is dialyzed at 4°C for 48 h against 0.1 M $NaHCO_3$ containing 10 mM NaN^3 and 10 m*M* benzamidine (Aldrich Chemicals). When 1 mg of mAb or pAb is biotinylated with the above indicated amounts of NHS-LC-biotin, molar biotin substitution ratios in Ab or 2, 12, and 25, respectively, are obtained. This method allows small amounts of proteins (100 μ g) to be reproducibly substituted with biotin.

IV. CONCLUSION

The SA biotin system offers many advantages for the flexible and inexpensive use of Abs in a variety of solid-phase applications. The SA biotin system circumvents problems associated with direct labeling of Abs with different signal generators and may prevent loss of AgCC when Ab must be immobilized on solid phases. The flexibility in the use of SA has been broadened by expressing SA in *E. coli*.[23] Not only is the time to produce SA shortened severalfold but SA could be used as a fusion protein which will greatly facilitate the purification of the coexpressed protein by affinity isolation.

In this review I have emphasized the importance of biotin substitution in Abs used in certain applications such as the PABC system and in affinity isolation. In most of the applications described, SA was used as the binding molecule for biotin. No systematic comparison between SA and avidin in these various applications was made. However, it is evident from the literature and from the limited experience of the author that avidin can be utilized instead of SA. SA may be used as a truncated form to achieve increased binding of proteins.[12]

Although most components of the biotin-SA system can be obtained commercially from numerous suppliers,[12] this article should provide some encouragement "to do it yourself", both with respect to the optimal biotinylation of molecules and to the production of "homemade" SA.

ACKNOWLEDGMENTS

I would like to thank my collaborators for their time spent inside and outside the lab to contribute to the original work and to Meir Wilchek and Ed Bayer (Weizman Institute Rehovot, Israel) for discussion and providing unpublished articles. The patience of Keetje v.d. Heuvel for typing the manuscript and Mark Sutherland for reading the manuscript is highly appreciated.

REFERENCES

1. **Köhler, G. and Milstein, C.,** Continuous cultures of fused cells secreting antibody of predefined specificity, *Nature,* 256, 495, 1975.
2. **Wilchek, M. and Bayer, E. A.,** The avidin-biotin complex in immunology, *Immunol. Today,* 5, 39, 1984.
3. **Gretch, D. R., Suter, M., and Stinski, M. F.,** The use of biotinylated monoclonal antibodies and Streptavidin affinity chromatography to isolate Herpes virus hydrophobic proteins or glycoproteins, *Anal. Biochem.,* 163, 270, 1987.

4. **Wilchek, M. and Bayer, E. A.,** The avidin-biotin complex in bioanalytical applications, *Anal. Biochem.*, 171, 1, 1988.
5. **Stapley, E. O., Mata, J. M., Miller, I. M., Demny, T. C., and Woodruff, H. B.,** Antibiotic MSD-235. I. Production by Streptomyces avidinii and Streptomyces lavendulae, *Antimicrob. Agents Chemother.*, 3, 20, 1963.
6. **Cazin, J., Jr., Suter, M., and Butler, J. E.,** Production of streptavidin in a synthetic medium, *J. Immunol. Methods,* 113, 75, 1988.
7. **Chaiet, L. and Wolf, F. J.,** The properties of streptavidin, a biotin-binding protein produced by streptomycetes, *Arch. Biochem. Biophys.*, 106, 1, 1964.
8. **Argarana, C. E., Kuntz, D., Birken, S., Axel, R., and Cantor, C. R.,** Molecular cloning and nucleotide sequence of the streptavidin gene, *Nucleic Acid Res.*, 14, 1871, 1986.
9. **Bayer, E. A., Ben-Hur, H., Gitlin, G., and Wilchek, M.,** An improved method for the single-step purification of streptavidin, *J. Biochem. Biophys. Methods,* 13, 103, 1986.
10. **Bayer, E. A., Ben-Hur, H., and Wilchek, M.,** Isolation and properties of Streptavidin, *Methods Enzymol.*, in press.
11. **Hofmann, K., Wood, S. W., Brinton, C. C., Montibeller, J. A., and Finn, F. M.,** Iminobiotin affinity columns and their application to retrieval of streptavidin, *Proc. Natl. Acad. Sci. U.S.A.*, 77, 4666, 1980.
12. **Bayer, E. A., Ben-Hur, H., Hiller, Y., and Wilchek, M.,** Postsecretory modifications of streptavidin, *Biochem. J.*, 259, 369, 1989.
13. **Suter, M. and Butler, J. E.,** The immunochemistry of sandwich ELISAs. II. A novel system prevents denaturation of capture antibodies, *Immunol. Lett.*, 13, 313, 1986.
14. **Suter, M., Cazin, J., Jr., Butler, J. E., and Mock, D. M.,** Isolation and characterization of highly purified streptavidin obtained in a two-step purification procedure from *Streptomyces avidinii* grown in a synthetic medium, *J. Immunol. Methods,* 113, 83, 1988.
15. **Suter, M., Butler, J. E., and Peterman, J. H.,** The immunochemistry of sandwich ELISAs. III. The stoichiometry and efficacy of the Protein-Avidin-Biotin Capture (PABC) system, *Mol. Immunol.*, 26, 221, 1989.
16. **Mock, D. M. and Dubois, D. B.,** A sequential solid-phase assay for biotin in physiological fluids that correlates with expected biotin status, *Anal. Biochem.*, 153, 272, 1986.
17. **Butler, J. E., Spradling, J. E., Suter, M., Dierks, S. E., Heyerman, H., and Peterman, J. H.,** The immunochemistry of sandwich ELISAs. I. The binding characteristics of immunoglobulins to monoclonal and polyclonal antibodies adsorbed on plastic and their detection by symmetrical and asymmetrical antibody-enzyme conjugates, *Mol. Immunol.*, 23, 971, 1986.
18. **Peterman, J. H.,** Factors which Influence the Binding of Antibody to Solid-Phase Antigens: Theoretical and Experimental Investigations, Ph.D. thesis, University of Iowa, Iowa City, 1986.
19. **Peterman, J. H., Tarcha, P .J., Chu, V. P., and Butler, J. E.,** The immunochemistry of sandwich-ELISAs. IV. The antigen capture capicity of antibody covalently attached to bromoacetyl surface-functionalized polystyrene, *J. Immunol. Methods,* 111, 271, 1988.
20. **Goding, J. W.,** *Monoclonal Antibodies: Principles and Practice,* Academic Press, London, 1983, 188.
21. **Updyke, T. V. and Nicolson, G. L.,** Immunoaffinity isolation of membrane antigens with biotinylated monoclonal antibodies and immobilized streptavidin matrices, *J. Immunol. Methods,* 73, 83, 1984.
22. **O'Shannessy, D. J., Dobersen, M. H., and Quarles, R. H.,** A novel procedure for labeling immunoglobulins by conjugation to oligosaccharide moieties, *Immunol. Lett.*, 8, 273, 1984.
23. **Sano, T. and Cantor, C. R.,** Expression of a cloned streptavidin gene in *Escherichia coli, Proc. Natl. Acad. Sci. U.S.A.*, 87, 142, 1990.
24. **Suter, M.,** unpublished observation.

Chapter 18

THE KINETICS OF SOLID-PHASE MICROTITER IMMUNOASSAYS

Burkart Franz and Michael Stegemann

TABLE OF CONTENTS

I. INTRODUCTION

The convenience of enzyme-linked immunosorbent assay (ELISA) technology resides in its solid-phase configuration; immobilization of either antigens or antibodies on the solid phase allows easy separation of bound and free reactants. However, the immunochemistry and kinetics of solid phase antigen-antibody reactions differ significantly from classical liquid-phase reactions involving haptens and antibodies in solutions.[1,2] First, the immobilized reactant can no longer diffuse and its contribution to the collision rate is eliminated. Second, the average nearest distance between antigen and antibody is greatly increased.

The kinetics of antibody binding to solid-phase antigen are of importance in the design and performance of a solid phase immunoassay. The standardization of ELISA tests would benefit by the establishment of conditions whereby time dependent reactions could reach equilibrium.

The aim of this study is to investigate the kinetics of binding of antibodies to an antigen adsorbed on microtiter wells. These empirical data should provide a method to facilitate attainment of equilibrium conditions within a reasonable period of time. Theoretical models to explain the characteristics of the binding of antibodies to antigens are discussed in Chapter 20.

II. MATERIAL AND METHODS

Monoclonal antibodies (mAbs) against bacterial type 1-fimbriae were chosen as a model for these investigations. The adhesiveness of pathogenic bacteria to host-cell surfaces is often correlated with the expression of extracellular fimbriae. The rigid structure of type 1-fimbriae is determined by the structural protein pilin, whereas the adhesiveness is mediated by the adhesin. Fimbriae are rather immunogenic and can be used as antigen for the differentiation of enterobacteriaceae.[3] Studies on the binding of fimbriae with their adhesin to solid phases suggest the structure and antigenicity of the pilin are maintained.[4]

The recombinant *Escherichia coli* strain HB101 pGG101 expressing type 1-fimbriae of an uropathogenic isolate of *Klebsiella pneumoniae* was used. The gene encoding the major fimbrial subunit was examined by nucleotide sequence analysis and four possible antigenic sites were determined.[5] Bacteria were grown in nutrient broth and fimbriae were concentrated by precipitation with $(NH_4)_2SO_4$ after mechanical detachment from the surface of the bacteria. Minor proteins, outer membrane proteins, and lipopolysacharides were separated from the fimbriae by preparative high-pressure liquid chromatography. The purity of the pilin was confirmed by sodium dodecyl sulfate-polyacrylamide gel electrophoresis and Western Blot analysis. Pure pilin had a molecular weight of 18 kDa.

The production of mAb in BALB/c mice was initiated by intraperitoneal immunization with 20 μg fimbrial protein. The animals were given three booster injections (100 μg of pure pilin) on three consecutive days. After 60 h the last booster injection was given and spleen cells were fused with the myeloma line NS-O/1.[6] Hybrid clones were screened for activity with purified pilin and isolated (phenol-water method) lipopolysacharides of *E. coli* strain HB101 using ELISA. Cloning was performed by limiting dilution. MAbs were purified by affinity chromatography using goat anti-mouse immunoglobulins linked to sepharose. The subisotypes of the mAbs were determined using specific antibodies (Sigma, München, Federal Republic of Germany). Four mAbs (6-C-11, 12-C-5, 4-H-4, and 8-H-4), all of the IgM isotype, were chosen to determine the maximum time required to reach equilibrium conditions.

Epitope analysis was performed by a competition ELISA using biotinylated mAbs. Each of the four antibodies described above was able to inhibit the binding of any of the other three antibodies by at least 75%. This finding is interpreted to mean that all four mAbs recognize the same epitopes or several very closely spaced epitopes on the pilin molecule.

Removawells (Dynatech, Denkendorf, West Germany) were coated with pilin (protein concentration 1 μ g/ml bicarbonate buffer) and used as the solid-phase antigen. The mAbs were iodinated with ^{125}Iodine using the method described by Markwell[7] using 0.5 μ Ci/μ g protein. Iodinated mAbs were purified on a sucrose gradient. The iodinated mAbs were added to antigen-coated microtiter wells in triplicates at concentrations of 300, 150, 75, 37.5, 18.3, 9.2, and 4.6 ng protein per 50 μ l phosphate buffered saline (PBS). The antigen-antibody reaction was allowed to proceed (see below for incubation times and conditions of shaking) at 20°C and the liquid phase of each well was then removed. The solid phase was rinsed and the amounts of free and bound radioactivity were determined in a LKB γ-counter. The proportion of [bound/(bound + free radioactivity)] was calculated.

All data were expressed in percent of maximal amount of antibody capable of binding. This amount was determined in a preliminary experiment by a fourfold consecutive transfer of the iodinated antibodies into another antigen-coated microtiter plate.

Three separate experiments were conducted using the iodinated mAbs following the general procedure described above.

Experiment 1 — The effect of mechanical shaking of microtiter plates on the binding of mAb 6-C-11 was tested. This mAb was incubated for 7.5, 15, 30, and 60 min at rotation speeds of 0, 400, 800, and 1200 rpm on a microplate shaker (IKA-Schüttler MTS 4; Janke and Kunkel, Erlangen, West Germany).

Experiment 2 — Differences in the kinetics of stable binding among mAbs 6-C-11, 12-C-5, 4-H-4, and 8-H-4 were investigated. MAbs were incubated for 7.5, 15, and 30 min and 1, 2, 4, 8, 16, and 32 h at rotation speeds of 0 and 1200 rpm on a microplate shaker.

Experiment 3 — The rate of dissociation of IgM antipilin was studied. MAb 6-C-11 was allowed to bind for 2 h at 1200 rpm. The liquid phase was then replaced by PBS containing 1, 2, and 4 times the amount of pilin used for coating of the microtiter plate. After 7.5, 15, and 30 min and again after 2, 4, 8, and 16 h, the proportion of bound radioactivity was determined. Furthermore, two experiments were performed to investigate other time-dependent processes often involved in multistep ELISA systems.

Experiment 4 — The effect of mechanical shaking of microtiter plates on the binding of mAb 6-C-11 to an uncoated microtiter plate was tested. The iodinated MAb was incubated for 7.5, 15, and 30 min and 1, 2, 4, 8, and 16 h at rotation speeds of 0, 400, 800, and 1200 rpm on a microplate shaker.

Experiment 5 — The effect of mechanical shaking of microtiter plates on the binding of streptavidin to a biotinylated protein adsorbed to a microtiter plate was tested. Streptavidin was produced and purified as described by Suter et al.[8] The purity of the streptavidin was confirmed by sodium dodecyl sulfate-polyacrylamide gel electrophoresis. The strepatavidin was iodinated as described above and the iodinated protein was incubated for 7.5, 15, and 30 min and 1, 2, 4, 8, and 16 h in a microtiter plate which had been coated with 1 μ g biotinylated human serum albumine per millilter bicarbonate buffer at rotation speeds of 0 and 1200 rpm on a microplate shaker.

III. RESULTS

The results of Experiment 1 are shown in Figure 1. The data indicate that shaking of the microtiter plate accelerates the stable binding of monoclonal IgM to antigen adsorbed on the well surface. A shaking speed of 1200 rpm increased the amount of antibody bound by at least a factor of 4.

Figure 2 presents the results of Experiment 2 and shows that the binding of monoclonal IgM to antigen adsorbed to an unshaken microtiter plates lasts 16 to 32 h. Rotation of microtiter plates at 1200 rpm reduced the time required to reach equilibrium conditions to 1 h.

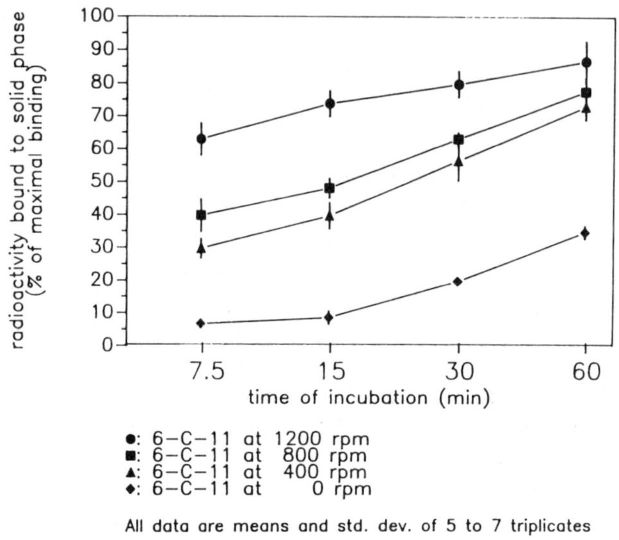

FIGURE 1. The influence of the speed of agitation on the binding kinetics of monoclonal antibody 6-C-11 to bacterial pilin adsorbed to microtiter plates.

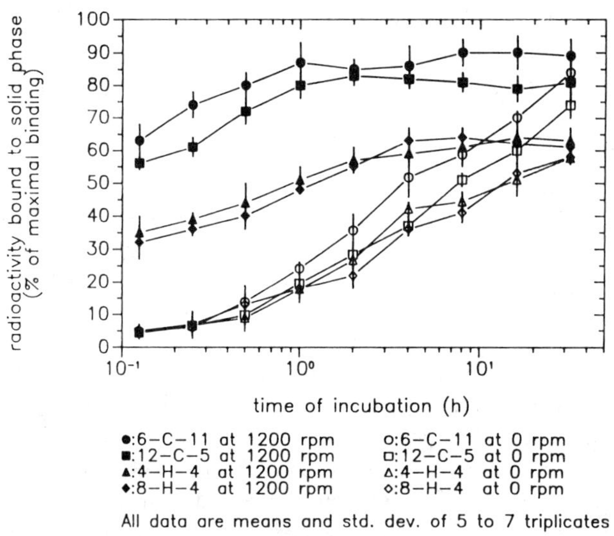

FIGURE 2. The influence of agitation on the binding kinetics of four monoclonal antibodies to bacterial pilin adsorbed to microtiter plates.

The results of Experiment 3 indicate that dissociation of mAb 6-C-11 from the solid phase pilin antigen could not be observed during the first 16 h even in the presence of free antigen.

Figure 3 presents the results of Experiment 4. Whereas both slow rotation speeds (400 and 800 rpm) seem to have little effect on the time necessary to reach equilibrium conditions, as far as the coating of a microtiter plate is concerned, at 1200 rpm this time is drastically reduced.

Figure 4 shows that the stable binding of streptavidin to a biotinylated immobilized protein is greatly accelerated by agitation at 1200 rpm. After 15 min almost 85% of the maximal amount capable of binding has bound to the solid phase.

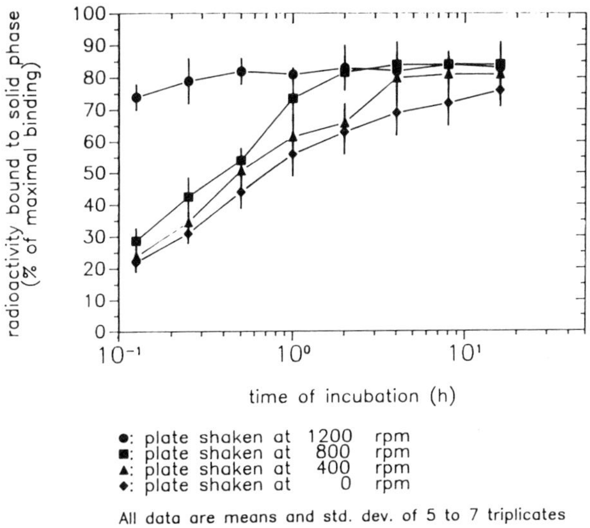

FIGURE 3. The influence of the speed of agitation on the binding kinetics of monoclonal antibody 6-C-11 to an uncoated microtiter plate.

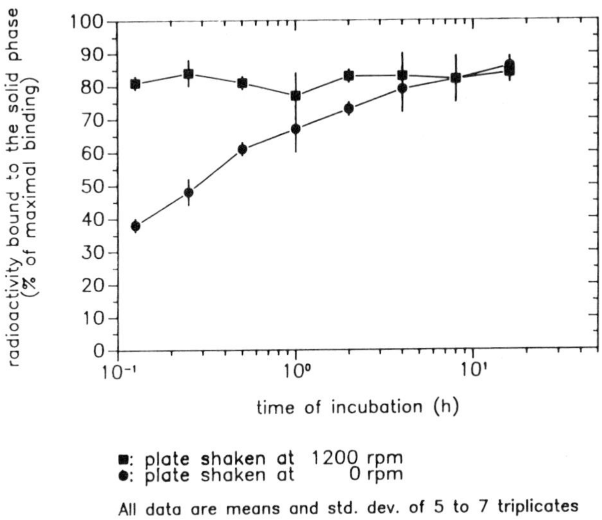

FIGURE 4. The influence of agitation on the binding kinetics of streptavidin to biotinylated protein adsorbed to microtiter plates.

IV. DISCUSSION

The effect of agitation on the equilibrium kinetics of the binding of antibodies to surface-immobilized antigen was investigated in this study. It was demonstrated that by agitation at high speed (1200 rpm) equilibrium conditions could be reached within 1 h. The time required to reach equilibrium without agitation was 16 to 32 h. This value seems to be compatible with the findings of others[9] concerning the binding kinetics of antibody-enzyme conjugates. These investigators showed that maximal binding of their conjugate occurs after 12 h.

The result obtained that 2 of the antibodies tested (6-C-11 and 5-C-12) have reached equilibrium at 85% and the other 2 (4-H-4 and 8-H-4) at 65% might be correlated to the affinity of the antibodies. In order to verify this hypothesis the affinity of all four antibodies was determined according to the thiocyanate elution method described by Macdonald et al.[10] It was found that the mAbs 6-C-11 and 5-C-12 can be eluted using a concentration of 1.84 and 2.08 mol/l thiocyanate, respectively, whereas the mAbs 4-H-4 and 8-H-4 can be eluted using as little as 0.35 and 0.44 mol/l, respectively (details not shown).

These data provide support for the concept of affinity dependence of the binding of antibodies to adsorbed antigens.

Furthermore, it could be shown that the binding of antibodies to an uncoated solid phase and the binding of streptavidin to a biotinylated protein adsorbed on a solid phase can be greatly accelerated by agitation.

In a solid phase system, immunologically active reactants, i.e., antigen and antibody, are not uniformly distributed. The distance between antibody and antigen is not the same for all antibodies in the solution: antibodies close to the solid phase collide and react faster with antigens on the solid phase than antibodies in solution in the middle of the microtiter well. Therefore, several compartments exist in the reaction vessel and the overall binding of antibodies with antigens consist of (1) the transport of antibodies to the solid-phase surface, (2) the binding between antibodies and antigens, (3) the dissociation of reactants, and (4) the transport of dissociated reactants back into the solution.

The first and the last step of this reaction are limited by the speed of diffusion of a molecule in an aqueous solution depending on its hydrodynamic radius according to the relationship (1).

$$D = k_B t/6\pi\eta r \tag{1}$$

where D is the diffusion constant, k_B is the Boltzmann constant, η is the viscosity of the fluid, and r is the hydrodynamic radius of the molecule.

In the absence of agitation, the movement of antibodies by diffusion is described by Fick's law of diffusion (2).

$$dS/dt = -DA\ dc/dx \tag{2}$$

where S is the number of diffusing molecules, D is the diffusion coefficient, c is the concentration gradient, and x is the direction in which the transport occurs.

The amount of antibodies bound to an antigen-coated solid phase is described by Equation (3).

$$d[AgAb]/dt = k_1[Ag_s][Ab] - k_2[AgAb] \tag{3}$$

where k_1 is the association and k_2 is the dissociation constant, [AgAb] is the molar concentration of the antigen-antibody complex, $[Ag_s]$ is the surface concentration of the solid phase antigen, and [Ab] is the antibody concentration.

A detailed examination of this model is presented by DeLisie and Metzger.[11] At low concentrations of solid-phase antigen, the rate of the bimolecular reaction would be the time-limiting process; diffusion is of less importance in determining the overall rate of the reaction. At high concentrations of solid-phase antigen, this reaction may occur so fast that the rate-limiting mechanism is the transport of antibodies to the solid phase.[12]

When the microtiter plate is vigorously agitated, enough energy for the transport of antibodies to the solid-phase antigen is provided so that diffusion is of less importance.

At high shaker-rotation speeds a second mechanism to facilitate the attainment of equilibrium might also be involved: the geometry of the fluid within the microtiter plate is be altered. Agitation at 1200 rpm results in a larger area of the surface of the microtiter well being exposed to the soluble reactants; the area covered by 50 μ l of PBS is increased by 50% (from

75 mm^2 to 115 mm^2; data obtained from high-speed photography), assuming a cylindrical geometry for the microtiter well. As the volume of the fluid in the well remains unchanged, the diffusion distance is reduced.

This could also contribute to the shorter time required for antibodies to reach equilibrium with solid-phase antigen because the reaction rate is proportional to the diffusion distance raised to the inverse third power.

We also performed an experiment to determine if the binding of monoclonal IgM to absorbed pilin was reversible. Our data indicate that no significant dissociation of antibodies from the solid phase could be observed. For liquid-phase reactions, it has been reported that the binding of antibodies to a hapten in solution is governed primarily by the dissociation constant rather than by the association constant.[13,14] This contradicts findings from studies on dissociation constants of reactants binding to cellular or artificial surfaces; it was reported[15] that the dissociation rate of antibodies from the surface of bacteria was slower than the dissociation rate of antibodies from antigens in solutions. These effects were only partly ascribed to multivalency, which was suspected by Crothers and Metzger,[16] because it was also demonstrated using F_{ab} fragments.

No dissociation of antibody from solid-phase antigen was seen in the absence of antigen.[17] This effect was suspected to be caused by the lack of mass transport away from the reaction volume.[18]

Collectively considered, these findings indicate that we are not yet capable of understanding the mechanism(s) of the binding of antibodies to antigens immobilized on a solid phase. Our failure to understand such reactions complicates the standardization of *in vitro* tests.

REFERENCES

1. **Kabat, E. A.,** Basic principles of antigen-antibody reactions, in *Methods of Enzymology,* Vol. 70, Colowick, S. P. and Kaplan, N. O., Eds., Academic Press, New York, 1980, 3.
2. **Karush, F.,** The affinity of antibody: range, variability and the role of multivalence, in *Comprehensive Immunology,* Vol. 5, Litman, G. W. and Good, R. A., Eds., Plenum Press, New York, 1978, 85.
3. **Orskov, I. and Orskov, F.,** Serology of *Escherichia coli, Prog. Allergy,* 33, 80, 1983.
4. **Brinton, C. C.,** The structure, function, synthesis and genetic control of bacterial pili and a molecular model for DNA and RNA transport in gram negative bacteria, *Trans. N.Y. Acad. Sci.,* 27, 1003, 1965.
5. **Gerlach, G. F. and Clegg, S.,** Characterization of two genes encoding antigenically distinct type-1 fimbriae of *Klebsiella pneumoniae, Gene,* 64, 231, 1988.
6. **Galfre, G. and Milstein, C.,** Preparation of monoclonal antibodies, strategies and procedures, in *Methods in Enzymology,* Vol. 73, Langone, J. J. and van Vunakis, H., Eds., Academic Press, Orlando, FL, 1981, 1.
7. **Markwell, M. A. K.,** A new solid-state reagent to iodinate proteins, *Anal. Biochem.,* 125, 427, 1982.
8. **Suter, M., Cazin, J., Butler, J. E., and Monck, D. M.,** Isolation and characterization of highly purified streptavidin obtained in a two-step purification procedure from *Streptomyces avidinii* grown in a synthetic medium, *J. Immunol. Methods,* 113, 83, 1988.
9. **Engvall, E., Jonsson, K., and Perlmann, P.,** Enzyme-linked immunosorbent assay. II. Quantitative assay of protein antigen, immunoglobulin G, by means of a enzyme-labeled antigen and antibody coated tubes, *Biochim. Biophys. Acta,* 251, 427, 1971.
10. **Macdonald, R. A., Hosking, C. S., and Jones, C. L.,** The measurement of relative antibody affinity by ELISA using thiocyanate elution, *J. Immunol. Methods,* 106, 191, 1988.
11. **DeLisie, C. and Metzger, H.,** Some physical chemical aspects of receptor-ligand interactions, *Immunol. Commun.,* 5, 417, 1976.
12. **Nygren, H. and Stenberg, M.,** Kinetics of antibody binding to solid phase immobilized antigen: influence of mass transport on the ELISA, *J. Colloid Interface Sci.,* 107, 560, 1985.
13. **Froese, A.,** Kinetic and equilibrium studies on 2,4-dinitrophenyl hapten antibody systems, *Immunochemistry,* 5, 253, 1968.
14. **Smith, T. W. and Skubitz, K. M.,** Kinetics of interactions between antibodies and haptens, *Biochemistry,* 14, 1496, 1975.

15. **Mason, D. W. and Williams, A. F.,** The kinetics of antibody binding to membrane antigens in solution and at the cell surface, *Biochem. J.,* 187, 1, 1980.
16. **Crothers, D. M. and Metzger, H.,** The influence of polyvalency on the binding properties of antibodies, *Immunochemistry,* 9, 341, 1972.
17. **Nygren, H., Werthén, M., and Stenberg, M.,** Kinetics of antibody binding to solid phase immobilised antigen. Effect of diffusion rate limitation and steric interaction, *J. Immunol. Methods,* 101, 63, 1987.
18. **Berzwsky, I. A. and Berkower, I. J.,** Antigen-antibody interaction, in *Fundamental Immunology,* Paul, W., Ed., Raven Press, New York, 1984, 595.

Chapter 19

RATE LIMITATION OF ANTIGEN-ANTIBODY REACTIONS: THEORETICAL AND PRACTICAL ASPECTS

Håkan Nygren and Manne Stenberg

TABLE OF CONTENTS

I. INTRODUCTION

The antigen-antibody reaction is essentially a reaction between a water-soluble macromolecule, the immunoglobulin, and its antigen epitope in a water environment. The forces involved in this reaction are van der Waals forces, hydrophobic interaction, electrostatic forces, and hydrogen binding.[1,2] These forces are generally operative in water solutions of macromolecules, but have been adapted by the structure of antibodies for the creation of sites specific for binding to their antigens. This specificity of antibodies is often exploited in their use as probes in immunoassays.

When interpreting results of immunoassays or trying to optimize assays, some assumptions have to be made regarding the mechanism of the reactions involved. In practice, the time and concentration dependence is most important. The aim of the present chapter is to review experimental studies on the rate limitation of antigen-antibody reactions and theoretical models currently used to explain the experimental results.

II. KINETICS OF ANTIGEN-ANTIBODY REACTIONS

A. THEORETICAL BACKGROUND

Studies on the kinetics of antigen-antibody reactions are undertaken to identify the rate limiting step in the bimolecular reaction

$$A + B \underset{k_{-0}}{\overset{k_0}{\rightleftharpoons}} \ldots \underset{k_{-1}}{\overset{k_i}{\rightleftharpoons}} (A \cdot B)_i \ldots \underset{k_{-n}}{\overset{k_n}{\rightleftharpoons}} AB \tag{1}$$

where A and B are the two reactants and AB is the final complex formed after n different intermediate states $(A \bullet B)_i$. The notations k_i are the different forward reaction rates, k_{-i} are the different reverse reaction rates. For multivalent reactants A or B the intermediate states $(A \bullet B)_i$ may represent states with different amounts of multiple binding. In the above reaction scheme the different reaction rates are most probably very different in magnitude. Normally, both the overall forward reaction rate and the overall reverse reaction rate are limited by one particular rate constant and the overall reaction scheme can be simplified to

$$A + B \underset{k_r}{\overset{k_f}{\rightleftharpoons}} AB \tag{2}$$

where k_f is the effective forward reaction rate constant and k_r is the effective reverse reaction rate constant. By considering the intermediate states as quasistationary, simplified expressions for the overall rate constants can be obtained. This procedure of reducing a complicated reaction scheme has been used in many studies; it has been used when analyzing the reaction of bivalent antibodies[3] as well as when studying the reaction of ligands with cell-surface receptors.[4,5]

Since we are dealing with very fast macromolecular reactions we must consider possible diffusion control in the reaction. One way to do this is to identify the initial intermediate state $(A \bullet B)_0$ as an encounter complex" which must first be formed before the reaction can continue. Formation of encounter complexes has been suggested in experimental studies in order to explain data on the forward reaction rate of antigen-antibody reactions and hapten-antibody reactions.[5-7] Let us study the simplified reaction scheme

$$A + B \underset{k_{-0}}{\overset{k_0}{\rightleftharpoons}} (A \cdot B)_0 \underset{k_{-1}}{\overset{k_1}{\rightleftharpoons}} AB \tag{3}$$

where the rate constant k_0 represents the diffusion rate constant with which the two reactants are appropriately positioned for reaction. It may involve both translational and rotational diffusion. Correspondingly, the rate constant k_{-0} represents the reverse diffusion rate constant with which are encounter complex can dissociate. The formation of an encounter complex is followed by a single reaction step with forward reaction rate k_1 and reverse rate k_{-1}.

By considering the concentration of the intermediate encounter complex as quasistationary ($d/dt\ ([(A \bullet B)_0]) \approx 0$) we get the simplified expressions for the overall rate constants k_f and k_r.

$$k_f = \frac{k_1 k_0}{k_1 + k_{-0}} \tag{4}$$

$$k_r = \frac{k_{-1} k_{-0}}{k_1 + k_{-0}} \tag{5}$$

The effective forward rate constant can be written as $k_f = a\, k_0$ in which we have introduced a "sticking coefficient" a which represents the reaction of successful encounter formations.

$$a = \frac{k_1}{k_1 + k_{-0}} \tag{6}$$

For $k_1 \gg k_{-0}$ we get $a \approx 1$ and diffusion limitation in the overall reaction with a forward reaction rate $k_f = k_0$. The overall reaction will be reaction limited if $k_1 \ll k_{-0}$ and the forward reaction rate is reduced by a factor $a < 1$ compared to the diffusion rate limited forward reaction rate k_0.

B. RATE LIMITATION OF ANTIGEN-ANTIBODY REACTIONS

1. Rate Limitation of Hapten-Antibody Reactions in Solution

The diffusion forward reaction rate k_0 depends mainly on the viscosity of the surrounding medium. It also depends on a geometrical factor which has a minimum for reactants A and B of equal size and uniform reactivity. For reactants of different sizes the diffusion rate constant increases and for reactants with limited reactive sites the diffusion reaction rate decreases. When these factors are considered, the diffusion limited reaction rate for a hapten-antibody reaction should be of the order of $1 \bullet 10^9\ M^{-1}\ s^{-1}$.[8-10] The fastest hapten-antibody reactions are a factor of two to ten slower than this value and it is generally accepted that these reactions are reaction limited and we can conclude that normally the sticking coefficient a is less than one.

Experiments performed to elucidate what factors can influence the reaction rate of hapten-antibody reactions have shown that the forward reaction rate depend on the chemical properties of the hapten. The reaction between dinitrophenyl-derivatives or fluorescein derivatives and antibodies proceeds with rate constants k_f of the order of $1 \bullet 10^7$ to $1 \bullet 10^8\ M^{-1}\ s^{-1}$.[11,12] The value of the rate constant of the reaction between cyclic AMP and antibodies is $7 \bullet 10^5\ M^{-1}\ s^{-1}$.[13] and the reaction between a carbohydrate hapten and antibodies proceeds with rate constants of the order of 10^5-$10^6\ M^{-1}\ s^{-1}$.[14] Chemical modification of dinitrophenyl derivatives changes the forward reaction rate by a factor of two to three.[7] The chemical modification of the hapten should not significantly alter its size and the forward diffusion rate constant k_0. This indicates that the change in reaction rate results from a change in the sticking coefficient, a.

Different monoclonal antibodies (mAbs) against fluorescein derivatives have been shown to differ in their forward reaction rate by as much as a factor of 60, but 4 out of 5 mAbs showed forward rates that varied within a factor of 10.[15] This variation is almost certainly the result of different sticking coefficients since the reactants involved have similar diffusion properties.

The ionic composition and pH of the reaction medium may have large affects on the forward rate of hapten-antibody reactions. The fluorescein-antifluorescein reaction measured by fluorescence polarization shows a reaction rate constant of 1.4 • 10^8 M^{-1} s^{-1} in phosphate buffer.[11] The reaction rate decreased to 1.3 • 10^7 M^{-1} s^{-1} in thiocyanate-containing medium, and further down to 1.3 • 10^5 M^{-1} s^{-1} in salicylate-containing medium.[11] Thus, the ion content of the reaction medium can modify the reaction rate by up to a factor of 1000. A change of pH of the reaction medium from pH 8.0 to 6.0 has been shown to alter the forward reaction rate of the fluorescein-antifluorescein reaction by a factor of 100 for 1 mAb but had little effect on the reaction rate of other mAbs.[15] The conditions in the reaction medium, e.g., pH and ionic strength, may influence the diffusion properties of macromolecules through the formation of aggregates. However, the finding that those ions which promote molecular association increase the reaction rate while those ions which promote dissociation decrease the reaction rate suggests that aggregation is not the cause of the changes seen. Most probably, ions have an effect on the sticking coefficient. In conclusion, the rate of hapten-antibody reactions in solution is reaction limited with a forward rate constant in the range of 6•10^8 to 1 • 10^5 M^{-1} s^{-1}. Most experiments performed show reaction rates of the order of 10^6 to 10^8 M^{-1} s^{-1} which is a factor of 10 to 1000 slower than the diffusion-limited rate.

2. Rate Limitation of Antibody Reactions with Macromolecular Antigens

The diffusion-limited forward reaction rate of macromolecules is influenced by the fact that only a small region of the molecules take part in the initial reaction (for a review of diffusion-controlled macromolecular interactions see Reference 16). Calculation of the diffusion rate constant for a reaction between antibodies and macromolecular antigens give values in the order of 1•10^7 M^{-1} s^{-1}.[10]

In experimental studies performed to analyze the forward reaction rate of reactions between antibodies and protein antigen, values of 1•10^5 to 1.4 •10^6 M^{-1} s^{-1} are usually obtained.[12,17,18] These values are a factor of 10 to 100 slower than the rate constant of a diffusion-limited reaction. Thus, antibodies react with approximately the same sticking coefficient with haptens and macromolecular antigens. Considering only the absolute values of the reaction rate constant leads to the conclusion that there is a higher activation barrier in the reaction of antibody with macromolecules than with haptens. However, if diffusion limitations are also accounted for there is no need to assume higher activation barriers to explain the experimental data. The difference in reaction rate of the DNP-anti DNP reaction in solution ($k_1 \approx 1 \cdot 10^8$ M^{-1} s^{-1}) and the same reaction with DNP immobilized at the surface of the bacteriophage Ø174 ($k_1 \approx 6 \cdot 10^6$ M^{-1} s^{-1})[12] is an experimental finding that does not necessarily result from different activation barriers for these reactions, but may well be explained by differences in diffusion orientation of the macromolecule which affects the diffusion forward rate constant k_0.

The ionic content of the reaction medium has been shown to influence the rate of the reaction between a protein antigen and antibodies. In tris-buffer, the reaction rate followed second order kinetics, but in phosphate buffer the reaction rate was scaled down by a factor $[Ab]^{0.6}$ which means that the reaction has a fractional rate order.[17] In a study on protein-immobilized haptens, fractional rate order was found for bivalent antibodies but not for their Fab fragments. Fractional reaction order was also found to correlate with the presence of nonchaotropic ions in the reaction medium, whereas the presence of chaotropic ions correlated with simple second-order rate kinetics.[6] This effect of chaotropic ions on the antigen-antibody reaction can only be explained by assuming secondary effects on the reaction between bivalent antibodies and protein-immobilized hapten. The authors suggest heterogeneity of the hapten-protein sites, conformation changes, or loss of water from the binding site as possible secondary effects.[6] A possible formation of aggregates may also be added to the list.

3. Rate Limitation of Antigen-Antibody Reactions at Surfaces

The diffusion forward reaction rate for bimolecular reactions where the sites operate independently at surfaces is further reduced by the limited radius of the reactive site and may well be of the order of 10^6 M^{-1} s^{-1}.[10] The rate of antibody reactions at cell surfaces, measured experimentally, is often in the order of 10^4 to 10^6 M^{-1} s^{-1} [19,20] which is close to diffusion rate limitation in some experiments but most often a factor of 10 to 100 slower than the diffusion forward reaction rate. Thus, there is no need to introduce other reaction mechanisms for the binding of antibodies to cell surfaces than those discussed above for the reaction of antibodies with haptens and macromolecules.

Immobilization of antigens at solid phases introduces possible cooperative effects on the antibody reaction. One well-documented effect is the depletion of antibodies in the solution close to the surface. This phenomenon occurs at high antigen density and makes the antigen-antibody reaction mass transport limited although the intrinsic bimolecular reaction at the surface may well be reaction rate limited (for a review see Reference 10).

Experimental measurements of antigen-antibody reactions at solid surfaces has shown that the initial reaction rate is often diffusion limited as a result of the depletion of antibodies in the reaction zone. At higher surface concentrations of bound antibodies, the reaction rate decreases and is often of the order of 10^3 M^{-1} s^{-1}.[21,22] Recently it was found that the decrease of the reaction rate was a continous process that could be described by assuming a time-dependent reaction rate constant[23] such that $k_1 = k(t) \propto t^{-h}$ where $0 \leq h \leq 1$.[24] The intrinsic forward rate constant for the antigen antibody reaction can be evaluated by use of the time dependence of the effective reaction rate coefficient. Initially, the forward reaction rate constant is $0.5 \bullet 10^5$ M^{-1} s^{-1};[23] this is of the same order of magnitude as forward reaction rates of homogeneous antibody reactions with protein antigens or at cell surfaces. This fact implies that the fractal kinetics of the solid phase reaction is the result of secondary effects, not a change in the intrinsic reaction rate. Thus, it becomes necessary to propose additional phenomena to explain the fractal kinetics of antigen-antibody reactions on solid phases.

Although the exact reaction mechanism of antibody-antigen reactions at interfaces is still unknown, the formalism for fractal reaction kinetics seem to be a fruitful way of describing experimental data. The time dependence of the forward reaction rate constant has been suggested to stem from nonrandomness in the distribution of molecules reacting at surfaces.[24] Theoretical models for the relation between the geometry of the distribution of reactant at the surface and the time dependence of the forward reaction rate constant are based on force-influenced brownian motion[25] or diffusion on fractal clusters.[24] Such models are still too etheric to allow direct comparison with the experiments presented above.

Experimentally, fractional reaction orders are not a new finding in immunochemistry. Oudin[26] found that the time dependence of antigen-antibody precipitation reactions was proportional to $t^{0.5}$. Levison and Dandliker[6,17] found that the rate of bivalent antibody reactions with protein antigen was proportional to $[Ab]^{0.5}$ or $[Ab]^{0.6}$ and Werthen et al.[23] found that the rate of antibody binding to surface-immobilized antigen was proportional to $t^{0.5}$ - $t^{0.7}$. A theoretical explanation to these experimental findings may be found in computer simulations of fractal kinetics.

III. PRACTICAL ASPECTS

The mass transport limitation of the initial antigen-antibody reaction at interfaces is often important since sensitive solid phase immunoassays measure the antigen-antibody reaction in this regime. Surprisingly, the dilution curves of an ELISA assay can often be explained by diffusion limitation. The assay measures the concentration of antibodies in the sample, and the diffusion limitation can be used as a convenient means of calibration.[10]

The finding that antigen-antibody reactions at interfaces seldom reach equilibrium within the time range used in experiments has important implications for the design of experiments and interpretation of the results of immunoassays. Competitive binding of reagents does not necessarily apply to equilibrium conditions and interpretation of such data becomes hazardous.

Fractal kinetics demonstrated for antigen-antibody reactions at interfaces also has unexpected consequences. The replacement of a time-independent rate constant used in the classical theory by a time-dependent reaction rate coefficient used in theories of fractal kinetics makes the reaction "history dependent".[24] Another consequence of the fractal reaction kinetics is the difficulty of attaining equilibrium due to the very long time constants involved. These possibilities must be considered when interpreting data from solid-phase immunoassays.

REFERENCES

1. **Karush, F.,** Immunologic specificity and molecular structure, *Adv. Immunol.,* 2, 1, 1962.
2. **van Oss, C. J. and Absolom, D. R.,** The physicochemical nature of the antigen-antibody interaction, in *Antibodies: Protective, Destructive and Regulatory Role,* Milgrom, Abeyounis, and Albini, Eds., S. Karger, Basel, 1985, 67.
3. **Perelson, A. S., Goldstein, B., and Rocklin, S.,** Optimal strategies in immunology. III. The IgM-IgG switch, *J. Math. Biol.,* 10, 209, 1980.
4. **Dembo, M., Goldstein, B., Sobotka, A. K., and Lichtenstein, L. M.,** Histamine release due to bivalent penicilloyl haptens: the relation of activation and desensitization of basophils to dynamic aspects of ligand binding to cell surface antigody, *J. Immunol.,* 122, 518, 1979.
5. **De Lisi, C.,** The effect of cell size and receptor density on ligand-receptor rate constants, *Mol. Immunol.,* 18, 507, 1981.
6. **Levison, S. A., Kierzenbaum, F., and Dandliker, W. B.,** Salt effects on antigen-antibody kinetics, *Biochemistry,* 9, 322, 1970.
7. **Haselkorn, D., Friedman, S., Givol, D., and Pecht, I.,** Kinetic mapping of the antibody combining site by chemical relaxation spectrometry, *Biochemistry,* 13, 2210, 1974.
8. **Wang, J. H.,** The role of diffusion in catalysis, *J. Phys. Chem.,* 59, 1115, 1955.
9. **Day, L. A., Sturtevant, J. M., and Singer, S .J.,** The kinetics of the reactions between antibodies to the 2,4 dinitrophenyl group and specific haptens, *Ann. N.Y. Acad. Sci.,* 103, 611, 1963.
10. **Stenberg, M. and Nygren, H.,** Kinetics of antigen-antibody reactions at solid-liquid interfaces, *J. Immunol. Methods,* 113, 3, 1988.
11. **Levison, S. A., Hicks, A. N., Portmann, A. J., and Dandliker, W. B.,** Fluorescence polarization and kinetic studies of antifluorescein antibody obtained at different stages of the immune response, *Biochemistry,* 14, 3778, 1975.
12. **Karush, F.,** The affinity of antibody: range variability and the role of multivalence, in *Immunoglobulins,* Litmann, G. W. and Good, R. A., Eds., Plenum Press, New York, 1978, 85.
13. **Denizot, F. C., Hirn, M. H., and Delange, M. A.,** Relationship between affinity and kinetics in a hapten-antibody reaction. Studies on anticyclic AMP antibodies, *Mol. Immunol.,* 16, 509, 1979.
14. **Zidovetski, R., Blatt, Y., and Pecht, I.,** A common mechanism of hapten binding to immunoglobulins and their heterologous chain recombinants, *Biochemistry,* 19, 2790, 1981.
15. **Kranz, D. M., Herron, J. N., and Voss, E. W.,** Mechanisms of ligand binding by monoclonal anti-fluorescyl antibodies, *J. Biol. Chem.,* 257, 6987, 1982.
16. **Berg, O. G. and von Hippel, P. H.,** Diffusion controlled macromolecular interaction, *Ann. Rev. Biophys. Chem.,* 14, 131, 1985.
17. **Levison, S. A. and Dandliker, W. B.,** Effect of phosphate ion on ovalbumin-antiovalbumin kinetics, as measured by fluorescence polarization and quenching techniques, *Immunochemistry,* 6, 253, 1969.
18. **Friguet, B., Djavadi-Ohaniance, L., and Goldberg, M. E.,** Polypeptide-antibody binding mechanism: conformational adaptation investigated by equilibrium and kinetic analysis, *Res. Immunol.,* 140, 355, 1989.
19. **Jerne, N. K., Henry, C., Nordin, A. A., Fuji, H., Koros, A. M. C., and Lefkovits, I.,** Plaque-forming cells: methodology and theory, *Transplant. Rev.,* 18, 130, 1974.
20. **Mason, D. W. and Williams A.F.,** Kinetics of antibody reactions and the analysis of cell surface antigens, in *Handbook of Immunology,* Vol. 1, Weir, D. M., Ed., Blackwell Scientific, Oxford, 1986, 38, 1.
21. **Nygren, H., Kaartinen, M., and Stenberg, M.,** Determination by ellipsometry of the affinity of monoclonal antibodies, *J. Immunol. Methods,* 92, 219, 1986.
22. **Nygren, H., Werthén, M., and Stenberg, M.,** Kinetics of antibody binding to solid-phase immobilized antigen. Effect of diffusion rate limitation and steric interaction, *J. Immunol. Methods,* 101, 63, 1987.

23. **Werthén, M., Stenberg, M., and Nygren, H.,** Theoretical analysis of the forward reaction of antibody binding to surface-immobilized antigen, *Prog. Colloid Polym. Sci.,* 82, 349, 1990.
24. **Kopelmann, R.,** Fractal reaction kinetics, *Science,* 242, 1620, 1988.
25. **Voss, R. F.,** Random fractals: characterization and measurement, in *Scaling Phenomena in Disordered Systems,* Pynn, R. and Skjeltorp, A., Eds., Plenum Press, New York, 1985, 11.
26. **Oudin, J.,** Graphic representation of specific precipitin reactions. Theoretical discussion, *Bull. Soc. Chim. Biol.,* 27, 254, 1945.

Chapter 20

A SUMMARY OF THE PRINCIPAL IMMUNOASSAY DATA ANALYSIS PACKAGES

Jeffrey H. Peterman

An important aspect of using solid-phase immunoassays (SPIA) is the method utilized to analyze the data from the assay. The general principles of such analysis are discussed elsewhere (see Chapter 3). There are many commercially available software packages for analyzing data from SPIA on personal computers. In developing data analysis software, the developer must balance between flexibility (which adds complexity) and simplicity. The developer must also balance between "user-friendly" data input/option screens and program performance (the former placing large demands on the computer's processor and memory). The compromises selected by the developers varies and this can be an important consideration when selecting a data analysis package. For example, a laboratory which performs a large number of very simple and well-established assays might be satisfied with a simple, fast, data analysis package. However, a laboratory performing varied, complex, poorly established assays, would need a flexible and versatile (i.e., complex) data analysis system. Many research laboratories have been unsatisfied with the commercially available software, finding that none offer the appropriate compromise for their uses. Under such circumstances, the best choice can be to have software custom developed.

The producers/distributors of some of the principal data analysis packages for ELISAs were contacted. They were asked whether their software supported each of the functions listed in the table below (Table 1). The table should not be considered comprehensive, as there are other suppliers of data analysis packages which were not contacted. Additionally, many of the functions can be performed by nondedicated software; for example, general purpose communications software (intended for use with modems, etc.) can be used to acquire the data from many plate readers, which can then be imported as ASCII files into spreadsheet or statistical software for analysis.

Information was obtained on the following aspects of the software.

Software — name of the software package. The names are generally registered trademarks of the software producer.

Company/phone numbers — Name and phone number of the principal producer/distributor. Some of the software packages are available from several distributors.

Price — Charge for the software as of January, 1990. The amount is subject to change and must be confirmed by the purchaser; many distributors offer substantial discounts in certain circumstances.

Computer — whether the software is available for the Apple Macintosh or the IBM PC. The specific minimum requirements (in terms of memory, graphics board, etc.) are not listed and should be obtained from the distributor.

Types of plot — Data may be plotted in the format of signal vs. concentration (nonlog), signal vs. log concentration (semi-log), log signal vs. log concentration (log-log), or logit signal vs. log concentration (logit-log). Some software packages will allow other types of plot, such as 1/signal vs. concentration.

Types of curve fit — The types of equation available for deriving the best curve to fit the data points are listed. Some software packages offer choices within each type of fit, such as third or fourth degree polynomial fits and some offer other types of fit which are not listed.

Kinetics function - Many software packages for ELISA systems allow the rate of substrate reaction to determined for each well. This value may then be used as the signal and has the

TABLE 1
Listing of Functions Performed by 12 SPIA Data Analysis Packages

	BioCalk	Delta-Soft	ELISA-AID[a]	ELISA+	ELISA-NALYSIS	Kineti-Calc	Ligand Data Calc	Micro-plate Manager	Microtek	Soft 2000	SOFT-max	Titersoft II
Software												
Company	Dynatech Laboratories, Inc.	Bio-Metalics Inc.	Robert Maciel Associates, Inc.	Meddata, Inc.	J. Peterman[b]	Bio-Tek Instruments, Inc.	David G. Rhoads Associates, Inc.	Bio -Rad	Denley Instruments Inc.	SLT Laboratory Instruments, Inc.	Molecular Devices Inc.	ICN Bio-medicals/ Flow labs, Inc.
Phone number	(800) 336-4543	(800) 899-1961	(617) 646-3627	(212) 260-7320	(301) 949-7127	(800) 851-5172	(215) 288-2626	(800) 824-6723	(800) 836-5391	(800) 828-4837	(8000) 835-5577	(800) 8683569
Price as of January 1990; $)	1750	1500	3000	1500	350	1375	295	1000	1500	1750	1500	981.75
Computer	P,M[c]	M	P[d]	M, M-U	P	P	P	P,M	P, M-U	P	P,M	P
Types of Plot												
Linear	Y[e]	Y	Y	Y	Y	Y	Y	Y	Y	Y	Y	Y
Semi-log	Y	Y	Y	Y	Y	Y	Y	Y	Y	Y	Y	Y
Log-log	Y	Y	N	Y	Y	Y	Y	Y	Y	Y	Y	Y
Logit-log	Y	Y	Y	Y[f]	Y	Y	Y	Y	Y	Y	Y	Y
Types of Curve Fit												
Linear regression	Y	Y	Y	Y	Y	Y	Y	Y	Y	Y	Y	Y
Polynomial	Y	N	N	Y	N	Y	Y	Y	N	Y	Y	Y
4-PLL	Y	Y	Y	Y	U	Y	Y	Y	Y	Y	Y	Y
Spline	Y	N	Y	N	N	Y	Y	N	N	Y	N	N

Linear interpolation	Y	Y	Y	Y	N	U	Y	N	Y	Y	N	N
Kinetics Function												
Calculate rate of absorbance change from multiple readings	Y	Y	N	U	Y[g]	Y	N	N[h]	Y	Y	Y	N
Plot kinetics for individual well	Y	Y	N	U	N	Y[h]	N	N[h]	Y	Y	Y	N
Parallelism Testing												
Setting sample as std	Y	Y	Y	Y	Y	Y	Y	Y	Y	Y	Y	Y
View multiple curves	Y	N	N	Y	Y	U	N	P[i]	N	N	N	N
Visual on sample conc data	Y	Y	Y	Y	Y	Y	Y	Y	Y	Y	Y	N
Statistical analysis of data	Y	N	P[j]	U	Y	N	N	P[i]	N	N	N	N
Device of Data Interface												
Import ASCII files	Y	Y	Y[k]	Y	Y	Y	Y	Y	Y	Y	Y	N
Data entry from keyboard	Y	Y	Y	Y	Y	Y	Y	Y-P,N-M	Y	Y	N	Y

TABLE 1 (continued)
Listing of Functions Performed by 12 SPIA Data Analysis Packages

	BioCalk	Delta-Soft	ELISA-AID[a]	ELISA+	ELISA-NALYSIS	Kineti-Calc	Ligand Data Calc	Micro-plate Manager	Microtek	Soft 2000	SOFT-max	Titersoft II
Supported devices[l]	4	3, 5, 6, 8	1—8[k]	1—8	3	N[n]	2	1	7	6	5	Y
Device controlled by computer	Y	Y	Y	Y	Y[o]	Y	N[n]	Y	N	N	N	N
Transfer data without computer control	Y	Y	Y	Y	Y	Y	N[n]	Y	N	N	N	Y

[a] A separate package specifically designed for radioimmunoassays (named RIA AID) is also available.
[b] This software has been written by, and is available from, the author of this chapter.
[c] P = IBM PC and compatibles, M = Apple Macintosh, Y-P, N-M = function available on PC version but not on Macintosh version, and MU = Macintosh version under development.
[d] This package is specifically designed to interact with mini- and mainframe computers with extensive communications and data tracking features.
[e] Y = has function, N = does not have function, U = function under development, P = has some aspects of function.
[f] The software uses a logit-log transformation internally, but this is plotted in a semi-log manner.
[g] The software can only calculate rates from pairs of absorbance values, unlike other software with this function, which can derive the rate based on regression analysis of data from multiple readings.
[h] A separate kinetics analysis package is available for both computer systems at a cost of $765.
[i] Once a sample has been assigned as a "standard", two standards may be displayed simultaneously and compared statistically.
[j] Custom feature available on request; may influence the final price of the software. The base package includes intraassay quality control analysis.
[k] The actual interface routines required for the software package must be specified by the purchaser and may influence the final price of the package. The base package includes an interface to one device as specified by the purchaser.
[l] Supported devices: 1 = Anthos; Denley Instrument, Inc.; 2 = Bio-Rad; 3 = Bio-Tek Instruments, Inc.; 4 = Dynatech Laboratories, Inc.; 5 = Flow Laboratories, Inc.; 6 = Molecular Devices, Inc.; 7 = SLT Laboratory Instruments; and 8 = Other devices.
[m] For this software, the interface to devices 1 and 7 requires "parallel to serial" conversion hardware.
[n] A "User Programmable Interface" version of Ligand Data Calc is available from the distributor for $400 more than the "Desk Top" (noninterfacing) version. Interfacing routines for this program are supplied for several instruments at no extra cost. Additional routines may be programmed either by the user or by the distributor.
[o] Reader control is only available for the Molecular Devices readers.

advantage of being less affected by variations in reaction time than is the raw absorbance value. Some packages generate the full kinetics curve for each well, allowing the user to ensure that the kinetics value is obtained while the reaction rate is constant.

Parallelism testing - Confirmation of parallelism between the date from a sample and the standard curve can be essential for the validity of sample estimation (see Chapters 3 and 9). Most software packages will allow the slope for each sample to be determined by changing the assay "template" such that the sample is set as a standard; this can be very cumbersome and even impractical. Some packages will allow multiple curves to be displayed at one time, such that the user can visually test for parallelism. Most packages will provide concentration data for each dilution of each sample corrected for dilution. For example, sample data which parallels the standard curve may be reported as 10 units for a 1/20 dilution, 11 for a 1/40 dilution, and 9.5 for a 1/80 dilution, while nonparallel sample date might be reported as 10 units for a 1/20 dilution, 15 units for a 1/40 dilution, and 20 for a 1/80 dilution. Ideally, the software performs a statistical test for parallelism on the data, comparing titrations of the sample with the standard curve.

Device/data interface - Data can be entered into analysis software by three methods: (1) by direct user entry from the keyboard, (2) from other software via a standard data file (e. g., an ASCII file), and (3) through an interface routine for a specific ELISA plate reader or other device. Devices such as plate readers do not export data in standard formats; formats can vary even between different models of plate readers from the same manufacturer. In order to directly acquire data from a device, the software must have the appropriate interface routines for that specific device. For example, a routine for the Flow Laboratories Titertek Multiskan Mark 1 will not work with any other plate reader — not even other readers from Flow Laboratories. Therefore, the software supplier should be contacted to confirm that the software will function with any particular device. Some plate readers may be controlled by computer with the appropriate software, adding capabilities not designed into the reader. Many of these readers will only send data to the computer if the computer takes complete control; this can reduce the functionality of the reader and require that the user move back and forth between the reader and the computer in order to read each plate. Some software packages can be set up to passively acquire data such that the user does not have to continuously interact with both the reader and computer.

It must be noted that software and SPIA reading devices (e. g., plate readers) are continually being modified, thus the data obtained within the table are subject to change and should be confirmed by contacting the software distributor. Therefore, the author of this table can make no guarantee as to its accuracy.

GLOSSARY

Absolute quantitation — The quantitation of antigen or antibody in gravimetric units of mass, e.g., mg/ml rather than in units of activity, e.g., titer.

Activation immunometric assay — A two-site immunoassay in which the analyte must be biologically altered or enzymatically cleaved either by *in vivo* or *in vitro* processes in order to expose unique antigenic determinants. *De novo* exposed antigenic determinants are then bound by either capture antibody or the detection antibody, depending on the design of the assay. Internal or modified antigenic epitopes that are generated at the time of the assay may be called neoantigens.

Active ester — An ester made with an alcohol whose structure allows it to be easily displaced by nucleophiles such as amines, forming a stable linkage. One common use of active esters is in the formation of soluble protein conjugates and in the solid-phase immobilization of proteinaceous molecules.

Affinity — The instrinsic attractiveness of one substance for another, e.g., antibody and antigen, which is typically expressed as the equilibrium constant (Keq or Ka) of the reaction. The instrinsic affinity constant is the Keq of the reaction between one antibody binding site and its univalent ligand.

Affinity column or affinity chromatography — A solid-phase preparative method which uses a specific receptor immobilized on an insoluble support to capture and subsequently purify a ligand from a liquid or merely deplete it from the liquid.

Ambient analyte immunoassay — A type of immunoassay in which analyte concentrations can be measured independently of both the sample volume and amount of antibody present.

Amplified immunoassay — See **Indirect enzyme immunoassay**.

Analyte — A term borrowed from clinical chemistry, and used in immunoassay to indicate the antigen or antibody to be measured.

Antibody binding site — The site within the variable region of antibodies which specifically binds to the specific groups, i.e., epitopes, of an antigen. In the context of idiotypic interactions, it may be referred to as the *paratope*.

Antibody occupancy principle — The immunoassay principle whereby the concentration of an analyte (or multiple analytes) is determined by the fractional occupancy of antibody binding sites.

Antigen capture capacity (AgCC) — A parameter which defines the functional properties of a capture antibody. Specifically, the percentage of the added antigen captured in the region of excess antibody. AgCC is a function of affinity and/or avidity, which is useful in situations in which reaction volumes, and consequently absolute concentrations, are difficult to determine.

Antigenicity and antigen — The ability of a substance to be recognized by the immune system. In solid-phase immunoassay, to be recognized by an antibody. The recognized substance is called *antigen*.

Antiglobulin — A generic term denoting an antibody which recognizes as its antigen another antibody.

Association rate constant (k_1) — The "forward" or "on" rate of a biomolecular ligand-receptor interaction that is governed by Mass Law principles.

Attentuated total reflection — A technique for the analysis of surfaces, usually employed with spectroscopies such as infrared or fluorescence. This technique uses a special fixture which directs the sampling beam at a critical reflection angle to the surface of the sample such that multiple reflections occur from only the first 1 to 2 μ m.

Avidin — A biotin binding glycosylated protein from egg white (see **Streptavidin**).

Avidity — The property of a molecular interaction which makes dissociation difficult. This can result from the instrinsic affinity of the interaction as well as from the complexity of the interaction as is the case when polymeric antibodies bind to multivalent antigen via multiple interactions or biomolecules bind to adsorptive surfaces at multiple contact points.

Biotin — Water-soluble vitamin of the B-complex (Vit. H) which binds to avidin and streptavidin.

Blocking — A process in solid-phase immunoassay, usually involving use of a protein solution or detergent, designed to prevent nonspecific interactions of the analyte or detection system with the solid-phase or other components of the system.

Blooming — Condition in which additives used in the formulation and/or processing of a plastic part migrate to the surface.

Bulk polymerization — Polymerization of a monomer in the absence of a solvent or suspending medium, initiated by a catalyst or heat.

Capture antibody — An immobilized antibody used to capture the antigen of interest from the test sample.

Chemiluminescence — A detection system for immunoassays in which the chemical generation of electronically excited states subsequently results in light emission.

Chimeric antibody — An antibody engineered using genes of different species so, for example, the binding sites and Fc-region are of different species origin.

Clonality — This property defines the number of different clones from which a population of biomolecules is derived. In the case of antibodies, monoclonals are derived from a single clone. All the antibodies from a particular clone are identical whereas polyclonals are derived from multiple clones and are structurally and functionally heterogeneous. Monoclonals may be derived from the fusion of one B-lymphocyte with a tumor B-cell line (hybridoma) or may be produced by a naturally occurring or induced B-cell tumor (plasmacytoma).

Colloid — A solid, liquid or gaseous substance made up of a very small (usually <2 μ m in any dimension) particles suspended in a non-solubilizing medium.

Competitive immunoassay — An immunoassay in which the detection of the analyte (ligand) is based on its ability to quantitatively inhibit the binding of a known amount of a standard ligand to a limiting amount of receptor, e.g., antibody. Both direct (labeled analyte) and indirect (labeled antibody) version are used for the quantitation of an antigen (*see* **Radioimmunoassay**).

Conjugate — A typically covalent complex of two different biomolecules. In EIA, complexes of enzyme and antibodies are called conjugates in which the enzyme is the signal generator. Fluorescent-labeled proteins or biotinylated proteins are also sometimes called conjugates.

Critical micelle concentration — The concentration of a soluble surface-active species above which association of the species occurs forming a micelle or "association colloid".

Detection system — The biomolecular assemblage of enzymes, antibodies, fluors, or radio-isotopes responsible for binding to the substance (analyte) to be measured. This assemblage then sends a signal (see signal generator) which is then physically measured.

Diffusion rate constant — The rate of association between biomolecules which is controlled by their respective diffusion constants and the solvent environment.

Direct EIA or direct RIA — An immunoassay in which the extent of reaction between antibody and antigen is directly measured because one or the other is attached to the signal generator (see **Signal generator**) used in the particular system.

Dissociation rate constant (k_2 or k_{-1}) — The reverse or "off" rate of a biomolecular ligand-receptor interaction that is governed by Mass Law principles. k_1/k_2 equals the equilibrium constant (see also **Affinity**).

Dot blot — A blotting method in which usually an antigen is directly spotted on a blotting membrane and subsequently detected by the application of specific antibody. Synonyms include dot immunoassay, dot immunoblot, dot spot, slot blot, immunodot, and antigen dot test.

Double reciprocal plot — See **Langmuir equation**.

Dynamic range or range — The titration range over which an immunoassay is capable of accurately measuring the concentration of the analyte.

Electrophoretic transfer — See **Transfer blot**.

ELISPOT — A modern version of a Jerne plaque assay in which potential antibody-secreting cells are incubated on antigen-coated solid-phases which capture the specific antibodies secreted by these cells and thus permit the number of secreting cells and their product to be determined. The solid phase may also be coated with capture antibody to determine, for example, the number of immunoglobulin-secreting cells, interleukin-secreting cells, etc.

Encounter complex — A reaction intermediate which theoretically forms during the attainment of equilibrium in biomolecular receptor-ligand interactions.

End-point titer method — A method of quantitation used in immunoassay in which the sample to be measured is titrated (diluted) until some fixed signal value is obtained.

Enzyme-immunoassay (EIA) — An immunoassay in which the signal generator is an enzyme.

Enzyme immunofiltration assay (EIFA) — Enzyme immunoassay using antigen- or antibody-bearing membranes as the solid phase and filtration of reactant liquids through the membrane to accelerate the reactions. Synonyms: enzyme-linked immunofiltration assay (ELIFA), immunoconcentration assay, immunofiltration assay, membrane filtration enzyme immunoassay, solid-phase immunofiltration assay.

Enzyme-linked immunoadsorbent assay (ELISA) — Generically equivalent to enzyme immunoassay (EIA). Specifically, ELISA denotes solid-phase heterogeneous EIA in which the substance to be measured is captured (bound) to a solid-phase reactant and separated from all unbound substances by washing of the solid phase.

Epitope — A biochemical structure on the solvent-exposed portion of a biomolecule which is specifically recognized by an antibody or antibody binding site. Antigenic determinant is synonymous.

Fab — A fragment of an antibody generated by proteolysis which contains one of the antibody binding sites.

Fc-region — That portion of an antibody molecule which is composed of the C-terminal domains of the heavy chains and is therefore at the opposite end of the molecule from the antibody binding sites.

Fluorescence — The property of a substance to adsorb light energy of a particular wavelength and to then emit light of lower energy at a longer wavelength.

Fluorescent immunoassay (FIA) — An immunoassay like RIA except that the signal generator produces a fluorescent signal (see **Fluorescence**).

Functional capture antibody concentration [CAb_t] — The concentration of capture antibody (see **Capture antibody**) or capture antibody binding sites empirically determined in solid-phase binding assays.

Hapten or haptenic — A substance unable to induce an immune response but which can be recognized by antibodies or cells as antigenic. Haptens are generally <1000 Da in size.

Heterogeneous immunoassay (HIA) — An immunoassay in which the receptor-ligand complexes (antibody-antigen complex) is separated from the reaction mixture to permit its detection. This is most conveniently done by immobilizing one principle reactant (receptor or ligand) on a solid phase. See **Solid-phase immunoassay**.

Heterologous and homologous — In serology and immunochemistry heterologous implies a difference of source of a reactant from that which serves as the point of reference. For example, antibodies made in goats to rabbit proteins are regarded as heterologous whereas those made in another rabbit would be homologous. A homologous immunological reaction is one between an antigen and the antibody made against it whereas a heterologous or "cross" reaction implies that either the antigen or antibody are not members of the original pair.

Heterologous interpolation — The process of interpolating unknown response data from a dose response curve that has been constructed with a specimen that contains a heterologous or different standard analyte. This term has been most extensively applied to assays in which the amount of antibody of a particular isotype, allotype, idiotype, or specificity is being quantitated and no reference sera exist with calibrated amounts of antibody specific for that antigen.

Homogeneous immunoassay — An immunoassay not requiring that the bound reactant be separated from the free reactant in order for it to be detected. Homogeneous immunoassays may not involve a solid-phase.

Hybridoma — See **Clonality**.

Hyperimmune — An organism that has been repeatedly exposed to antigen and as a result, expresses a mature immune response typically of high activity, i.e., affinity, to the immunizing antigen.

Idiotope and idiotype — An epitope found in the variable regions of antibodies. These are often associated with sequences which determine the specificity of an antibody and therefore may be diagnostic markers for certain specificities. An idiotype is an antibody possessing a particular constellation of idiotopes which uniquely define it from others.

Immunoblot and immunoblotting — A broad term describing a type of solid-phase immunoassay involving adsorption of macromolecules to a blotting membrane followed by immunochemical analysis. See **Dot blot** and **Transfer blot**.

Immunodominant — A particular chemical grouping on an antigen, e.g., protein, virus or bacterium, which elicits the production of proportionally large amounts and/or high affinity antibodies to that grouping.

Immunofiltration assay — See **Enzyme immunofiltration assay**.

Immunogenicity — The ability of a substance to induce an immune response in an animal.

Immunomagnetic isolation — The use of magnetized microparticles or beads coated with antibody to purify cells or antigens.

Immunoradiometric assay (IRMA) — A competitive immunoassay in which the detection system is a radioactive antibody instead of a radioactive antigen as is common to classical radioimmunoassay.

Immunosensors — A type of signal generator composed of solid-phase probes coated with specific antibodies which upon reaction with analyte results in transduction of a measurable electronic or optical signal.

Indirect enzyme- or radioimmunoassay — A solid-phase immunoassay in which the primary antigen-antibody reaction is detected using an antiglobulin (see **Antiglobulin**) conjugated to a signal generator (see **Signal generator**). An indirect enzyme immunoassay can be amplified through the use of more complex signal generators.

Injection molding — The process by which a resin is plasticized by heat and mechanical energy and forced under pressure into the cavities of a mold.

Inverse emulsion polymerization — Typically, a polymerization of a water soluble monomer in a water-in-oil emulsion, where the monomer is dissolved in water droplets and dispersed in an oil phase with the aid of a water-in-oil emulsifier. Oil-soluble or water-soluble initiators are used to give a colloidal dispersion of water-swollen polymer particles in oil.

Isomerizing conformers — The mixture of conformations which exist for small biomolecules, e.g., peptides, in solution.

Isotype — The class or subclass of an antibody as defined by the type of heavy chain constant region it possesses.

Latex — A colloidal suspension in which the insoluble particles are composed of a natural- or synthetic-associated polymer molecule or molecules.

Latex polymerization — Typically, a polymerization of a water insoluble monomer predispersed as an oil-in-water emulsion, usually with the aid of an oil-in-water emulsifier. The reaction is so designed that nucleation of particle formation takes place in the continuous phase or in the monomer-swollen micelles. The emulsion droplets serve only as reservoirs for monomer and not a loci of polymerization. The technique is applicable to formation of particles in a size range of about 10 to 2000 nm. The method is often referred to as "emulsion polymerization" which is somewhat a misnomer.

Langmuir isotherm equation — A means of graphically expressing the interaction of receptor and ligand. This is often called a "double reciprocal plot" and a derivation, which permits estimation of functional antibody concentration ($[Ab_t]$), is called a Steward-Petty plot.

Ligand — The molecule specifically recognized by a receptor and in binding studies, typically the one whose concentration is varied to permit the interaction to be studied. In solid-phase systems it is usually the soluble component of the interaction.

Linear binding region (LBR) — A region on binding plots in which the percentage bound remains constant, i.e., linear, when the plots are expressed as percentage of the ligand (analyte) bound vs. ligand added. In this region the receptor concentration is in excess.

Logit-log — A transformation used to convert the sigmoidal nature of the titration curve of an immunoassay to a straight line which is then applicable to linear regression analyses.

Membrane filtration enzyme immunoassay — See **Enzyme immunofiltration assay**.

Microparticle — A small particle of <1 μ m diameter composed of some plastic with or without a magnetic core. Microparticles, as opposed to "beads", behave as a colloidal suspension. *See* **Colloid**.

Microspot immunoassay — An occupancy-based immunoassay in which the capture antibody is restricted to a microspot on a solid-phase and analyte concentration is determined from the ratio of two signals emitted by the capture and detection antibodies, respectively.

Microtiter and microtiter plate — Microtiter refers to serological or immunochemical tests which are performed in 85 × 128-mm plastic trays containing 96 6.5-mm diameter wells which can hold a maximum of 300 μ l.

Molding additives — Substances added to polymer resins, or applied to the mold surface, to facilitate the manufacturing of a plastic part.

Monoclonal antibody (mAb) — See **Clonality**.

Multianalyte immunoassay — A configuration of solid-phase immunoassay in which a number of antibodies with specificity for different analytes are attached to microparticles, affinity columns, or other surfaces, to permit the simultaneous capture and subsequent measurement of different analytes in the test sample.

Multipoint analyses — The practice of measuring the concentration of a substance in a test sample at various dilutions of the sample.

Nitrocellulose (NC) — The most popular membrane used for immunoblotting which is manufactured as a pure cellulose-nitrate ester or as a cellulose nitrocellulose acetate-mixed ester. Various porosities are available.

Noncompetitive immunoassay — An immunoassay in which detection of the analyte is direct as opposed to indirect, which is the case for classical competitive radioimmunoassay. See **Sandwich assay**.

Noncompetitive immunometric assay — See **Sandwich assay**.

Nonspecific binding — Binding of a label or labeled conjugate during solid-phase immunoassay by nonimmune-specific means. This phenomenon can lead to high background signals or false positive results. See also **Blocking**.

Number average molecular weight — The arithmatic mean molecular weight.

Parallelism — The ability of test and reference samples to produce parallel titration curves in an immunoassay.

Plasmacytoma — See **Clonality**.

Protein avidin biotin capture (PABC) system — A configuration which permits proteins to be immobilized on a solid phase not involving either adsorptive forces or covalent bonding.

Polyvinylidene difluoride (PVDF) — A type of immunoblotting membrane that is chemically distinct from nitrocellulose, e.g., Immobilon P®.

Polydispersity ratio — In polymer chemistry, the ratio of weight average divided by number average measurement, which can be used to describe the heterogeneity of a given sample. A value of 1 indicates a monodisperse system.

Precision profile — The precision of an assay in measuring an analyte as a function of its position on the standard curve. Precision is expressed as the coefficient of variation of replicate measurements.

Prozone — A phenomenon in immunoassays in which the addition of high concentrations of ligand *decreases* rather than *increase* the signal generated. Prozone is also called a "hook" in some literature.

Polyclonal antibody (pAb) — See **Clonality**.

Paratope — See **Antibody binding site**.

Precision — The agreement of replicate measurements within an assay or between assays within a laboratory or between laboratories. A measure of reproducibility.

Radioimmunoassay (RIA) — Generically speaking any immunological reaction which is measured using a radioactive reporter substance or signal generator. Specifically, RIA most often refers to classical competitive, labeled-antigen immunoassays used to quantitate antigens of medical importance.

RAST — Radioallergosorbent test, in which IgE antibodies in the sera of patients to allergens immobilized on membranes, are measured.

Receptor — A biomolecule with specific affinity for another biomolecule as, for example, an antibody for an antigen. In solid-phase immunoassay the receptor is usually the solid-phase capture reactant.

Receptor excess — A situation in which the ratio between the receptor concentration and the ligand concentration is high enough that the free receptor concentration approximates the total receptor concentration.

Reference equation method — A method o estimatingf sample concentration in which an equation is first obtained for the best curve fit to a plot of a reference standard. The concentrations of the samples are then estimated using this equation together with the binding data from individual sample dilutions.

Reference standard (RS) — A solution of purified protein or antibody or a mixture of proteins or antibodies, such as serum, that has been designated (or shown) to contain (1) a certain concentration of the protein or antibody or (2) a certain antibody activity toward the antigen of interest. RSs are selected and stored in such a manner that they maintain their activity over a long period of time.

Reflectance photography — A method for determining the intensity of a colored spot usually on a solid-phase blotting membrane for which an equation relating intensity to concentration can be written.

Reporter substance — A biomolecule which, through its unique chemical properties, generates a fluorescent signal, a radioactive signal, or produces a color reaction as an indication of its presence.

Resin — A polymeric base material used to formulate or produce an intermediate or finished product.

Sandwich assay — A configuration of solid-phase immunoassay in which the analyte is captured by one antibody and detected by another. Also called the noncompetitive two-site immunoradiometric assay. The analyte (antigen) must be multivalent. Both symmetrical (both antibodies of same specificity) and asymmetrical (antibodies of different specificity) configurations are used.

Saturation analyses — The practice of progressively increasing ligand concentration until the receptor sites are saturated. Mathematical equations can then be applied to calculate receptor concentration and equilibrium constants.

Scatchard plot — A means of graphically expressing the interaction of ligand and receptor such that an estimate can be made of the degree of heterogeneity, the equilibrium constant, and the receptor (e.g., antibody) valence.

Schiff base — An imine, which forms as the reaction product of an aldehyde and an amine. It should be chemically reduced to form a stable linkage, unless the carbon-nitrogen double bond is stabilized by conjugation with aryl groups or other double bonds in the species (conjugated in the chemical sense).

Sensitivity — The minimum detectable amount or activity that can be accurately measured by an immunoassay. Sensitivity is often expressed as the limits of detection (LOD).

Serology — The study of factors in serum, usually antibodies or antigens, which are considered of diagnostic value in disease.

Signal generator — See reporter substance.

Size exclusion chromatography — A chromatographic technique whereby separation of molecules occurs by size.

Slot blot — See **Dot blot**.

Solid-phase — Any insoluble surface that can serve as a support for protein or other biomolecules.

Solid-phase immunoassay (SPI or SPIA) — An immunoassay in which one of the principle reactants (antibody or antigen; receptor or ligand) is immobilized on an insoluble interface. SPI is a heterogeneous immunoassay.

Spacer arms — Molecules or chains of molecules having two or more reactive functionalities, used to conjugate molecules to other molecules or to surfaces.

Specific activity — A measure of the activity of a protein per microgram of that protein. Activity could be in terms of radioactivity, enzyme activity, or antibody activity.

Specific antibody immunoassay (SpAbI) — An immunoassay configured so that the substance to be measured (analyte) is a specific antibody. Some literature calls these "antigen-specific" immunoassays.

Standard curve — A curve or straight line produced by mathematically fitting a curve to the data from the reference standard. The equation for this curve is used in the Reference equation method.

Steric stabilization — A process of increasing the free energy of interaction between surfaces surrounded by a liquid (and hence causing repulsion) by anchoring solvent-soluble oligomeric or polymeric chains to the surface.

Steward-Petty plot — See **Langmuir equation**.

Sticking coefficient — A coefficient which expresses the degree to which the dissociation of an encounter complex influences the association rate of a receptor-ligand reaction (*see* **Encounter complex**).

Streptavidin (SA) — 70-kDa protein produced by *S. avidinii* that binds biotin with igh affinity.

Suspension polymerization — Typically, polymerization of an oil-soluble monomer dispersed into water as an emulsion with the aid of an oil-in-water emulsifier. In this method, the main loci of polymerization are the emulsion droplets. Typical particle dimensions range from about 2 to 1000 μ m.

Titer — The reciprocal of the highest dilution of sample which gives a certain preestablished level of activity in a particular assay.

Transfer blotting — See **Western blot**.

Weight average molecular weight — The arithmatic mean of the square of the molecular weights divided by the arithmatic mean of the weights.

Western blot — A transfer blotting technique in which proteins are transferred from agarose or polyacrylamide gels onto blotting membranes and subsequently detected, usually with specific antibodies or special strains.

X-ray photoelectron spectroscopy (XPS) — An analytical technique sensitive to the first 1000 Å of a surface, which uses X-rays to cause the emission of photoelectrons, the energy of which is indicative of both the original element and its chemical environment.

Index

INDEX

A

D

E

F

G

H

I

K

L

M

N

O

P

Q

R

S

T

U

V

W

X

Z